Applied Mathematics and Engineering Science Texts

SERIES EDITOR

PROFESSOR ALAN JEFFREY
*University of Newcastle upon Tyne
and University of Delaware*

Applied Mathematics and Engineering Science Texts

NONLINEAR ORDINARY DIFFERENTIAL EQUATIONS

R. GRIMSHAW
University of New South Wales
Australia

BLACKWELL SCIENTIFIC PUBLICATIONS

OXFORD LONDON EDINBURGH

BOSTON MELBOURNE

© 1990 by
Blackwell Scientific Publications
Editorial offices:
Osney Mead, Oxford OX2 0EL
25 John Street, London WC1N 2BL
23 Ainslie Place, Edinburgh EH3 6AJ
3 Cambridge Center, Suite 208
  Cambridge Massachusetts 02142, USA
107 Barry Street, Carlton
  Victoria 3053, Australia

First published 1990

Set by the Pi-squared Press, Nottingham
Printed and bound in Great Britain
by Hartnolls Ltd, Bodmin,
Cornwall

DISTRIBUTORS

Marston Book Services Ltd
PO Box 87
Oxford OX2 0DT
(*Orders*: Tel: 0865 791155
         Fax: 0865 791927
         Telex: 837515)

USA
Publishers' Business Services
PO Box 447
Brookline Village
Massachusetts 02147
(*Orders*: Tel: (617) 524-7678)

Canada
Oxford University Press
70 Wynford Drive
Don Mills
Ontario M3C 1J9
(*Orders*: Tel: (416) 441-2941)

Australia
Blackwell Scientific Publications
(Australia) Pty Ltd
107 Barry Street
Carlton, Victoria 3053
(*Orders*: Tel: (03) 347-0300)

British Library
Cataloguing in Publication Data

Grimshaw, R.
    Nonlinear ordinary differential equations.
    1. Nonlinear ordinary differential equations
    I. Title   II. Series
    515'.352

    ISBN 0-632-02708-8
    ISBN 0-632-02709-6 pbk

Library of Congress
Cataloging-in-Publication Data

Grimshaw, R.
    Nonlinear ordinary differential equations /
    R. Grimshaw
        p.   cm.—(Applied mathematics and
    engineering science texts)
    Includes bibliographical references.
    ISBN 0-632-02708-8
    ISBN 0-632-02709-6 pbk
    1. Differential equations, Nonlinear.
    I. Title.   II. Series
    QA372.G75   1990
    515'.352—dc20

CONTENTS

PREFACE

Ordinary differential equations are crucial to many areas of science and technology. The subject has grown in parallel with the development of mathematics from the time of Newton and Leibnitz and, while modern mathematics now encompasses a vast range of topics, differential equations have retained their central role. That this is the case is undoubtedly because of the wide variety of applications, ranging from the traditional areas of physics and engineering to the more recent use of differential equations in biology, chemistry, ecology and economics. As a discipline differential equations has retained a high degree of vitality and nowhere is this more evident today than in the current fascination with chaos. Indeed, the recent discovery that simple-looking differential equations are capable of describing complicated and chaotic behaviour is a timely reminder that even in traditional areas there is still much to be discovered. Much of current interest in differential equations is due to the advent of high-speed computing facilities, and the consequent ability to shift emphasis from the classical study of linear systems to the fascinating problems encountered in the study of nonlinear systems. Hence the emphasis in this text on nonlinear differential equations.

The material in this book is an outgrowth of the lectures given to third- and fourth-year undergraduate students in the Universities of Melbourne and New South Wales. Like these lecture courses, the text is designed to be a bridge between elementary courses on ordinary differential equations and advanced graduate courses on topics such as chaotic phenomena, bifurcation theory or Hamiltonian dynamics. For background, students taking a course based on this text should have some familiarity with the elementary theory of first-order differential equations and second-order ordinary differential equations with constant coefficients. They would also be expected to have a good grounding in calculus and linear algebra. Courses which provide this kind of background are normally available in the first and second years of undergraduate programs in UK, USA or Australian universities. Given this, the text is self-contained, and is designed to take the student from basic elementary notions to the point where the exciting and fascinating developments in the modern theory of nonlinear differential equations can be understood and appreciated. While the emphasis in this text is on developing an understanding of the subtle and sometimes unexpected properties of nonlinear systems, and simultaneously introducing practical analytical techniques to analyze nonlinear phenomena, I have endeavoured at the same time to provide a coherent and rigorous framework so that the development of the material in the text can be seen as logical and structured.

In using this text as a basis for an undergraduate lecture course of 24 to 28 lectures, I have usually omitted the detailed proofs of chapter 1, and instead preferred to emphasize the significant implications of the existence and uniqueness theorems. Depending on the background of the students it may also be possible to omit much of chapter 2 on the basic theory of linear differential equations, although, of course, this material should be included if the students have not already encountered it elsewhere. These two chapters have been included in the text to make it as self-contained as possible. The core of the lecture course is based on chapters 3 through 8, with a possible selection of topics from the more advanced material in chapters 9 to 11. In a lecture course I have tended to omit the details of the more technical and lengthy proofs in keeping with an applied approach to the subject, but have nevertheless included these proofs in the text for the sake of completeness. I have also at times used material in chapters 9 to 11 as the basis for an advanced course at the level of the final undergraduate year, or at the beginning graduate level. While the text does not include any discussion of numerical methods *per se*, as I believe this topic is now sufficiently important to warrant a separate lecture course, and text, of its own, I would strongly recommend that, in a lecture course, the increasingly important role that numerical integration now plays in the study of nonlinear differential equations be emphasized both by the use of graphic displays and through student assignments. Indeed, the interplay between numerical discovery and underlying theoretical concepts can constitute an exciting learning environment for the student.

Each chapter in the text is designed to be as self-contained as possible, and where the theoretical development is based on results from a previous chapter, these are referred to explicitly. Consequently, and to avoid an overly cumbersome system, the numbering system for equations is self-contained within each chapter. Thus, for instance, (3.7) is the seventh numbered equation in section 3 of the chapter in which it appears. References to numbered equations in other chapters are given by quoting explicitly the equation number and chapter as well. However, theorems are numbered by the chapter within which they appear, so that, for instance, theorem 6.4 is the fourth theorem in chapter 6. Each chapter contains a selection of problems, most of which have been tested in my lecture courses. Some brief solutions to these problems are included and, as well, detailed working of some related problems is contained in the body of the text.

Many people have contributed to the development of my interest and enthusiasm for nonlinear differential equations and, hence, directly or indirectly, to the material for this text. They are too numerous to mention by name, but my greatest debt is to the students who have endured my lectures and, over the years, contributed to the material for this text.

ROGER GRIMSHAW
University of NSW, Sydney
November 1989

CHAPTER ONE
INTRODUCTION

1.1 Preliminary notions

Ordinary differential equations involve an independent variable, t, and a dependent variable, x, which is to be a function of t so that $x = x(t)$. We shall denote the derivative of x with respect to t by x' so that

$$x' = \frac{\mathrm{d}x}{\mathrm{d}t}.$$

In this section x will be a scalar variable, although later, in section 1.2 and elsewhere in this text, we shall allow x to be a vector. The simplest general form for a differential equation that we can pose is

$$x' = f(x, t),$$

where $f(x, t)$ is a specified function of x and t. This is said to be a first-order differential equation where the terminology order refers to the highest derivative of x which appears in the equation. The physical interpretation of the variables x and t depends of course on the physical context from which the differential equation arises. However, it is often the case that t corresponds to the time and then the differential equation describes the evolution of some dynamical process as t increases.

A simple example of a first-order differential equation is

$$x' = \mu x,$$

which can be used to describe the growth of a population, when the growth rate is assumed to be proportional to the population itself, the factor of proportionality being the constant μ. This equation has the solution

$$x = C e^{\mu t},$$

which contains an arbitrary constant, C, and describes the exponential growth of $x(t)$. In general, the solution of any first-order differential equation will contain an arbitrary constant. Hence an extra condition is needed to characterize a solution completely, and often this will be the initial condition

$$x(t_0) = x_0,$$

where x_0 and t_0 are specified. For the exponential equation discussed above, we see that $x_0 = C e^{\mu t_0}$ and so

1

$$x = x_0 \exp \mu(t - t_0).$$

For specified values of x_0 and t_0 the solution is now unique. Further we note that the solution $x(t)$ as well as depending on t also depends on the initial data x_0 and t_0, and on the parameter μ, so that we should write $x(t; \mu, x_0, t_0)$.

Higher-order equations will also need to be considered. For instance, the general form for a second-order differential equation is

$$x'' = f(x, x', t),$$

where

$$x'' = \frac{d^2 x}{dt^2}$$

and f is a specified function of the three variables x, x' and t. By analogy with the first-order equation we expect the general solution to contain two arbitrary constants and, hence, require two initial conditions to characterize it completely. For instance, consider the equation for a simple harmonic oscillator:

$$x'' + \omega^2 x = 0,$$

where ω is a real constant describing the frequency of the oscillator. The general solution is

$$x = C \cos \omega t + D \sin \omega t,$$

containing two arbitrary constants C and D, and describing sinusoidal oscillations with frequency ω and period $T = 2\pi/\omega$. Appropriate initial conditions are

$$x(t_0) = x_0, \qquad x'(t_0) = x_1,$$

where x_0, x_1 and t_0 are specified, and the solution which satisfies these conditions is

$$x = x_0 \cos \omega(t - t_0) + \frac{x_1}{\omega} \sin \omega(t - t_0).$$

Extrapolating, the general form for an nth order equation is

$$x^{(n)} = f(x, x', \ldots, x^{(n-1)}, t),$$

where

$$x^{(r)} = \frac{d^r x}{dt^r} \quad (r = 1, 2, \ldots, n).$$

Now f is a specified function of the $n+1$ variables $x, x', \ldots, x^{(n-1)}$ and t, and we expect the general solution to contain n arbitrary constants.

The two examples discussed so far are examples of linear equations, in which the right-hand side is a linear function of the dependent variable. However, many equations encountered in practice are nonlinear, and it is the study of these which is the main concern of this text. An elementary example of a non-linear equation is

$$x' = x^2,$$

which has the general solution

$$x = (C-t)^{-1},$$

where C is an arbitrary constant. Note that this solution is singular when $t = C$. For instance, with the initial condition $x(0) = 1$ the solution is $x = (1-t)^{-1}$, and we see that the solution for this initial condition is only defined for $0 \leqslant t < 1$ and develops a singularity as $t \to 1$. The presence of this singularity cannot readily be predicted by an examination of the equation alone. Indeed, if the initial condition is instead $x(0) = -1$, the solution is $x = -(1+t)^{-1}$ and is defined for all $t \geqslant 0$, although it is, of course, singular at $t = -1$. Nonlinear differential equations are often characterized by subtle and unexpected behaviour of the solutions, and in this respect can be contrasted with linear differential equations for which there is a well-developed general theory.

1.2 First-order systems

When constructing the general theory of differential equations it is convenient to consider a system of n first-order equations for n unknown functions $x_1(t), \ldots, x_n(t)$. The standard form is

$$x_i' = f_i(x_1, \ldots, x_n, t) \quad (i = 1, \ldots, n). \tag{2.1}$$

Here each f_i is a specified function of $n+1$ variables $x_1, \ldots, x_n$ and t. The simplicity of the concept of a first-order system is most apparent when the system (2.1) is written in vector notation. Thus we introduce the **vector** x with components x_i $(i = 1, \ldots, n)$ so that

$$x = \begin{bmatrix} x_1 \\ \vdots \\ x_n \end{bmatrix}.$$

The derivative of a vector-valued function $x(t)$ is x', and is defined to be the vector whose components are x_i' $(i = 1, \ldots, n)$, so that

$$x' = \begin{bmatrix} x_1' \\ \vdots \\ x_n' \end{bmatrix}.$$

Similarly we define the vector-valued function $f(x, t)$, a specified function of the vector x and t, to be the vector whose components are $f_i(x, t)$ $(i = 1, \ldots, n)$, so that

$$f(x,t) = \begin{bmatrix} f_1(x,t) \\ \vdots \\ f_n(x,t) \end{bmatrix}.$$

It follows at once that the first-order system (2.1) can be written in the compact form

$$x' = f(x,t). \tag{2.2}$$

If, in (2.2), $f = f(x)$ is explicitly independent of t, then (2.2) is said to be auto-nomous; otherwise it is said to be non-autonomous.

In developing the theory of (2.2) we shall need to use a convenient measure of the magnitude of x. This is the **norm**

$$|x| = \sum_{i=1}^{n} |x_i|.$$

We generally prefer to use this norm rather than the more conventional Euclidean length

$$\|x\| = \sqrt{\sum_{i=1}^{n} |x_i|^2}.$$

They are, of course, equivalent since it is readily shown that

$$\|x\| \leqslant |x| \leqslant \sqrt{n}\|x\|.$$

Throughout this text notions of continuity, differentiability, and so on, are understood to be defined with respect to this norm, unless explicitly stated otherwise.

We are now in a position to define more precisely what is meant by the statement that $x(t)$ is a solution of the system of differential equations (2.2). Let I be an open interval, $a < t < b$, of the real line, and let D be a domain (an open, connected set) in the space of vectors x. We shall suppose that $f(x,t)$ is a continuous function of x and t for the product domain in which $x \in D$ and $t \in I$. The solution $x(t)$ of (2.2) is required to be differentiable for $t \in I$, that is, $x(t)$ is continuous and has a continuous derivative on I. It is sometimes con-venient to state results for closed intervals I, $a \leqslant t \leqslant b$ or for intervals of the form $a < t \leqslant b$ or $a \leqslant t < b$. In each of these cases the definition of con-tinuity, or differentiability, is extended to include continuity, or differentiability, from the right at a, or from the left at b, as the case may require. For any such interval I the definition of a solution of (2.2) follows.

Definition: $x(t)$ is a **solution** of (2.2) for $t \in I$ if $x(t)$ is differentiable on I and
    (i) $x(t) \in D$ for $t \in I$;
    (ii) $x'(t) = f(x(t),t)$ for $t \in I$.

However, as we saw in section 1.1 there are generally many solutions of a differential equation. In order to be able to classify solutions, and to have a selection mechanism which can identify a particular solution, we adjoin to (2.2) the initial condition

$$x(t_0) = x_0, \qquad (2.3)$$

where $t_0 \in I$ and $x_0 \in D$ are specified constants. The initial-value problem for (2.2) can now be formulated as follows.

Initial-value problem: For some interval I containing t_0 and domain D containing x_0, to find a solution $x(t)$ of (2.2) satisfying the initial condition (2.3).

The **existence** of a solution to the initial-value problem, without any further restriction on the function $f(x,t)$, is guaranteed by the Cauchy–Peano theorem, which we state here without proof (see, for instance, Coddington and Levinson, 1955, p. 6; Hartman, 1973, p. 10; or Birkhoff and Rota, 1978, p. 166).

Theorem 1.1 (Cauchy–Peano): If $f(x,t)$ is continuous for $|t-t_0| \leq \alpha$ and $|x-x_0| \leq \beta$, and if $|f(x,t)| \leq M$ there, then there exists a solution to (2.2) and (2.3) for $|t-t_0| \leq \delta$, where $\delta = \min\{\alpha, \beta/M\}$.

This theorem, while establishing the existence of at least one solution, gives no information on the number of solutions to be expected. Nevertheless, on geometrical and intuitive grounds we expect that the initial-value problem will have a unique solution. Indeed, experience with elementary examples such as those discussed in section 1.1 suggests that the general solution of (2.2) will contain n arbitrary constants, and one method of determining those constants might be to specify the n components of x_0 at some particular t_0. More convincingly, (2.2) can be interpreted geometrically as specifying a direction field at each point (x,t) and, hence, the initial condition (2.3) determines a unique direction field at the point (x_0, t_0). This notion forms the basis of the Cauchy–Euler construction of a simple approximate solution. Let the interval $t_0 \leq t \leq t_0 + \alpha$ be divided into N parts by the sequence of points t_r $(r = 1, \ldots, N)$, where

$$t_0 < t_1 < \ldots < t_N = t_0 + \alpha.$$

Then for the initial condition (2.3) form the corresponding sequence of points x_r $(r = 1, \ldots, N)$, where

$$x_r = x_{r-1} + f(x_{r-1}, t_{r-1})(t_r - t_{r-1}) \quad (r = 1, \ldots, N).$$

Each x_r is given recursively in terms of x_{r-1}. Then construct the approximate solution $x^N(t)$ of (2.2) as follows,

$$x^N(t) = x_{r-1} + f(x_{r-1}, t_{r-1})(t - t_{r-1}) \quad (\text{for } t_{r-1} \leq t < t_r; r = 1, \ldots, N).$$

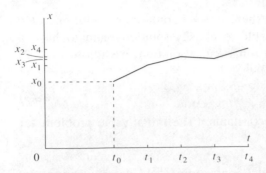

Fig. 1.1 Cauchy–Euler construction of an approximate solution of equation (2.2)

The construction is shown schematically in figure 1.1. This approximate solution satisfies the initial condition (2.3) and is continuous for all t $(t_0 \leqslant t \leqslant t_0 + \alpha)$ and is differentiable except at the set of points t_r $(r = 1, \dots, N)$. This simple construction is the basis of many numerical schemes for the approximate integration of (2.2), although in practice the numerical schemes currently in favour are considerably more accurate and sophisticated. The Cauchy–Euler construction can also form the basis of the proof of the Cauchy–Peano existence theorem by considering the limit $N \to \infty$, although the proof generally establishes only the convergence of a subsequence of the sequence of approximate solutions $x^N(t)$ to an exact solution of (2.2) and (2.3).

However, these intuitive arguments need rigorous substantiation as the following simple example demonstrates. Suppose that $n = 1$ and that we temporarily write x for the scalar component x_1. Then consider

$$x' = |x|^{2/3}, \qquad x(0) = 0.$$

Here $f(x, t) = |x|^{2/3}$ is a continuous function of x for all x (and all t), but

$$x = \tfrac{1}{27} t^3 \quad \text{and} \quad x \equiv 0,$$

are both solutions of the initial-value problem for all t. Indeed

$$x = \begin{cases} \tfrac{1}{27}(t-b)^3 & (b < t < \infty) \\ 0 & (a \leqslant t \leqslant b) \\ \tfrac{1}{27}(t-a)^3 & (-\infty < t < a) \end{cases}$$

is differentiable and satisfies the differential equation for all constants a and b such that $a \leqslant b$. If also $a \leqslant 0 \leqslant b$, then each member of this family of solutions satisfies the initial condition as well. In this instance the initial-value problem has a two-parameter family of solutions. It is also interesting to note that in this case the Cauchy–Euler construction will yield only the identically zero solution. We conclude from this, and similar anomalous examples, that for the initial-value problem to have a unique solution some further restrictions must be placed on $f(x, t)$. This will be taken up in section 1.3.

Moreover, Hadamard has pointed out that for a differential equation such as (2.2) to provide a meaningful representation of a physical system it is not only necessary for the solution of the initial-value problem to exist, it should also be unique and depend continuously on the initial data, (x_0, t_0), and on any external parameters which may be contained in the function $f(x, t)$. When these conditions are satisfied, we say that the initial-value problem is **well-set**. Solutions without these properties are essentially useless in a physical context as they cannot correspond in a sensible way to physical measurements. Hence we shall, in section 1.3, spend some time developing theorems which ensure that the differential equations are well-set in this sense.

However, before taking up these issues in section 1.3, we note that our selection of the first-order system (2.2) as the standard form of differential equation to be considered does not exclude application of the theory to higher-order equations. In general, any higher-order system of equations can be reduced to an equivalent first-order system. To illustrate the procedure consider the single nth-order equation for the unknown function $u(t)$:

$$u^{(n)} = g(u, u', \dots, u^{(n-1)}, t), \tag{2.4}$$

where g is a specified function of $u, u', \dots, u^{(n-1)}$ and t. Now define

$$x_1 = u, \qquad x_2 = u', \qquad \dots, \qquad x_n = u^{(n-1)}, \tag{2.5}$$

and form the equivalent first-order system

$$x_1' = x_2, \qquad x_2' = x_3, \qquad \dots, \qquad x_{n-1}' = x_n, \qquad x_n' = g(x_1, x_2, \dots, x_n, t). \tag{2.6}$$

This clearly has the same form as (2.1) or (2.2), and we conclude that any results obtained for the first-order system (2.2) have their counterparts for the nth-order equation (2.4). To complete the correspondence we note that corresponding to the initial condition (2.3), the transformation (2.5) determines the appropriate initial conditions for (2.4), which are

$$u(t_0) = u_1, \qquad u'(t_0) = u_2, \qquad \dots, \qquad u^{(n-1)}(t_0) = u_n. \tag{2.7}$$

Then x_0 in (2.3) is the vector with components $u_1, u_2, \dots, u_n$.

1.3 Uniqueness and existence theorems

We saw in section 1.2 that some further requirement of the function $f(x, t)$ in (2.2) is needed, in addition to the clearly necessary continuity conditions, if we are to be assured that the initial-value problem (i.e. (2.2) and (2.3)) has a unique solution. The most commonly used such requirement is the **Lipschitz condition**, which we formulate as follows.

Lipschitz condition: There exists a constant L such that

$$|f(x, t) - f(y, t)| \leq L|x - y|$$

uniformly for all $x, y \in D$ and $t \in I$.

The constant L is called the Lipschitz constant and, of course, is independent of x, y and t but may, in general, depend on the interval I and the domain D. We shall see that imposition of the Lipschitz condition is not only sufficient to guarantee uniqueness of the initial-value problem, but will also be useful in proving existence theorems and in establishing other properties of the solution of (2.2).

In seeking to establish whether or not a given function $f(x,t)$ satisfies a Lipschitz condition one can, of course, appeal directly to the definition above. However, this is often quite tedious and difficult and the following lemma provides a simple condition which suffices for most purposes.

Lemma: If $f(x,t)$ is a continuous function of x and t in the product domain for which $x \in D$ and $t \in I$, and the partial derivative $\partial f/\partial x$ exists and is bounded for all x in the convex domain D and all $t \in I$, then f satisfies a Lipschitz condition with the Lipschitz constant

$$L = \max_{D,I} \left| \frac{\partial f}{\partial x} \right|.$$

Here a convex domain D is one for which the line segment joining any two points in D lies entirely in D. The partial derivative $\partial f/\partial x$ is the $n \times n$ matrix

$$\frac{\partial f}{\partial x} = \left[\frac{\partial f_i}{\partial x_j} \right],$$

and the norm of the matrix is defined to be

$$\left| \frac{\partial f}{\partial x} \right| = \sum_{i=1}^{n} \sum_{j=1}^{n} \left| \frac{\partial f_i}{\partial x_j} \right|.$$

The proof of the lemma follows directly from the mean-value theorem, which here states that

$$f(x,t) - f(y,t) = \frac{\partial f}{\partial x}(\xi,t)(x-y),$$

where ξ is some point on the line segment joining x and y. The notation is condensed since ξ, as well as depending on x, y and t, also depends on i (the component of f) and j (the component of $x-y$). Note that the right-hand side here is the matrix product of the matrix $\partial f/\partial x$ with the vector $x-y$. Since the domain D is convex, the point ξ lies in D and it follows that

$$|f(x,t) - f(y,t)| \leq \left| \frac{\partial f}{\partial x}(\xi,t) \right| |x-y|.$$

The Lipschitz condition follows immediately, with the Lipschitz constant being that stated in the lemma.

It is interesting to note here that the function $f(x,t) = |x|^{2/3}$ (for the case $n = 1$, where we again temporarily write x for the scalar component x_1) satisfies a Lipschitz condition in any bounded closed region which excludes $x = 0$, but cannot satisfy a Lipschitz condition in any domain which includes $x = 0$ since $|x|^{2/3} \geqslant M|x|$ for any constant M as $|x| \to 0$. In section 1.2 we showed that the initial-value problem

$$x' = |x|^{2/3}, \qquad x(0) = 0,$$

did not have a unique solution, and this can now be attributed to the failure of the Lipschitz condition at $x = 0$. This example also illustrates the fact that, although the function $f(x,t)$ is in general a nonlinear function of x, the Lipschitz condition implies that $f(x,t)$ can always be bounded by a linear function of x, and it is essentially this property which makes the Lipschitz condition so useful.

As a preliminary to proving existence and uniqueness theorems, it is useful to reformulate the initial-value problem (2.2) and (2.3) as an equivalent integral equation,

$$x(t) = x_0 + \int_{t_0}^{t} f(x(s),s) \, \mathrm{d}s. \tag{3.1}$$

Indeed, if $x(t)$ is a solution of (2.2) with the initial condition (2.3), then a simple integration of (2.2) leads immediately to (3.1). Conversely, if $x(t)$ is a continuous function of t which satisfies (3.1) for $t \in I$, then clearly $x(t_0) = x_0$. Also $f(x(t),t)$ is then a continuous function of t and, since its indefinite integral is differentiable, (3.1) shows that $x(t)$ is differentiable, and differentiation leads immediately to (2.2). In terms of the integral equation (3.1), the initial-value problem can be reformulated as follows.

Initial-value problem: For some interval I containing t_0 and domain D containing x_0, to find a continuous function $x(t)$ which satisfies (3.1) for $t \in I$.

It is significant here that it is sufficient to find only a continuous function which satisfies (3.1), as satisfaction of (3.1) implies that $x(t)$ is then also differentiable. It is partly this fact which makes (3.1) more useful to work with than the original differential equation.

We shall begin by establishing the uniqueness of the initial-value problem. As a preliminary, we shall prove the following lemma.

Gronwall's lemma: Let $r(t)$ be continuous for $|t - t_0| \leqslant \delta$ and satisfy the inequalities

$$0 \leqslant r(t) \leqslant \epsilon + \delta \left| \int_{t_0}^{t} r(s) \, \mathrm{d}s \right|$$

for some non-negative constants ϵ and δ. Then

$$0 \leqslant r(t) \leqslant \epsilon \exp\{\delta |t - t_0|\}.$$

Proof: It is sufficient to consider the case $t \geq t_0$. The proof for the case $t \leq t_0$ is analogous. Let

$$R(t) = \int_{t_0}^{t} r(s) \, ds.$$

Then $R(t)$ is continuous and differentiable, and $R' = r$. Hence the inequality takes the form

$$0 \leq R' \leq \epsilon + \delta R.$$

Now multiply both sides by the non-negative function $\exp\{-\delta(t - t_0)\}$, so that

$$0 \leq [R \exp\{-\delta(t - t_0)\}]' \leq \epsilon \exp\{-\delta(t - t_0)\}.$$

Integration of this inequality from t_0 to t then shows that

$$0 \leq R \exp\{-\delta(t - t_0)\} \leq \frac{\epsilon}{\delta}[1 - \exp\{-\delta(t - t_0)\}],$$

or

$$0 \leq R \leq \frac{\epsilon}{\delta}[\exp\{\delta(t - t_0)\} - 1].$$

But,

$$0 \leq r = R' \leq \epsilon + \delta R,$$

or

$$0 \leq r \leq \epsilon \exp\{\delta(t - t_0)\},$$

which completes the proof.

As we shall see, Gronwall's lemma is extremely useful, not only in establishing the uniqueness theorems to follow below, but also in establishing a number of other results which appear later in this text.

    First we show that Gronwall's lemma can be used to establish the following theorem.

Theorem 1.2: Let $f(x, t)$ satisfy a Lipschitz condition with constant L for $x \in D$ and $|t - t_0| \leq \delta$. Let $x(t)$ and $y(t)$ be solutions of the initial-value problem for $|t - t_0| \leq \delta$ such that

$$x(t_0) = x_0, \qquad y(t_0) = y_0,$$

where $x_0, y_0 \in D$. Then

$$|x(t) - y(t)| \leq |x_0 - y_0| \exp\{L|t - t_0|\}. \tag{3.2}$$

Proof: In the integral equation formulation (3.1) it follows that

$$x(t) = x_0 + \int_{t_0}^{t} f(x(s), s) \, ds, \qquad y(t) = y_0 + \int_{t_0}^{t} f(y(s), s) \, ds.$$

Subtracting, we see that

$$x(t) - y(t) = x_0 - y_0 + \int_{t_0}^{t} [f(x(s), s) - f(y(s), s)] \, ds.$$

Taking the norm of both sides and applying the Lipschitz condition, it follows that

$$0 \leqslant |x(t) - y(t)| \leqslant |x_0 - y_0| + \left| \int_{t_0}^{t} L |x(s) - y(s)| \, ds \right|.$$

From Gronwall's lemma, with $r(t) = |x(t) - y(t)|$, $\epsilon = |x_0 - y_0|$ and $\delta = L$, the result follows.

Putting $x_0 = y_0$ in theorem 1.2 we see that $x(t) = y(t)$ for all $|t - t_0| \leqslant \delta$, thus establishing the uniqueness of the initial-value problem, whenever $f(x, t)$ satisfies a Lipschitz condition. This result is sufficiently important to be stated as a separate theorem.

Theorem 1.3: Let $f(x, t)$ satisfy a Lipschitz condition for $x \in D$ and $|t - t_0| \leqslant \delta$. Then the initial-value problem has a unique solution, that is, there is at most one continuous function $x(t)$ which satisfies (3.1).

Theorem 1.2 also shows that the solutions of the initial-value problem are continuous in the initial data, since it follows from (3.2) that if $y_0 \rightarrow x_0$, then $y(t) \rightarrow x(t)$ uniformly for all $|t - t_0| \leqslant \delta$. Thus, provided the Lipschitz condition holds, the initial-value problem is well-set. However, we shall not state this continuity result as a separate theorem here as a more general result will be proved in section 1.4. It is also significant to note here that continuity with respect to x_0 is generally limited to some bounded interval $|t - t_0| \leqslant \delta$, and that the exponential growth on the right-hand side of (3.2) cannot, in general, be improved. These comments are particularly relevant when we come to discuss questions of stability in chapter 4. The following simple example demonstrates these observations. Thus consider

$$x' = x, \qquad x(0) = x_0,$$

so that

$$x = x_0 e^t.$$

Here we have $n = 1$ and $f = x$, and we have again temporarily written x for the scalar component x_1. Clearly the Lipschitz constant $L = 1$ here and is the best choice for the inequality (3.2). Also, while there is continuity with respect to x_0 over any bounded interval, solutions corresponding to the initial values x_0 and y_0 diverge as $t \to \infty$ whenever $x_0 \neq y_0$, no matter how small the value of $|x_0 - y_0|$.

Next we shall take up the question of existence of a solution to the initial-value problem. While the Cauchy–Peano theorem (see section 1.2) gives an acceptable existence result, the use of the Lipschitz condition allows a relatively simple, constructive existence proof to be given, which also admits a number of useful extensions. The construction involves the iteration of the operator on the right-hand side of the integral equation, and is generally known as Picard iteration.

Theorem 1.4 (Picard iteration): Let $f(x, t)$ be continuous for

$$|t - t_0| \leq \alpha, \qquad |x - x_0| \leq \beta,$$

and satisfy a Lipschitz condition with constant L in this region. Let $|f(x, t)| \leq M$ there and let $\delta = \min\{\alpha, \beta/M\}$. Then the initial-value problem (3.1) has a unique solution for $|t - t_0| \leq \delta$.

Proof: Uniqueness of the solution has already been established in theorem 1.3. The proof of the existence of a solution proceeds in a number of steps.

(i) First we define the following sequence of iterates:

$$x^0(t) = x_0,$$

$$x^1(t) = x_0 + \int_{t_0}^{t} f(x^0(s), s) \, ds,$$

$$\vdots$$

$$x^{N+1}(t) = x_0 + \int_{t_0}^{t} f(x^N(s), s) \, ds. \tag{3.3}$$

The method of constructing these iterates is clear. The first approximation, x^0, is formed by neglecting the integral term in (3.1). The second approximation, x^1, is then formed by using x^0 in the integral term as a correction term. This process is then repeated. However, before proceeding, we must show that each iterate stays within the domain of definition of $f(x, t)$, which is clearly necessary if the next iterate is to be defined. We show that the iterates are defined for $|t - t_0| \leq \delta$ and satisfy the inequalities

$$|x^{N+1}(t) - x_0| \leq \beta \quad (\text{for } |t - t_0| \leq \delta). \tag{3.4}$$

First, it is trivial that

$$|x^0(t) - x_0| = 0 \leq \beta \quad (\text{for } |t - t_0| \leq \delta).$$

As the induction hypothesis, assume that

$$|x^r(t) - x_0| \leq \beta \quad (\text{for } |t - t_0| \leq \delta; r = 0, 1, \ldots, N).$$

Then, from (3.3),

$$|x^{N+1}(t) - x_0| \leq \left| \int_{t_0}^{t} |f(x^N(s), s)| \, ds \right|$$

$$\leq M|t - t_0|$$

$$\leq \beta \quad (\text{for } |t - t_0| \leq \delta).$$

This completes the induction.

(ii) Next we show that the sequence of iterates is uniformly and absolutely convergent. From (3.3) it follows that

$$x^{N+1}(t) - x^N(t) = \int_{t_0}^{t} \{f(x^N(s), s) - f(x^{N-1}(s), s)\} \, ds.$$

Taking the norm of both sides and using the Lipschitz condition, we get

$$|x^{N+1}(t) - x^N(t)| \leq L \left| \int_{t_0}^{t} |x^N(s) - x^{N-1}(s)| \, ds \right|. \tag{3.5}$$

This sequence of inequalities can be solved as follows. First, it follows from part (i) that

$$|x^1(t) - x^0(t)| \leq M|t - t_0|,$$

where we recall that $x^0(t) = x_0$. Then the induction hypothesis is that

$$|x^r(t) - x^{r-1}(t)| \leq ML^{r-1} \frac{|t - t_0|^r}{r!} \quad (r = 1, \ldots, N).$$

Hence it follows from (3.5) that

$$|x^{N+1}(t) - x^N(t)| \leq \frac{ML^N}{N!} \left| \int_{t_0}^{t} |s - t_0|^N \, ds \right|$$

$$= ML^N \frac{|t - t_0|^{N+1}}{(N+1)!}.$$

Now $x^N(t)$ can be written as the partial sum

$$x^0(t) + \sum_{r=1}^{N} \{x^r(t) - x^{r-1}(t)\} = x^N(t),$$

and the series on the left-hand side is dominated, term by term, by the series

$$x_0 + \sum_{r=1}^{N} ML^{r-1} \frac{|t-t_0|^r}{r!}.$$

As $N \to \infty$, this series is uniformly and absolutely convergent to

$$x_0 + \frac{M}{L}[\exp\{L|t-t_0|\} - 1].$$

Hence, by the comparison test, $x^N(t)$ converges uniformly for $|t-t_0| \leq \delta$ to a limit function, $x(t)$ say.

(iii) From (3.3), using induction, it can be seen that each iterate $x^N(t)$ is a continuous function of t for $|t-t_0| \leq \delta$. Hence, since $x^N(t)$ converges uniformly to $x(t)$, it follows that $x(t)$ is a continuous function of t for $|t-t_0| \leq \delta$. Similarly, it follows from (3.4) as $N \to \infty$ that

$$|x(t) - x_0| \leq \beta \quad \text{(for } |t-t_0| \leq \delta\text{)}.$$

It remains to show that $x(t)$ satisfies (3.1). This is established by letting $N \to \infty$ in (3.3). On the left-hand side $x^{N+1}(t) \to x(t)$, while on the right-hand side

$$\int_{t_0}^{t} f(x^N(s), s) \, ds \to \int_{t_0}^{t} f(x(s), s) \, ds \quad \text{(as } n \to \infty\text{)}.$$

This last result is established as follows. Using the Lipschitz condition, it follows that

$$\left| \int_{t_0}^{t} \{f(x(s), s) - f(x^N(s), s) \, ds \right| \leq \left| \int_{t_0}^{t} L|x(s) - x^N(s)| \, ds \right|$$

$$\leq L\delta \max_{|t-t_0| \leq \delta} |x(t) - x^N(t)|,$$

and the right-hand side now tends to zero as $N \to \infty$.

This completes the proof of theorem 1.4. A significant and useful feature of this proof is that the iterates $x^N(t)$ converge uniformly to $x(t)$, and we shall see later how this is utilized. The proof is also constructive in that it provides a method of obtaining approximate solutions to any required degree of accuracy. However, from this point of view, it is rarely of practical value as the .iterates usually converge too slowly to be useful. It should also be noted that theorem 1.4 is only a local existence theorem in that, whatever the original interval of definition, $|t-t_0| \leq \alpha$, in general the solution is only guaranteed to exist in a smaller interval, $|t-t_0| \leq \delta$, where $\delta \leq \alpha$. The following simple example illustrates this. Thus consider

$$x' = tx^3, \qquad x(0) = 1,$$

where $n = 1$ and we are again temporarily writing x for the scalar component

x_1. Here $f(x,t) = tx^3$ and is a continuous function for all x and t, and satisfies a Lipschitz condition in any bounded region. Hence the interval of definition can be as large as desired, but the solution is $x = (1-t^2)^{-1/2}$ and is only defined for $|t| < 1$. However, in chapter 2, we shall show that linear systems (i.e. $f(x,t)$ is a linear function of x) are exceptional in that there is then a global existence theorem.

Up to this point we have, in the main, been considering the solutions of (3.1) for fixed values of x_0 and t_0. However, in general, we can regard the solution as a function of x_0 and t_0 as well, and write

$$x = x(t; x_0, t_0).$$

We have already noted that it is a consequence of theorem 1.2 that x is a continuous function of x_0. A similar argument may be developed to show that x is a continuous function of both x_0 and t_0. However, we shall not develop this approach any further here, as a more general result will be proved in section 1.4. Finally, we note that if t_0 is held fixed, but x_0 is chosen arbitrarily, subject only to the restriction that $x_0 \in D$, then we have in effect constructed a solution which contains n arbitrary constants, namely the n components of x_0. Further, if x_0 is mapped in a one-to-one way into another set of n arbitrary constants, then we have constructed a general solution of (2.2). This is defined to be a solution which contains n arbitrary constants. Note that the uniqueness theorem prevents us from allowing t_0 to vary arbitrarily simultaneously with x_0 and, hence, there can be no more than n arbitrary constants. However, although the concept of a general solution is sometimes useful, we should recall again that theorem 1.4 is only a local existence theorem and, hence, we must normally expect a general solution to exist in only a certain interval about the point t_0, the length of the interval being dependent on the arbitrary constants. To illustrate this, consider

$$x' = tx^3, \qquad x(0) = x_0,$$

where $n = 1$ and we are again temporarily writing x for the scalar component x_1. The solution is $x = x_0(1 - x_0^2 t^2)^{-1/2}$ and is only defined for $|t| < x_0^{-1}$. Here x_0 can be regarded as an arbitrary constant and, as x_0 varies, the solutions fill the entire x–t plane. This general solution is shown in figure 1.2. Nevertheless, for each particular x_0, the corresponding unique solution is defined only over an interval whose size depends on x_0. However, linear systems form an important special case and we shall show in chapter 2 that general solutions then exist globally.

1.4 Dependence on parameters, and continuation

In many practical problems the dynamical system which the differential equation (2.2) describes contains external parameters, as well as the dependent variable

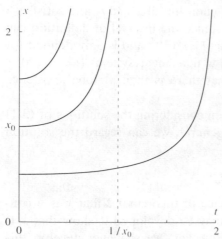

Fig. 1.2 Plots of the solution of $x' = tx^3$, $x(0) = x_0$, for different values of x_0.

x. We let the vector μ, with m components $\mu_1, \ldots, \mu_m$, represent these parameters, and replace (2.2) with

$$x' = f(x, t; \mu), \qquad (4.1)$$

where μ varies over a domain D_μ and $f(x, t; \mu)$ is required to be a continuous function of x, t and μ for $x \in D$, $t \in I$ and $\mu \in D_\mu$. The initial condition for (4.1) is again (2.3), and theorems 1.3 and 1.4 show that, for each fixed μ, there exists a unique solution which now, however, depends on μ as well. It is clearly desirable now to be able to determine the behaviour of the solution with respect to the parameter μ and, in particular, whether continuity of $f(x, t; \mu)$ with respect to μ is sufficient to guarantee that the solution is also continuous with respect to μ. More generally, the solution of the initial-value problem for (4.1) will depend on x_0 and t_0 as well, and we write

$$x = x(t; \mu, x_0, t_0).$$

Our aim now is to establish conditions under which x is a continuous function of t, μ, x_0 and t_0. When this is the case the solution of the initial-value problem is **well-set** in the Hadamard sense (see the discussion in section 1.2). First, we must extend the notion of the Lipschitz condition as follows.

Lipschitz condition: There exists a constant L such that

$$|f(x, t; \mu) - f(y, t; \mu)| \leq L|x - y|$$

uniformly for all $x, y \in D$, $t \in I$ and $\mu \in D_\mu$.

The Lipschitz constant L is independent of x, y, t **and** μ.

In order to incorporate the dependence of the solution on x_0 and t_0, as well as on μ, we adopt the device of allowing x_0 and t_0 to depend on μ so that $x_0 = x_0(\mu)$ and $t_0 = t_0(\mu)$ are continuous functions of μ. This is not a restriction as it can always be achieved by simply extending the number of parameters to include x_0 and t_0. Hence we now write $x = x(t;\mu)$, where it is understood that the dependence on μ may include the dependence of the solution on x_0 and t_0. Then the counterpart of theorem 1.4 for (4.1) is as follows.

Theorem 1.5: Let $f(x,t;\mu)$ be continuous for

$$|t-\tau| \le \alpha, \qquad |x-\xi| \le \beta, \qquad |\mu-\mu_0| \le \gamma$$

and satisfy a Lipschitz condition with constant L in this region. Let $x_0(\mu)$ and $t_0(\mu)$ be continuous functions of μ for $|\mu-\mu_0| \le \gamma$ such that $x_0(\mu_0) = \xi$ and $t_0(\mu_0) = \tau$. Then there exists a $\delta > 0$ and $\epsilon > 0$ such that the initial-value problem

$$x' = f(x,t;\mu), \qquad x(t_0(\mu)) = x_0(\mu)$$

has a unique solution for $|t-\tau| \le \delta$ and $|\mu-\mu_0| \le \epsilon$ which is a continuous function of t and μ.

Proof: As for (2.2) we first convert the initial-value problem for (4.1) into an equivalent integral equation,

$$x(t;\mu) = x_0(\mu) + \int_{t_0(\mu)}^{t} f(x(s;\mu),s;\mu)\, ds.$$

The proof is analogous to that of theorem 1.4. Thus we define the iterates

$$x^0(t;\mu) = x_0(\mu), \qquad \ldots, \qquad x^{N+1}(t;\mu) = x_0(\mu) + \int_{t_0(\mu)}^{t} f(x^N(s;\mu),s;\mu)\, ds.$$

The first step is to choose regions over which t and μ can vary such that each iterate remains within the domain of definition of $f(x,t;\mu)$. Let

$$\delta = \tfrac{1}{2}\min\{\alpha, \beta/M\},$$

where M is such that $|f(x,t;\mu)| \le M$ for $|t-\tau| \le \alpha$, $|x-\xi| \le \beta$ and $|\mu-\mu_0| \le \gamma$. Then choose ϵ ($0 < \epsilon \le \gamma$) so that, for $|\mu-\mu_0| \le \epsilon$, $|t_0(\mu)-\tau| < \tfrac{1}{2}\delta$ and $|x_0(\mu)-\xi| < \tfrac{1}{4}\beta$. It follows that, if $|t-\tau| < \delta$ then

$$|t-t_0(\mu)| \le |t-\tau| + |t_0(\mu)-\tau| \le \tfrac{3}{2}\delta.$$

Next, just as in the proof of theorem 1.4, it may be shown

$$|x^{N+1}(t;\mu) - x_0(\mu)| \le M|t-t_0(\mu)| \le \tfrac{3}{4}\beta \quad (\text{for } |t-t_0(\mu)| \le \tfrac{3}{2}\delta).$$

But then, for $|t-\tau| \le \delta$, it follows that

$$|x^{N+1}(t;\mu) - \xi| \le |x^{N+1}(t;\mu) - x_0(\mu)| + |x_0(\mu) - \xi| \le \beta.$$

Thus each iterate is well-defined for $|t-\tau| \le \delta$ and $|\mu-\mu_0| \le \epsilon$. Next, it can be shown, exactly as in theorem 1.4, that

$$|x^{N+1}(t;\mu) - x^N(t;\mu)| \leqslant ML^N \frac{|t - t_0(\mu)|^{N+1}}{(N+1)!}$$

and, hence, the sequence $x^{N+1}(t;\mu)$ is uniformly convergent to a limit function $x(t;\mu)$. The key new result here is that the continuity of $f(x,t;\mu)$ in x, t and μ, and the continuity of $x_0(\mu)$ and $t_0(\mu)$, ensures that each iterate is a continuous function of t and μ for $|t - \tau| \leqslant \delta$ and $|\mu - \mu_0| \leqslant \epsilon$. Hence, since the convergence is uniform, $x(t;\mu)$ is also a continuous function of t and μ. The rest of the proof, that $x(t,\mu)$ solves the initial-value problem, now follows exactly as in theorem 1.4. Note that the key feature of this extension of theorem 1.4 is that uniform convergence preserves continuity.

A number of other results may also be proved which establish conditions under which the solution may have some further degree of smoothness, such as differentiability with respect to t and μ (see, for instance, Coddington and Levinson, 1955, pp. 22–32). We shall give only one such result here.

Theorem 1.6: Let $f(x,t;\mu)$ be analytic for $|t - t_0(\mu)| \leqslant \alpha$, $|x - x_0(\mu)| \leqslant \beta$ and $|\mu - \mu_0| \leqslant \gamma$ and satisfy a Lipschitz condition with constant L in this region. Let $x_0(\mu)$ and $t_0(\mu)$ be analytic functions of μ for $|\mu - \mu_0| \leqslant \gamma$ such that $x_0(\mu_0) = \xi$ and $t_0(\mu_0) = \tau$. Then there exists a $\delta > 0$ and $\epsilon > 0$ such that the initial-value problem

$$x' = f(x,t;\mu), \qquad x(t_0(\mu)) = x_0(\mu)$$

has a unique solution for $|t - \tau| \leqslant \delta$ and $|\mu - \mu_0| \leqslant \epsilon$ which is an analytic function of t and μ.

Here a function is said to be analytic with respect to a set of variables if it has a convergent power-series expansion in those variables. We shall not give the proof here (but see, for instance, Coddington and Levinson, 1955, pp. 32–37; or Birkhoff and Rota, 1978, pp. 158–160), but note that it is analogous to that of theorem 1.5 and depends on the property that uniform convergence preserves analyticity. Indeed, in the statement of theorem 1.5, we have simply replaced continuous with analytic to obtain the statement of theorem 1.6, and the corresponding proofs are likewise analogous. Theorem 1.6 will be important in subsequent chapters as it will enable us to construct approximate solutions by power-series expansions with respect to a parameter, a procedure first used extensively by Poincaré.

To conclude this section we consider the question of the continuation of a solution and, for this purpose, we omit any possible dependence on the parameter μ. We noted at the conclusion of theorem 1.4 that the existence of a solution is guaranteed by this theorem generally only for some interval about the initial point. In other words, it is only a **local** existence theorem, establishing the existence of a local solution. However, we shall show that this local solution can be extended into a **global** solution which is defined up to the boundary of the domain of definition of $f(x, t)$.

Theorem 1.7: Let $f(x, t)$ be continuous for x in the domain D and t in the open interval I, and satisfy a Lipschitz condition there. Then, for any point $x_0 \in D$ and $t_0 \in I$, the initial-value problem

$$x' = f(x, t), \qquad x(t_0) = x_0$$

has a unique solution $x(t)$ which is defined for $t_0 \leqslant t < T$ $(T \leqslant \infty)$ and is such that, if $T < \infty$, then either $|x(t)| \to \infty$ as $t \to T$ or $(x(t), t)$ approaches the boundary of the product domain, (D, I), as $t \to T$.

Proof: First, we can conclude from theorem 1.4 that there exists a unique solution $x(t)$ to the initial-value problem for $t_0 \leqslant t \leqslant t_1$, where $t_1 \in I$ and $x_1 = x(t_1) \in D$. We may now choose (x_1, t_1) as a new initial condition and apply theorem 1.4 again to show that $x(t)$ may be continued uniquely to the interval $t_1 \leqslant t \leqslant t_2$, where $t_2 \in I$ and $x_2 = x(t_2) \in D$. In this manner we may generate a sequence of points $t_0, t_1, \ldots, t_R$ and $x_0, x_1, \ldots, x_R$, where $x_R = x(t_R)$, $t_R \in I$, $x_R \in D$ and the solution $x(t)$ exists and is unique for $t_0 \leqslant t \leqslant t_R$. Since $t_0 < t_1 < \ldots < t_R$, by the monotonic-sequence theorem, there exists T $(\leqslant \infty)$ such that $t \to T$ as $R \to \infty$. Hence the solution $x(t)$ exists and is unique for $t_0 \leqslant t < T$.

If T is finite (i.e. $T < \infty$), consider the sequence x_R as $R \to \infty$. Either $|x_R| \to \infty$, or the sequence is bounded. In the former case, we conclude that $|x(t)| \to \infty$ as $t \to T$. In the latter case, the sequence (x_R, t_R) is bounded, and hence, by the Bolzano–Weierstrass theorem contains a limit point (X, T), where both X and T are finite. It remains to show that (X, T) must be on the boundary of the product domain (D, I).

Suppose, on the contrary, that (X, T) lies inside the boundary of the product domain (D, I). Then there exist $\delta > 0$ and $\epsilon > 0$ such that $|t - T| \leqslant \delta$ is in I and $|x - X| \leqslant \epsilon$ is in D. Further, there exists $M > 0$ such that $|f(x, t)| \leqslant M$ for $|t - T| \leqslant \delta$ and $|x - X| \leqslant \epsilon$. Since (X, T) is a limit point of $(x(t), t)$, we may choose T_1 and T_2 so that $|T_i - T| \leqslant \delta$ and $|x(T_i) - X| \leqslant \epsilon$ $(i = 1, 2)$, where, of course, $t_0 \leqslant T_i < T$. But (3.1) holds for $t_0 \leqslant t < T$ and, hence,

$$|x(T_1) - x(T_2)| = \left| \int_{T_1}^{T_2} f(x(s), s) \, ds \right|$$

$$\leqslant M |T_1 - T_2|.$$

Thus $|x(T_1) - x(T_2)| \to 0$ uniformly as $T_1, T_2 \to T$ and, by the Cauchy criterion for convergence, it follows that $x(t)$ converges as $t \to T$. Since $x(t)$ is continuous, it follows that $x(t) \to X$ as $t \to T$. But we can now use (X, T) as a new initial condition and, by theorem 1.4, the solution can be continued uniquely for

$t > T$, which contradicts the result that T is a limit point for the sequence $t_0, t_1, \ldots, t_R$.

Theorem 1.7 shows that the solution $x(t)$ to the initial-value problem either exists for all t ($t_0 \leq t < \infty$), or there exists T ($< \infty$) such that the solution exists only for $t_0 \leq t < T$. In this latter case, either $|x(t)| \to \infty$ as $t \to T$, or $(x(t), t)$ approaches the boundary of the domain of definition of $f(x, t)$. The following examples illustrate that all three cases can occur. For all of the examples $n = 1$, and we temporarily write x for the scalar component x_1. First consider

$$x' = x, \qquad x(0) = 1,$$

which has the solution $x = e^t$. Here the solution exists for all t ($0 \leq t < \infty$). Next consider

$$x' = x^2, \qquad x(0) = 1,$$

which has the solution $x = (1 - t)^{-1}$. Here $T = 1$, the solution exists for $0 \leq t < 1$ and $|x| \to \infty$ as $t \to 1$. The final two examples are cases when the solution approaches the boundary of the domain of definition of $f(x, t)$. Consider

$$x' = \frac{x}{t} - \frac{1}{2\sqrt{|t|}}, \qquad x(-1) = 0,$$

which has the solution $x = -t - \sqrt{|t|}$. Here $t_0 = -1$ and the domain of definition of $f(x, t)$ is all x and $-\infty < t < 0$. In this case $T = 0$, $X = 0$ and the solution, although remaining finite, is not differentiable as $t \to 0$. Finally consider

$$x' = -\frac{1}{x}, \qquad x(0) = 1,$$

which has the solution $x = \sqrt{1 - 2t}$. Here $t_0 = 0$ and the domain of definition of $f(x, t)$ is $x > 0$ and all t. In this case $T = \frac{1}{2}$, $X = 0$ and, as above, the solution is not differentiable at the limit point.

Problems

1 Suppose that here x is a 2-vector with components x_1, x_2.
   (a) Show that

$$f(x, t) = \begin{bmatrix} 1 + x_1 \\ x_2^2 \end{bmatrix}$$

satisfies a Lipschitz condition when x lies in any bounded domain D (e.g. $|x| < M$, where M is a constant), but cannot satisfy a Lipschitz condition for all x.
   (b) Show that

$$f(x, t) = \begin{bmatrix} \sqrt{|x_1|} \\ x_2 \end{bmatrix}$$

is continuous for all x, but does not satisfy a Lipschitz condition in any domain D which contains $x = 0$.

2 For the first-order system

$$x_1' = x_2, \qquad x_2' = -\sin x_1 - x_2|x_2| + \cos t,$$

show that the right-hand side satisfies a Lipschitz condition in the domain $|x| \leq \beta$ for $|t| \leq \alpha$, where α and β are arbitrary but finite numbers. Use theorem 1.4 to deduce that the initial-value problem

$$x_1(0) = 0, \qquad x_2(0) = 0$$

has a unique solution for $|t| \leq \delta$, and obtain an estimate for δ. By allowing α and β to be as large as possible, attempt to improve your estimate of δ.

3 Letting x be a scalar variable (i.e. $n = 1$), show that the initial-value problem

$$x' = \sqrt{|x|}, \qquad x(0) = 0,$$

has the two solutions

$$x = 0, \quad \text{and} \quad x = \tfrac{1}{4}t^2 \operatorname{sign} t.$$

Generalize, and show that this initial-value problem has a two-parameter family of solutions. Comment in relation to theorem 1.4.

4 Write the single nth-order equation

$$u^{(n)} = g(u, u', \dots, u^{(n-1)}, t)$$

as a first-order system and, hence, establish the counterpart of theorem 1.4 for the initial-value problem.

5 (Extended Gronwall's lemma) Let $r(t)$ and $v(t)$ be continuous for $|t - t_0| \leq \delta$, where $v(t) \geq 0$, and suppose that

$$0 \leq r(t) \leq \epsilon + \delta \left| \int_{t_0}^{t} v(s) r(s) \, ds \right|.$$

Show that

$$r(t) \leq \epsilon \exp \left\{ \delta \left| \int_{t_0}^{t} v(s) \, ds \right| \right\}.$$

6 In each of the following initial-value problems x is a scalar variable (i.e. $n = 1$). In each case verify that the specified function is the solution and comment in relation to theorem 1.7.

(a) $x = \exp\{-\tfrac{1}{2}t^2\}$ solves

$$x' = -tx, \qquad x(0) = 1.$$

Here x exists for all t.

(b) $x = (1 - 2t)^{-1/2}$ solves

$$x' = x^3, \qquad x(0) = 1.$$

Here $x \to \infty$ as $t \to \tfrac{1}{2}$.

(c) $x = (1-t)^{\frac{1}{3}}$ solves

$$x' = -\frac{1}{3x^2}, \qquad x(0) = 1.$$

Here $x \to 0$ (the boundary of the domain of definition of the right-hand side) as $t \to 1$.

(d) $x = \sqrt{1-t}$ solves

$$x' = \frac{x}{2(t-1)}, \qquad x(0) = 1.$$

Here $x \to 0$ as $t \to 1$ (the boundary of the domain of definition of the right-hand side).

CHAPTER TWO
LINEAR EQUATIONS

2.1 Uniqueness and existence theorem for a linear system

Although this text is largely concerned with nonlinear equations, we shall find it necessary to have at our disposal some key results for linear equations, and these form the subject matter of this chapter. The first-order system (2.2) is said to be a linear system whenever $f(x, t)$ is a linear function of x. Thus the general form for a first-order linear system is

$$x' = A(t)x + g(t), \tag{1.1}$$

where $A(t)$ is an $n \times n$ matrix function of t, and $g(t)$ is a vector function of t. In component form,

$$x_i' = \sum_{j=1}^{n} A_{ij}(t)x_j + g_i(t) \quad (i = 1, \ldots, n), \tag{1.2}$$

where $A_{ij}(t)$ are the elements $A(t)$, $g_i(t)$ are the components of $g(t)$ and we recall that x_i are the components of the vector x.

We shall assume that $A(t)$ and $g(t)$ are continuous functions of t on the closed interval I, $a \leqslant t \leqslant b$. Let the right-hand side of (1.1) be denoted by $f(x, t)$, so that here

$$f(x, t) = A(t)x + g(t).$$

Hence $f(x, t)$ is a continuous function of x and t for all x and $t \in I$. Further, since $A(t)$ is a continuous function of t on the closed interval I, there exists a constant M such that $|A(t)| \leqslant M$ for $t \in I$. It follows that $f(x, t)$ satisfies a Lipschitz condition with constant M for all x and $t \in I$. Indeed

$$f(x, t) - f(y, t) = A(t)(x - y),$$

and so

$$|f(x, t) - f(y, t)| \leqslant M |x - y|.$$

Here we should recall that the norm of the matrix is defined to be

$$|A| = \sum_{i=1}^{n} \sum_{j=1}^{n} |A_{ij}|,$$

and it can readily be shown that

$$|Ax| \leqslant |A||x|. \tag{1.3}$$

Thus $f(x,t)$, as defined by the right-hand side of (1.1), satisfies all the conditions needed for the validity of theorems 1.2, 1.3 and 1.4 (section 1.3) and, hence, we can conclude that there exists a unique solution to the initial-value problem for (1.1), where the initial condition is, for $t_0 \in I$,

$$x(t_0) = x_0. \tag{1.4}$$

Note that there is now no restriction on x_0, as here $f(x,t)$ is defined for all x. However, for the linear system (1.1) a stronger result is available, and the following global existence theorem holds.

Theorem 2.1: Let $A(t)$ and $g(t)$ be continuous functions of t for $t \in I$, $a \leqslant t \leqslant b$. Then there exists a unique solution to the initial-value problem (1.4) for (1.1) for $t \in I$.

Proof: The uniqueness part of this theorem is an immediate consequence of theorem 1.3 (section 1.3) since we have established above that $f(x,t)$, as defined by the right-hand side of (1.1), satisfies all the conditions of that theorem. The significant new aspect here which distinguishes this theorem from the more general case of theorem 1.4 (section 1.3) is that existence is established over the whole interval I. Thus, in contrast to that result, this theorem establishes a global result.

The proof of existence proceeds along the same lines as the proof of theorem 1.4 of (section 1.3), with the important proviso that step (i) in the proof is no longer needed. First, we define the sequence of iterates (3.3) as before, where now $f(x,t)$ is defined by the right-hand side of (1.1). Since this $f(x,t)$ is defined for all x, each iterate in (3.3) is now defined for all $t \in I$. Hence step (i) in the proof of theorem 1.4 (section 1.3) is not necessary and can be omitted. We proceed immediately to steps (ii) and (iii) of that proof. This part is unchanged, and we can show that the iterates converge uniformly to a solution of the initial-value problem for all $t \in I$.

Thus for the linear equation (1.1) solutions are defined over the whole interval I where $A(t)$ and $g(t)$ are continuous. Solutions can fail to exist, that is, the solution ceases to be differentiable, only for those values of t where $A(t)$ and $g(t)$ fail to be continuous. Hence, these points can be identified by inspection of the equation. In particular, if $A(t)$ and $g(t)$ are continuous for all t $(-\infty < t < \infty)$, then solutions exist for all t. Also, we can now construct a general solution by fixing $t_0 \in I$ and allowing x_0 to vary arbitrarily without restriction. Hence, for the linear equation (1.1), general solutions, containing n arbitrary constants, will exist for all $t \in I$.

Next we note that theorems 1.5 and 1.6 (section 1.4) have their counterparts for the linear system (1.1). Thus if $A(t;\mu)$ and $g(t;\mu)$ are continuous (respectively analytic) functions of t and the parameter μ, then the solutions of (1.1) are likewise continuous (respectively analytic) functions of t and μ, the result now holding for all $t \in I$.

2.2 Homogeneous linear systems

The first-order system (1.1) is said to be homogeneous whenever $g(t)$ is identically zero. Thus the general form for a first-order homogeneous linear system is

$$x' = A(t)x, \tag{2.1}$$

where $A(t)$ is an $n \times n$ matrix function of t, continuous for $t \in I$, $a \leq t \leq b$. The homogeneous system (2.1) has two important properties. The first is that the identically zero function, $x(t) = 0$ for all $t \in I$, is a solution of (2.1), and is the unique solution such that $x(t_0) = 0$ for any $t_0 \in I$. The second is that, if $x^1(t)$ and $x^2(t)$ are solutions of (2.1), then so is

$$x(t) = c_1 x^1(t) + c_2 x^2(t)$$

for any two scalar constants c_1 and c_2. Both properties are readily established by inspection and are a consequence of the fact that the right-hand side of (2.1) is a homogeneous, linear function of x.

Thus the set of all solutions of (2.1) on an interval I form a vector space. In order to determine the dimension of this space we shall need the following definition.

Definition: A set of m functions $x^1(t), \ldots, x^m(t)$ are **linearly dependent** on an interval I if there exist scalar constants $c_1, \ldots, c_m$, not all zero, such that

$$c_1 x^1(t) + \ldots + c_m x^m(t) = 0 \quad \text{(for } t \in I).$$

Otherwise they are **linearly independent**.

Note that the definition must hold for all $t \in I$. Thus, for instance, with $n = 2$, the vectors

$$\begin{bmatrix} 1 \\ t \end{bmatrix}, \quad \begin{bmatrix} t^2 \\ 1 \end{bmatrix}$$

are linearly independent vector functions of t for t in any interval I, but at $t = 1$ both reduce to

$$\begin{bmatrix} 1 \\ 1 \end{bmatrix}$$

and, hence, are linearly dependent 2-vectors for this particular value of t. With this definition of linear independence we can show that theorem 2.1 (section 2.1) implies the following result.

Theorem 2.2: The set of all solutions of (2.1) on an interval I form an n-dimensional vector space.

Proof: We have already shown above that the set of all solutions form a vector space. Next we show that there exist n linearly independent solutions. Let $d^1, \ldots, d^n$ be n linearly independent n-vectors and, using theorem 2.1 (section 2.1), let $x^r(t)$ be the unique solution of (2.1) such that

$$x^r(t_0) = d^r \quad (r = 1, \ldots, n),$$

where $t_0 \in I$. Then $x^1(t), \ldots, x^n(t)$ are linearly independent on I. For suppose that the contrary holds and there exist constants $c_1, \ldots, c_n$, such that

$$c_1 x^1(t) + \ldots + c_n x^n(t) = 0 \quad (\text{for } all \ t \in I).$$

Then, putting $t = t_0$, it follows that

$$c_1 d^1 + \ldots + c_n d^n = 0.$$

But, since $d^1, \ldots, d^n$ are linearly independent n-vectors, it follows that $c_1 = 0$, $\ldots$, $c_n = 0$. This establishes that the dimension of the vector space is at least n.

To show that the dimension is exactly n, we next demonstrate that any solution $x(t)$ of (2.1) can be written as a linear combination of $x^1(t), \ldots, x^n(t)$. Indeed, given $x(t)$, let $x_0 = x(t_0)$, where $t_0 \in I$. Then, since $d^1, \ldots, d^n$ are linearly independent n-vectors, there exists a unique set of constants $c_1, \ldots, c_n$ such that

$$x_0 = c_1 d^1 + \ldots + c_n d^n.$$

Now consider

$$c_1 x^1(t) + \ldots + c_n x^n(t).$$

By construction, this is a solution of (2.1) and satisfies the initial condition $x(t_0) = x_0$. But theorem 2.1 (section 2.1) states that there is a unique solution to the initial-value problem and, hence, it must be identically equal to $x(t)$. We have shown that

$$x(t) = c_1 x^1(t) + \ldots + c_n x^n(t)$$

for all $t \in I$. In effect, theorem 2.1 allows us to transfer linear independence of the initial conditions at t_0 to the solution for all $t \in I$.

Theorem 2.2 shows that the general solution of (2.1) is

$$x(t) = c_1 x^1(t) + \ldots + c_n x^n(t), \tag{2.2}$$

where $x^1(t), \ldots, x^n(t)$ are n linearly independent solutions and $c_1, \ldots, c_n$ are n arbitrary constants. As an illustration, consider the example

$$x_1' = x_2, \qquad x_2' = -x_1, \tag{2.3}$$

where $n = 2$ and the 2×2 matrix A is, for all t,

$$\begin{bmatrix} 0 & 1 \\ -1 & 0 \end{bmatrix}.$$

Methods for solving systems such as (2.3) will be developed in section 2.5. It can be shown that two linearly independent solutions are

$$x^1 = \begin{bmatrix} \cos t \\ -\sin t \end{bmatrix}, \qquad x^2 = \begin{bmatrix} \sin t \\ \cos t \end{bmatrix}$$

and the general solution is given by (2.2) with $n = 2$, or

$$\begin{aligned} x_1 &= c_1 \cos t + c_2 \sin t, \\ x_2 &= -c_1 \sin t + c_2 \cos t. \end{aligned}$$

It is also instructive to consider the example

$$x_1' = \frac{2t^2 x_1 - 2t x_2}{t^3 - 1}, \qquad x_2' = \frac{-x_1 + t^2 x_2}{t^3 - 1}, \tag{2.4}$$

where again $n = 2$ and the 2×2 matrix A is, for all $t \neq 1$,

$$\frac{1}{t^3 - 1} \begin{bmatrix} 2t^2 & -2t \\ -1 & t^2 \end{bmatrix}.$$

Here $A(t)$ is continuous for all $t \neq 1$, and two linearly independent solutions are

$$x^1 = \begin{bmatrix} 1 \\ t \end{bmatrix}, \qquad x^2 = \begin{bmatrix} t^2 \\ 1 \end{bmatrix},$$

which are defined for all t, even though $A(t)$ is not defined at $t = 1$. The general solution is

$$\begin{aligned} x_1 &= c_1 + c_2 t^2, \\ x_2 &= c_1 t + c_2. \end{aligned}$$

A convenient notation for the homogeneous linear system (2.1) is to let a set of n solutions form an $n \times n$ matrix as follows.

Definition: Let $x^1(t), \dots, x^n(t)$ be n solutions of (2.1) on an interval I, and put

$$X(t) = [x^1(t), \dots, x^n(t)], \tag{2.5}$$

where $X(t)$ is an $n \times n$ matrix solution of

$$X' = AX. \tag{2.6}$$

If $x^1, \dots, x^n$ are also linearly independent, then X is a **fundamental matrix** and, if $X(t_0) = E$, the unit matrix, then $X(t)$ is the **principal** fundamental matrix. Further

$$W(t) = \det X(t) \tag{2.7}$$

is called the **Wronskian**.

The property (2.6) follows immediately from the definition (2.5) and (2.1). Further, if $X(t)$ is a fundamental matrix solution of (2.6), then so is $X(t)C$ for any non-singular constant matrix C. Indeed, let

$$Y(t) = X(t)C.$$

Then $Y(t)$ is non-singular, and

$$Y' = X'C = AXC = AY.$$

Note that the columns of Y are linear combinations of the columns of X. Also the general solution (2.2) can be written in the form

$$x(t) = X(t)c, \tag{2.8}$$

where c is an arbitrary n-vector, with components $c_1,\ldots,c_n$. The following theorem gives some useful results for the Wronskian.

Theorem 2.3: Let $W(t)$ be defined by (2.5), (2.6) and (2.7). Suppose that, for some $t_0 \in I$, $W(t_0) = 0$. Then $x^1(t),\ldots,x^n(t)$ are linearly dependent and $W(t) = 0$ for all $t \in I$. Conversely, if $W(t_0) \neq 0$ for some $t_0 \in I$, then $x^1(t),\ldots,x^n(t)$ are linearly independent and $W(t) \neq 0$ for all $t \in I$. Further

$$W(t) = W(t_0)\exp\left\{\int_{t_0}^{t} \operatorname{tr} A(s)\ \mathrm{d}s\right\}. \tag{2.9}$$

Proof: Of course both parts of the theorem follow from (2.9). However, we shall give an independent proof and then establish (2.9). Let $x^r(t_0) = d^r$. Then if $W(t_0) = 0$ it follows that $d^1,\ldots,d^n$ are linearly dependent n-vectors and, hence, there exist constants $c_1,\ldots,c_n$, not all zero, such that

$$c_1 d^1 + \ldots + c_n d^n = 0.$$

Now consider, for $t \in I$,

$$x(t) = c_1 x^1(t) + \ldots + c_n x^n(t).$$

By construction, this is a solution of (2.1) and satisfies the initial condition $x(t_0) = 0$. But theorem 2.1 (section 2.1) implies that $x(t) = 0$ for all $t \in I$ is the unique solution with this initial condition and, hence, $x^1(t),\ldots,x^n(t)$ are linearly dependent, so that $W(t) = 0$ for all $t \in I$. Conversely, if $W(t_0) \neq 0$, then $d^1,\ldots,d^n$ are linearly independent n-vectors and, just as in the proof of theorem 2.2, it follows that $x^1(t),\ldots,x^n(t)$ are linearly independent, so that $W(t) \neq 0$ for all $t \in I$.

To establish (2.9), we first observe that, as $t \to t_0$,

$$X(t) = X(t_0) + (t - t_0) X'(t_0) + o(t - t_0),$$

or

$$X(t) = X(t_0) + (t - t_0) A(t_0) X(t_0) + o(t - t_0),$$

where we have used (2.6) to calculate $X'(t_0)$. But then, using the definition (2.7) for $W(t)$, it follows that

$$W(t) = W(t_0) \det[E + (t - t_0) A(t_0)] + o(t - t_0)$$

or

$$W(t) = W(t_0)\{1 + (t - t_0)\operatorname{tr} A(t_0)\} + o(t - t_0),$$

where $\operatorname{tr} A$ for an $n \times n$ matrix A is the sum of the diagonal elements. Now we can also write

$$W(t) = W(t_0) + (t - t_0) W'(t_0) + o(t - t_0).$$

Hence, on taking the limit $t \to t_0$, we see that

$$W'(t_0) = W(t_0)\operatorname{tr} A(t_0).$$

But t_0 can be any point in I and, hence, for all t in I,

$$W' = W \operatorname{tr} A.$$

Integration with respect to t now gives (2.9).

As an illustration, consider the example (2.3). In this case

$$X = \begin{bmatrix} \cos t & \sin t \\ -\sin t & \cos t \end{bmatrix}, \qquad W = 1.$$

Here X is a fundamental matrix and, since $X(0) = E$, is the principal fundamental matrix. In this example $\operatorname{tr} A = 0$ and so the Wronskian is a constant, equal to its value at $t = 0$. By contrast, for example (2.4),

$$X = \begin{bmatrix} 1 & t^2 \\ t & 1 \end{bmatrix}, \qquad W = 1 - t^3,$$

and X is a fundamental matrix for $t \neq 1$, since $W \neq 0$ for $t \neq 1$. Indeed, X is a principal fundamental matrix since $X(0) = E$. Here $\operatorname{tr} A = 3t^2(t^3 - 1)^{-1}$ and the Wronskian is not constant. We see that $W = 0$ when $t = 1$, at which point X is a singular matrix. Of course, this does not contradict theorem 2.3 since $A(t)$ is not defined at $t = 1$, and, hence, this point must be excluded from the interval I on which theorem 2.3 holds.

2.3 Inhomogeneous linear systems

We now return to consider (1.1) when $g(t)$ is not necessarily zero, in which case (1.1) is said to be inhomogeneous. Let x_H be any solution of the homogeneous equation (2.1) and let x_P be a particular solution of (1.1). Then

$$x = x_H + x_P \tag{3.1}$$

is also a solution of the inhomogeneous equation (1.1). Indeed

$$x' = x'_H + x'_P,$$
$$= Ax_H + Ax_P + g,$$
$$= Ax + g.$$

Further, choosing x_H to be the general solution of (2.1) (e.g., given by (2.8)), it follows that (3.1) is the general solution of (1.1).

Hence, to solve the inhomogeneous equation (1.1) it is sufficient to find a particular solution x_P. To achieve this, suppose that x is a solution of (1.1) and put

$$x = Xy,$$

where X is a fundamental matrix for the homogeneous equation (2.1), and so satisfies (2.6). Then

$$x' = X'y + Xy',$$
$$= AXy + Xy',$$
$$= Ax + Xy'.$$

Comparing this with (1.1) it follows that

$$Xy' = g,$$

or

$$y = c + \int_a^t X^{-1}(s)g(s)\,\mathrm{d}s,$$

where c is an arbitrary constant vector, and we recall that the equation (1.1) is defined over the interval $a \leqslant t \leqslant b$. It follows that the general solution of (1.1) is

$$x = X(t)c + X(t) \int_a^t X^{-1}(s)g(s)\,\mathrm{d}s. \tag{3.2}$$

Here the first term, Xc, can be recognized as x_H, the general solution of the homogeneous equation (2.1), and the second term is a particular solution x_P. Note that here we have chosen the lower terminal of the integral to be $t = a$ so that $x_P(a) = 0$, but clearly any other choice is permissible. This method of constructing a particular solution x_P is sometimes called 'variation-of-parameters', since, in effect, we have replaced the constant vector c in (2.8), the general solution of the homogeneous equation, by the new variable y. The expression (3.2) shows that to solve the inhomogeneous equation (1.1) it is sufficient to have available a fundamental matrix for the homogeneous equation (2.1).

To aid in interpreting x_P we define

$$G_0(t,s) = \begin{cases} 0 & (t \leqslant s \leqslant b), \\ X(t)\,X^{-1}(s) & (a \leqslant s < t). \end{cases} \tag{3.3}$$

Then the particular solution x_P in (3.2) can be written as

$$x_P(t) = \int_a^b G_0(t,s)\,g(s)\ \mathrm{d}s. \tag{3.4}$$

Here $G_0(t,s)$ is called a Green's matrix for (1.1). It satisfies the matrix equation (2.6) as a function of t for each s, provided that $t \neq s$. As $t \to s$ from above and below we see that

$$\left[G_0(t,s)\right]_{t=s-}^{t=s+} = E,$$

where we recall that E is the unit matrix. Thus $G_0(t,s)$ is discontinuous at $t = s$ and the discontinuity is E. Also $G_0(t,s)$ satisfies the initial condition

$$G_0(a,s) = 0 \quad (\text{for } a < s \leqslant b).$$

Indeed $G_0(t,s)$ is the unique solution of the matrix equation (2.6) with this prescribed initial condition and the specified discontinuity at $t = s$. It can be interpreted as the response to a point impulse at $t = s$, and the particular solution (3.4) can then be regarded as the superposition of this response with $g(s)$. Further, this interpretation can be described in terms of the Dirac delta-function $\delta(t-s)$, since $G_0(t,s)$ as a function of t, for each fixed s, satisfies the matrix equation

$$G' = AG + \delta(t-s)\,E.$$

Indeed any solution of this equation is a Green's matrix, $G(t,s)$, and the general solution is

$$G(t,s) = X(t)\,C(s) + G_0(t,s),$$

where $C(s)$ is an arbitrary function of s. In the discussion above we chose $C(s) = 0$ so that the prescribed initial condition could be satisfied, but other choices are possible. Clearly $G(t,s)$, like $G_0(t,s)$, is discontinuous at $t = s$ and the discontinuity is again E. Also if $G_0(t,s)$ is replaced by $G(t,s)$ in (3.4), then the resulting expression is again a particular solution of (1.1), although, of course, the result is a different particular solution to (3.4).

Chapter 2

2.4 Second-order linear equations

Consider the single second-order linear equation for the unknown function $u(t)$

$$u'' + a_1(t) u' + a_2(t) u = f(t), \tag{4.1}$$

where $a_1(t)$, $a_2(t)$ and $f(t)$ are continuous functions of t on the closed interval I ($a \leq t \leq b$). Using the procedure described in section 1.2 (see (2.5) and (2.6)), this may be converted to an equivalent first-order system. Thus let

$$x_1 = u, \qquad x_2 = u' \tag{4.2}$$

and form the equivalent first-order system

$$x_1' = x_2, \qquad x_2' = -a_1(t) x_2 - a_2(t) x_1 + f(t). \tag{4.3}$$

This is the linear system of the form (1.1), with $n = 2$ and the matrix $A(t)$ and vector $g(t)$ given by

$$A(t) = \begin{bmatrix} 0 & 1 \\ -a_2 & -a_1 \end{bmatrix}, \qquad g(t) = \begin{bmatrix} 0 \\ f \end{bmatrix}. \tag{4.4}$$

Clearly $A(t)$ and $g(t)$ are continuous functions of t for $t \in I$. Hence all the results of section 2.1, 2.2 and 2.3 can be translated into corresponding results for (4.1).

The initial-value problem for (4.1) is derived from (1.4) and (4.2),

$$u(t_0) = u_0, \qquad u'(t_0) = u_1. \tag{4.5}$$

It then follows from theorem 2.1 (section 2.1) that there is a unique solution to the initial-value problem for $t \in I$. Next, from the theory of section 2.2 and section 2.3 applied to the system (4.3), it follows that the general solution of (4.1) is

$$u = c_1 u^1(t) + c_2 u^2(t) + u_P(t), \tag{4.6}$$

where $u^1(t)$ and $u^2(t)$ are two linearly independent solutions of the homogeneous equation

$$u'' + a_1(t) u' + a_2(t) u = 0 \tag{4.7}$$

and $u_P(t)$ is a particular solution of the inhomogeneous equation (4.1). Then, using the transformation (4.2) and the definition (2.5), we see that the fundamental matrix for the equivalent first-order system is

$$X = \begin{bmatrix} u^1 & u^2 \\ u^{1\prime} & u^{2\prime} \end{bmatrix}. \tag{4.8}$$

The Wronskian of (4.8) is $\det X$ (see (2.7)) and hence is given by

$$W = u^1 u^{2\prime} - u^2 u^{1\prime}. \tag{4.9}$$

Theorem 2.3 (section 2.2) then shows that $u^1(t)$ and $u^2(t)$ are linearly independent if and only if $W(t) \neq 0$ for all $t \in I$. Also, since here $\mathrm{tr}\, A = -a_1$ (see (4.4)), it follows that (2.9) becomes

$$W(t) = W(t_0)\exp\left\{ -\int_{t_0}^{t} a_1(s) \; ds \right\}. \tag{4.10}$$

The particular solution $u_P(t)$ can be obtained from (3.3) and (3.4). Thus, from (4.8) we see that

$$X^{-1} = \frac{1}{W}\left[\begin{array}{cc} u^{2'} & -u^2 \\ -u^{1'} & u^1 \end{array} \right]. \tag{4.11}$$

Using (3.3) and (3.4) we are now in a position to calculate the particular solution $u_P(t)$ in terms of u^1, u^2 and f. We note that here the vector g is given by (4.4) and hence is non-zero only in the last entry, while u_P corresponds to the first component of x_P (see (4.8)). It follows that

$$u_P(t) = \int_{a}^{b} U_0(t,s)f(s) \; ds,$$

where $U_0(t,s)$ is the $(1,2)$-element of $G_0(t,s)$ and is called a Green's function for (4.1). It is given by

$$U_0(t,s) = \left\{ \begin{array}{ll} 0 & (t < s \leq b), \\ \dfrac{1}{W(s)}[u^2(t)u^1(s) - u^1(t)u^2(s)] & (a \leq s < t). \end{array} \right.$$

This result is an immediate consequence of the definition (3.3) and the expressions (4.8) and (4.11). Finally, substituting this expression for $U_0(t,s)$ into the expression for $u_P(t)$ we find that

$$u_P(t) = \int_{a}^{t} \frac{f(s)}{W(s)}[u^2(t)u^1(s) - u^1(t)u^2(s)] \; ds. \tag{4.12}$$

It can now be readily verified by direct differentiation that $u_P(t)$ is indeed a particular solution of (4.1). Note that

$$u_P'(t) = \int_{a}^{t} \frac{f(s)}{W(s)}[u^{2'}(t)u^1(s) - u^{1'}(t)u^2(s)] \; ds$$

and that $u_P(t)$ is the unique solution of (4.1) which satisfies the initial conditions $u_P(a) = u_P'(a) = 0$.

By way of illustration, consider the example

$$u'' + u = f(t) \tag{4.13}$$

for which the equivalent first-order system for the homogeneous equation is (2.3). Two linearly independent solutions of the homogeneous equation are

$$u^1 = \cos t, \quad u^2 = \sin t,$$

and a simple calculation shows that W (4.9) is 1. The expression (4.16) for $u_P(t)$ simplifies to

$$u_P(t) = \int_a^t f(s)\sin(t-s)\ ds$$

and the general solution of (4.13) is

$$u = c_1 \cos t + c_2 \sin t + \int_a^t f(s)\sin(t-s)\ ds. \tag{4.14}$$

Note that the general solution to the homogeneous equation

$$u'' + u = 0$$

is

$$u = c_1 \cos t + c_2 \sin t,$$

which is periodic with period 2π, i.e. $u(t+2\pi) = u(t)$ for all t. Subsequently in this text we shall need to consider under what conditions the general solution of the inhomogeneous equation (4.13) is also a periodic function, given that $f(t)$ is periodic with period 2π. The following theorem gives the required necessary and sufficient conditions for this to be the case.

Theorem 2.4: If $f(t)$ is periodic with period 2π, so that

$$f(t+2\pi) = f(t)\quad \text{(for all } t),$$

then the general solution of (4.13) is also periodic with period 2π if and only if

$$-\int_0^{2\pi} f(s)\sin s\ ds = \int_0^{2\pi} f(s)\cos s\ ds = 0. \tag{4.15}$$

Proof: The general solution of (4.13) is periodic if and only if

$$u(t+2\pi) = u(t)\quad \text{(for all } t).$$

This implies that

$$u(2\pi) = u(0),\qquad u'(2\pi) = u'(0), \tag{4.16}$$

where we first put $t = 0$ and, secondly, differentiate with respect to t and then put $t = 0$.

Conversely, if the conditions (4.16) hold, then the initial-value problem posed at $t = 2\pi$ is identical to that posed at $t = 0$. Further, since $f(t+2\pi) = f(t)$, it follows from the existence and uniqueness theorem 2.1 that the initial conditions at $t = 2\pi$ generate the same solution as the initial conditions at $t = 0$, provided t is replaced by $t+2\pi$. Hence $u(t)$ is periodic.

Next we apply the conditions (4.16) to the general solution (4.14), where, without loss of generality, we may choose $a = 0$. The result (4.15) follows immediately. Importantly for developments later in this text the conditions in (4.15) correspond term-by-term to the two conditions (4.16), since it is readily seen that

$$u(2\pi) - u(0) = -\int_0^{2\pi} f(s)\sin s \ ds, \qquad u'(2\pi) - u'(0) = \int_0^{2\pi} f(s)\cos s \ ds.$$
(4.17)

An alternative proof proceeds directly from the general solution (4.14) since clearly $u(t)$ is periodic if and only if

$$\int_t^{t+2\pi} f(s)\sin(t-s) \ ds = 0.$$

That this condition is equivalent to (4.15) is left as an exercise, although it should be noted that the lemma in theorem 3.1 of section 3.1 will be useful.

Returning to the general theory, a similar development can be carried out for the single nth-order linear equation

$$u^{(n)} + a_1(t)u^{(n-1)} + \ldots + a_n(t)u = f(t),$$
(4.18)

since the transformation (2.5) (section 1.2) converts (4.18) into an equivalent first-order linear system of the form (1.1) for a vector x with n components. The details are left to the exercises.

2.5 Linear equations with constant coefficients

Here we consider the special case of (1.1) when the coefficient matrix is independent of t. Such equations occur often in applications, and also will form the basis for much of our development of the theory of nonlinear equations in this text. Hence it will be useful to summarize some useful results here. Linear equations with constant coefficients are notable because, in contrast to the general situation for differential equations, it is in principle possible to construct solutions explicitly. First we consider the homogeneous equation (2.1), where we now write

$$x' = Ax,$$
(5.1)

since the coefficient matrix A is now a constant matrix and hence is defined for all t. Thus, from theorem 2.1, these solutions are also defined for all t.

The solution procedure is to substitute

$$x = e^{\lambda t}u$$
(5.2)

into (5.1), where u is a constant vector. This will be a solution if

$$Au = \lambda u, \tag{5.3}$$

and hence λ is an eigenvalue of A and u is the corresponding eigenvector. There are n eigenvalues $\lambda_1, \ldots, \lambda_n$, which are the zeros of

$$\det[A - \lambda E] = 0, \tag{5.4}$$

and we let $u^1, \ldots, u^n$ be the corresponding eigenvectors. Thus we have obtained the solutions to (5.1),

$$x^1 = e^{\lambda_1 t} u^1, \qquad \ldots, \qquad x^n = e^{\lambda_n t} u^n. \tag{5.5}$$

When $u^1, \ldots, u^n$ are linearly independent eigenvectors, these n solutions are linearly independent and can be used to form a fundamental matrix (2.5). This is the generic case and occurs, in particular, whenever the eigenvalues are distinct.

When the eigenvalues are real-valued, this procedure will yield a family of real-valued solutions. However, complex-valued eigenvalues may also occur and, in this case, we need to modify the procedure to obtain real-valued solutions. Since A is a real-valued matrix, complex-valued eigenvalues will occur in complex-conjugate pairs. Thus if λ is an eigenvalue, and u the corresponding eigenvector, it is a consequence of (5.3) that $\bar{\lambda}$ is also an eigenvalue with $\bar{u}$ the corresponding eigenvector. The solution procedure described above remains valid so that $ue^{\lambda t}$ and $\bar{u}e^{\bar{\lambda}t}$ are both solutions of (5.1), but are now complex-valued solutions. To obtain two real-valued solutions we first add this pair of solutions and then subtract them after multiplying by i. This procedure is equivalent to taking the real and imaginary parts of $ue^{\lambda t}$. Thus let

$$\lambda = \alpha + i\beta, \qquad u = v + iw, \tag{5.6}$$

where α, β, v and w are real-valued. It follows that

$$e^{\lambda t} = e^{\alpha t}(\cos \beta t + i \sin \beta t).$$

Hence we can conclude that

$$\operatorname{Re}(ue^{\lambda t}) = e^{\alpha t}(\cos \beta t \, v - \sin \beta t \, w), \qquad \operatorname{Im}(ue^{\lambda t}) = e^{\alpha t}(\sin \beta t \, v + \cos \beta t \, w),$$
$$\tag{5.7}$$

are both real-valued solutions of (5.1), and are linearly independent if $\beta \neq 0$. Thus, for every complex-conjugate pair of eigenvalues, a pair of solutions of the form (5.7) replaces the corresponding pair of complex-valued solutions in (5.5), and we obtain a set of n linearly independent real-valued solutions.

The algebraic results which underlie the procedures described above can also be used to transform equation (4.1) into a canonical form. Thus let

$$S = [u^1, \ldots, u^n]. \tag{5.8}$$

Then it follows from (5.3) that

$$AS = S\Lambda, \quad \text{where } \Lambda = \text{diag}[\lambda_1, \ldots, \lambda_n]. \tag{5.9}$$

Here Λ is the matrix which has $\lambda_1, \ldots, \lambda_n$ on its main diagonal and zeros elsewhere. When $u^1, \ldots, u^n$ are linearly independent eigenvectors, S is a nonsingular matrix, and we introduce the linear transformation

$$x = Sy, \tag{5.10}$$

where x is a solution of (5.1). Then

$$x' = Sy'$$

and

$$x' = Ax = ASy = S\Lambda y,$$

so that

$$y' = \Lambda y. \tag{5.11}$$

Equation (5.11) is the canonical form of (5.1) and can be solved directly, since (5.11) is equivalent to

$$y_1' = \lambda_1 y_1, \qquad \ldots, \qquad y_n' = \lambda_n y_n.$$

These equations are uncoupled and can be solved immediately. It follows that a fundamental matrix for (5.11) is

$$Y = \text{diag}[e^{\lambda_1 t}, \ldots, e^{\lambda_n t}].$$

When complex-valued eigenvalues occur this procedure must be modified if y is to be a real-valued variable. Suppose then that λ and $\bar{\lambda}$ are eigenvalues of A, with eigenvectors u and $\bar{u}$, and let $\lambda = \alpha + i\beta$ and $u = v + iw$ (see (5.6)). It is then a consequence of (5.3) that

$$Av = \alpha v - \beta w, \qquad Aw = \beta v + \alpha w. \tag{5.12}$$

The transformation matrix S is again defined by (5.8) with the proviso that each complex-conjugate pair of eigenvectors u and $\bar{u}$ in S is replaced by the corresponding pair of real vectors v and w. To illustrate this, consider the case $n = 2$, so that A is a 2×2 matrix with eigenvalues λ and $\bar{\lambda}$. Let

$$S = [v, w].$$

Then it follows from (5.12) that

$$AS = S\Omega, \quad \text{where } \Omega = \begin{bmatrix} \alpha & \beta \\ -\beta & \alpha \end{bmatrix}. \tag{5.13}$$

Again we introduce the linear transformation (5.10),

$$x = Sy,$$

where x is a solution of (5.1). Since S is real-valued, y is also a real-valued variable. Also v and w are linearly independent, and S is non-singular if $\beta \neq 0$. Then, using a similar procedure to that described above, we find that

$$y' = \Omega y. \tag{5.14}$$

This is the canonical form for a system of two equations when A has complex-valued eigenvalues. In component form it reduces to

$$y_1' = \alpha y_1 + \beta y_2, \qquad y_2' = -\beta y_1 + \alpha y_2. \tag{5.15}$$

Unlike the canonical form when the eigenvalues are real, these equations are coupled. Nevertheless it is readily shown that a fundamental matrix is

$$Y = e^{\alpha t} \begin{bmatrix} \cos \beta t & \sin \beta t \\ -\sin \beta t & \cos \beta t \end{bmatrix}.$$

In the general case, when A is an $n \times n$ matrix, the canonical form is again (5.11), but the expression (5.9) for Λ is replaced by a matrix whose only non-zero elements are diagonal entries, corresponding to real eigenvalues, and 2×2 sub-matrices of the form Ω centred on the main diagonal, corresponding to each complex-conjugate pair of eigenvalues.

As an illustration, let us consider the following two examples. First, suppose that

$$\begin{aligned} x_1' &= -3x_1 + 2x_2, \\ x_2' &= x_1 - 2x_2. \end{aligned}$$

The matrix A is

$$A = \begin{bmatrix} -3 & 2 \\ 1 & -2 \end{bmatrix},$$

and the eigenvalues, found from (5.4), are $\lambda_1 = -1$ and $\lambda_2 = -4$. The corresponding eigenvectors are

$$u^1 = \begin{bmatrix} 1 \\ 1 \end{bmatrix}, \qquad u^2 = \begin{bmatrix} -2 \\ 1 \end{bmatrix},$$

and two linearly independent solutions are

$$x^1 = e^{-t} \begin{bmatrix} 1 \\ 1 \end{bmatrix}, \qquad x^2 = e^{-4t} \begin{bmatrix} -2 \\ 1 \end{bmatrix}.$$

The general solution is

$$x_1 = c_1 e^{-t} - 2c_2 e^{-4t},$$

$$x_2 = c_1 e^{-t} + c_2 e^{-4t}.$$

The fundamental matrix corresponding to x^1 and x^2 is

$$X = \begin{bmatrix} e^{-t} & -2e^{-4t} \\ e^{-t} & e^{-4t} \end{bmatrix}.$$

Next, suppose that

$$x_1' = 5x_1 + 10x_2,$$
$$x_2' = -x_1 - x_2.$$

The matrix A is

$$\begin{bmatrix} 5 & 10 \\ -1 & -1 \end{bmatrix},$$

and the eigenvalues, found from (5.4), are $\lambda_1 = 2+i$ and $\lambda_2 = 2-i$. This is a complex-conjugate pair with $\alpha = 2$ and $\beta = 1$. The eigenvector corresponding to λ_1 is

$$u = \begin{bmatrix} 3+i \\ -1 \end{bmatrix}$$

and, hence, the real and imaginary parts are

$$v = \begin{bmatrix} 3 \\ -1 \end{bmatrix}, \qquad w = \begin{bmatrix} 1 \\ 0 \end{bmatrix}.$$

A complex-valued solution is

$$x = e^{2t}(\cos t + i \sin t) \begin{bmatrix} 3+i \\ -1 \end{bmatrix}.$$

Taking the real and imaginary parts we obtain two linearly independent real-valued solutions:

$$x^1 = e^{2t} \begin{bmatrix} 3\cos t - \sin t \\ -\cos t \end{bmatrix}, \qquad x^2 = e^{2t} \begin{bmatrix} 3\sin t + \cos t \\ -\sin t \end{bmatrix}.$$

The general solution is

$$x_1 = e^{2t}\{c_1(3\cos t - \sin t) + c_2(3\sin t + \cos t)\},$$
$$x_2 = e^{2t}\{-c_1 \cos t - c_2 \sin t\}.$$

The fundamental matrix corresponding to x^1 and x^2 is

$$X = e^{2t} \begin{bmatrix} 3\cos t - \sin t & 3\sin t + \cos t \\ -\cos t & -\sin t \end{bmatrix}.$$

The discussion so far has assumed that $u^1, \ldots, u^n$, real- or complex-valued, are linearly independent eigenvectors. This is the generic case and the one which occurs most often in practice. Nevertheless, we shall discuss briefly the procedure to be followed when this assumption fails. Suppose then that λ is an eigenvalue of A with algebraic multiplicity m (i.e. its multiplicity as a solution of (5.4)), while there are p linearly independent eigenvectors $u^1, \ldots, u^p$ corresponding to this eigenvalue (p is called the geometric multiplicity). Then, in general, $1 \leqslant p \leqslant m$. The generic case discussed above occurs when $p = m$, but here we shall suppose that $p < m$. Then there are p linearly independent solutions of (5.1) corresponding to the eigenvalue λ, each solution having the form (5.1), where u is one of the eigenvectors $u^1, \ldots, u^p$. Next, to increase the quota of solutions from p to m we replace the trial solution (5.2) with

$$x = e^{\lambda t}(v + tu). \tag{5.16}$$

Substitution into (5.1) then gives

$$\lambda(v + tu) + u = A(v + tu),$$

and, hence, this will be a solution if

$$Au = \lambda u \quad \text{and} \quad Av = \lambda v + u. \tag{5.17}$$

Eliminating u,

$$(A - \lambda E)^2 v = 0. \tag{5.18}$$

Suppose now that this equation has q linearly independent solutions for v: $v^1, \ldots, v^q$, such that $(A - \lambda E)v = u \neq 0$. Each vector v is called a generalized eigenvector of A. Note that u is always an eigenvector of A. Hence we have obtained an additional q linearly independent solutions of (5.1), each solution being of the form (5.16). It can be shown that $1 \leqslant q \leqslant m - p$, although the proof of this requires deep and intricate results from the theory of linear algebra which are beyond the scope of this text. If $p + q = m$, then we have obtained a full quota of solutions corresponding to the eigenvalue λ. However, if $p + q < m$ we must proceed further and replace (5.16) with the trial solution

$$x = e^{\lambda t}(w + tv + \tfrac{1}{2}t^2 u). \tag{5.19}$$

Substitution into (5.1) then gives

$$Au = \lambda u, \qquad Av = \lambda v + u, \qquad Aw = \lambda w + v. \tag{5.20}$$

Thus u is an eigenvector, v is a generalized eigenvector satisfying (5.18) and w is also a generalized eigenvector, now satisfying

$$(A - \lambda E)^3 w = 0. \tag{5.21}$$

This has r linearly independently solutions, where $1 \leqslant r \leqslant m - p - q$, and, hence, we have generated an additional r linearly independent solutions of (5.1). If $p + q + r < m$, the process is repeated. A basic theorem of linear algebra states that this procedure terminates. Indeed it can be shown that there exist m linearly independent generalized eigenvectors u satisfying

$$(A - \lambda E)^s u = 0 \tag{5.22}$$

for some integer s, and no smaller integer, where $1 \leqslant s \leqslant m - p + 1$. Of course, included amongst the solutions of (5.22) are the eigenvectors satisfying (5.3) and the generalized eigenvectors satisfying (5.18) and (5.21). The proof of this result is beyond the scope of this text. However, we can use the result to conclude that corresponding to each eigenvalue λ with multiplicity m, there exist m linearly independent solutions of (5.1) of the form

$$x^1 = e^{\lambda t}p^1(t), \qquad \ldots, \qquad x^m = e^{\lambda t}p^m(t), \tag{5.23}$$

where $p^1(t), \ldots, p^m(t)$ are m linearly independent vector-valued polynomials in t of degree $s-1$. When λ is complex-valued, $\bar{\lambda}$ is also an eigenvalue, and $2m$ real-valued linearly independent solutions can be obtained by taking the real and imaginary parts of (5.23).

To illustrate the procedure, consider the following example

$$\begin{aligned} x_1' &= x_1 + 2x_2 - x_3, \\ x_2' &= x_2 \quad , \\ x_3' &= -x_2 + 2x_3. \end{aligned}$$

The matrix A is

$$A = \begin{bmatrix} 1 & 2 & -1 \\ 0 & 1 & 0 \\ 0 & -1 & 2 \end{bmatrix},$$

and the eigenvalues are $\lambda_1 = 2$ and $\lambda_2 = \lambda_3 = 1$. The eigenvector corresponding to λ_1 is

$$u^1 = \begin{bmatrix} 1 \\ 0 \\ -1 \end{bmatrix}$$

and the corresponding linearly independent solution is

$$x^1 = e^{2t}u^1.$$

For the eigenvalues λ_2 and λ_3 the multiplicity $m = 2$, but there is just a single eigenvector (i.e. $p = 1$):

$$u^2 = \begin{bmatrix} 1 \\ 0 \\ 0 \end{bmatrix}$$

and the corresponding linearly independent solution is

$$x^2 = e^t u^2.$$

To find a third solution, we use the trial substitution (5.16) and find that (5.17), with $\lambda = 1$, becomes

$$\begin{bmatrix} 0 & 2 & -1 \\ 0 & 0 & 0 \\ 0 & -1 & 1 \end{bmatrix} v = \begin{bmatrix} 1 \\ 0 \\ 0 \end{bmatrix}.$$

This has the solution

$$v^2 = k \begin{bmatrix} 1 \\ 0 \\ 0 \end{bmatrix} + \begin{bmatrix} 0 \\ 1 \\ 1 \end{bmatrix},$$

where k is an arbitrary constant. Clearly the term proportional to k just produces the eigenvector u^2, and so we may set $k = 0$ without loss of generality. Thus the third linearly independent solution is

$$x^3 = e^t(v^2 + tu^2).$$

The general solution is

$$\begin{aligned} x_1 &= c_1 e^{2t} + (c_2 + c_3 t)e^t, \\ x_2 &= c_3 e^t, \\ x_3 &= -c_1 e^{2t} + c_3 e^t. \end{aligned}$$

In this particular example, this solution can also be obtained directly, as the equation for x_2 is uncoupled from those for x_1 and x_3, and, hence, x_2 can be found immediately. Then x_3 can be found as the equation for x_3 does not depend on x_1 and, with x_2 and x_3 known, x_1 can be found. We leave it as an exercise to verify that this leads to the solution obtained above.

An interesting alternative procedure for solving (5.1) is through the exponential of the matrix A. This is defined by

$$e^A = E + \sum_{m=1}^{\infty} \frac{A^m}{m!}. \tag{5.24}$$

The series is absolutely convergent using the matrix norm (1.2). It can then be shown that

$$X = e^{At} \tag{5.25}$$

is the unique solution of the matrix equation

$$X' = AX$$

such that $X(0) = E$. Indeed

$$e^{At} = E + \sum_{m=1}^{\infty} \frac{t^m A^m}{m!}, \tag{5.26}$$

where the series is uniformly and absolutely convergent in the matrix norm (1.2). Term-by-term differentiation then gives the required result. The general solution of (5.1) is then

$$x = e^{At} c,$$

where c is an arbitrary constant vector. While the expression (5.25) is apparently appealingly simple, it is not so useful in practice as explicit evaluation of (5.25) through (5.26) is a very inefficient procedure. Instead (5.25) is best evaluated through a transformation of the form

$$A = SBS^{-1},$$

where S is a non-singular matrix. It can readily be shown from (5.24) that

$$e^A = Se^B S^{-1}.$$

For instance, if A has eigenvalues $\lambda_1, \ldots, \lambda_n$ and n linearly independent eigenvectors $u^1, \ldots, u^n$, then we choose S to be given by (5.8) and use the transformation (5.9) to show that

$$e^{At} = Se^{\Lambda t}S^{-1}, \quad \text{where } e^{\Lambda t} = \text{diag}[e^{\lambda_1 t}, \ldots, e^{\lambda_n t}].$$

We leave it as an exercise to establish the second line of this result from (5.26). When an eigenvalue of A has an algebraic multiplicity greater than the geometric multiplicity, the evaluation of e^{At} is much more complicated, and we shall omit the details (but see, for instance, Coddington and Levinson, 1955, pp. 75–78, or Birkhoff and Rota, 1978, pp. 321–325).

Finally we turn to a consideration of the inhomogeneous equation

$$x' = Ax + g(t). \tag{5.27}$$

From (3.2) the general solution of (5.27) is

$$x = X(t)c + X(t)\int_a^t X^{-1}(s)g(s)\, ds,$$

where $X(t)$ is a fundamental matrix for (5.1). We now show that

$$X(t)X^{-1}(s) = X(t-s)X^{-1}(0)$$

and, hence, the general solution of (5.27) is

$$x = X(t)c + \int_a^t X(t-s)X^{-1}(0)g(s)\, ds. \tag{5.28}$$

To show this, we put

$$Y(t,s) = X(t)X^{-1}(s)$$

and observe that, considered as a function of t, for all s,

$$Y' = AY \quad \text{and} \quad Y(s,s) = E.$$

But also, let

$$Z(t,s) = X(t-s)X^{-1}(0).$$

Again, considered as a function of t, we see that

$$Z' = AZ \quad \text{and} \quad Z(s,s) = E.$$

Here, it is crucial that, because A is a constant matrix, the solutions of (5.1) are translation-invariant, so that

$$X'(t-s) = AX(t-s),$$

for any value of s. Because the solution of the initial-value problem

$$X' = AX, \qquad X(0) = E,$$

is unique, it follows that $Z = Y$ for all t and s. The result (5.28) will be of considerable use to us in subsequent parts of this text.

Problems

1 Find a fundamental matrix and the Wronskian for each of the following linear systems.
(a) $x_1' = x_2, x_2' = x_1$.
(b) $x_1' = \sin t x_2, x_2' = -x_2$.
(c) $x_1' = \dfrac{tx_1 - x_2}{t^2 - 1}, x_2' = \dfrac{-x_1 + tx_2}{t^2 - 1}$ (for $t \neq \pm 1$).
Explain why here the Wronskian is zero at $t = \pm 1$. [*Hint*: Find an equation for x_1 alone.]

2 Find the particular solution which vanishes at $t = 0$ for each of the following linear systems and, in each case, identify the Green's matrix $G_0(t, s)$.
(a) $x_1' = a(t)x_1 + b(t)$.
(b) $x_1' = x_2 + g_1(t), x_2' = -x_1 + g_2(t)$. [*Hint*: See (2.3).]
(c) $x_1' = x_2 + g_1(t), x_2' = x_1 + g_2(t)$. [*Hint*: See Problem 1(a).]

3 Show that the single nth-order linear homogeneous equation

$$u^{(n)} + a_1(t)u^{(n-1)} + \ldots + a_n(t)u = 0,$$

where $a_1(t), \ldots, a_n(t)$ are continuous functions of t for $a \leqslant t \leqslant b$, is equivalent to a first-order linear homogeneous system

$$x' = A(t)x,$$

and find the matrix A. Develop the analogues of theorems 2.2 and 2.3, and, hence, show that the general solution $u(t)$ is an arbitrary linear combination of n linearly independent solutions $u^1(t), \ldots, u^n(t)$. Obtain an expression for the Wronskian in terms of $u^1, \ldots, u^n$.

4 Show that the single nth-order linear inhomogeneous equation

$$u^{(n)} + a_1(t)u^{(n-1)} + \ldots + a_n(t)u = f(t),$$

where $a_1(t), \ldots, a_n(t)$ and $f(t)$ are continuous functions of t for $a \leqslant t \leqslant b$, is equivalent to a first-order linear inhomogeneous system of the form

$$x' = A(t)x + g(t),$$

and find the matrix A and the vector g. Use the analogue of (3.2) to show that a particular solution is

$$u_P(t) = \int_a^b U_0(t,s)f(s) \, ds,$$

where $U_0(t,s)$ is a Green's function, given by the properties:

(a) $U_0(t,s)$ satisfies the homogeneous equation for $t \neq s$;

(b) $U_0, \dfrac{\partial U_0}{\partial t}, \ldots, \dfrac{\partial^{(n-2)} U_0}{\partial t^{(n-2)}}$ are continuous at $t = s$;

(c) $\left[\dfrac{\partial^{(n-1)} U_0}{\partial t^{(n-1)}} \right]_{t=s-}^{t=s+} = 1.$

5 Find the general solution of the equation

$$u'' + \omega^2 u = f(t),$$

where ω is a real constant. [*Hint*: See (4.13) and (4.14).]

6 Find the general solution of

$$x' = Ax$$

in each of the following cases. Also find the fundamental matrix $X(t)$ such that $X(0) = E$.

(a) $A = \begin{bmatrix} 2 & 1 \\ -3 & 6 \end{bmatrix}$, (b) $A = \begin{bmatrix} -2 & 1 \\ -1 & -2 \end{bmatrix}$,

(c) $A = \begin{bmatrix} 0 & 1 \\ -1 & 2 \end{bmatrix}$. (d) $A = \begin{bmatrix} -3 & 4 \\ -2 & 1 \end{bmatrix}$,

(e) $A = \begin{bmatrix} 0 & -1 & 2 \\ 0 & 1 & 0 \\ 1 & 1 & -1 \end{bmatrix}$, (f) $A = \begin{bmatrix} 0 & 1 & 0 \\ 4 & 3 & -4 \\ 1 & 2 & -1 \end{bmatrix}$.

7 The canonical form for the linear system

$$x' = Ax,$$

where A is a 2×2 matrix with complex eigenvalues λ and $\bar{\lambda}$, and corresponding complex eigenvectors u and $\bar{u}$, is

$$x_1' = \alpha x_1 + \beta x_2,$$
$$x_2' = -\beta x_1 + \alpha x_2,$$

where $\lambda = \alpha + i\beta$ $(\beta \neq 0)$. (See (5.12) to (5.15).) Obtain the general solution and find the fundamental matrix $X(t)$ such that $X(0) = E$.

Also show that if z is the complex variable

$$z = x_1 - ix_2,$$

then

$$z' = \lambda z, \qquad \bar{z}' = \bar{\lambda}\bar{z}.$$

Explain this in terms of the complex-valued transformation

$$x = Sy,$$

where $S = [u, \bar{u}]$.

8 Find the general solution of

$$x_1' = x_2 + f_1(t), \qquad x_2' = -x_1 + f_2(t).$$

If $f_1(t)$ and $f_2(t)$ are both periodic with period 2π, show that the conditions for x_1 and x_2 to be periodic with period 2π are

$$\int_0^{2\pi} \{f_1(s)\cos s - f_2(s)\sin s\} \; ds = 0 \quad \text{and} \quad \int_0^{2\pi} \{f_1(s)\sin s + f_2(s)\cos s\} \; ds = 0.$$

CHAPTER THREE
LINEAR EQUATIONS WITH
PERIODIC COEFFICIENTS

3.1 Floquet theory

The general form for a linear homogeneous system with periodic coefficients is

$$x' = A(t)x, \tag{1.1}$$

where

$$A(t+T) = A(t) \quad \text{(for all } t). \tag{1.2}$$

Thus the coefficient matrix is periodic with a period T. We shall suppose that T is minimal, that is, the relation (1.2) does not hold for any smaller value of T. Note that $A(t)$ is also periodic with period kT for any integer k ($k = \pm 1, \pm 2, \ldots$) since it is readily shown from (1.2) that $A(t+kT) = A(t)$. Equations of the form (1.1) are of interest in their own right in that they describe dynamical systems with intrinsic periodicities. An example of such a system will be given in section 3.2. However, our interest in this text arises because equations of the form (1.1) occur in the study of the stability of periodic motion, and the results obtained will be exploited later (see in particular chapters 4, 7 and 8).

Although the coefficient matrix in (1.1) is periodic, in general the solutions are not periodic, as the following simple example demonstrates. Supposing that $n = 1$ and x is a scalar variable, let

$$x' = (1+\sin t)x.$$

Here the period T is 2π, and the general solution is

$$x = c\exp(t-\cos t),$$

where c is an arbitrary constant. Clearly the solution is not periodic and, in fact, $|x| \to \infty$ as $t \to \infty$. This example is typical and, as we shall see, illustrates the kind of behaviour to be expected. The following theorem lays the groundwork for the development of the general theory of (1.1).

Theorem 3.1: Let $X(t)$ be a fundamental matrix for (1.1). Then $X(t+T)$ is also a fundamental matrix, and there exists a non-singular constant matrix B such that

$$X(t+T) = X(t)B \quad \text{(for all } t). \tag{1.3}$$

47

Also,

$$\det B = \exp\left\{\int_0^T \operatorname{tr} A(s)\, ds\right\}.$$ (1.4)

Proof: Since $X(t)$ is a fundamental matrix, it follows that (see (2.6) of section 2.2)

$$X'(t) = A(t)X(t).$$

Let $Y(t) = X(t+T)$. Then

$$Y'(t) = X'(t+T)$$

$$= A(t+T)X(t+T)$$

$$= A(t)Y(t),$$

where the last line is a consequence of (1.2) and the definition of $Y(t)$. Thus $Y(t)$ is also a fundamental matrix and, hence, has the form $X(t)B$ for some constant non-singular matrix B (see section 2.2). This establishes (1.3). To prove (1.4) we use theorem 2.3 to deduce that (see (2.9) of section 2.2)

$$W(t) = W(t_0)\exp\left\{\int_0^t \operatorname{tr} A(s)\, ds\right\},$$

where $W(t) = \det X(t)$ is the Wronskian of $X(t)$ (see (2.7) of section 2.2). Hence

$$W(t+T) = W(t_0)\exp\left\{\int_{t_0}^t \operatorname{tr} A(s)\, ds + \int_t^{t+T} \operatorname{tr} A(s)\, ds\right\}.$$

But the result (1.3) shows that

$$W(t+T) = W(t)\det B$$

and, combining these two results, it follows that

$$\det B = \exp\left\{\int_t^{t+T} \operatorname{tr} A(s)\, ds\right\}.$$

The result (1.4) then follows since $\operatorname{tr} A(t)$ is a periodic function of t with period T. In establishing this last step it is useful to note the following lemma.

Lemma: Let $f(t)$ be a periodic function of t with period T, so that

$$f(t+T) = f(t), \quad \text{(for all } t\text{)}.$$ (1.5)

Then

$$\int_t^{t+T} f(s)\, ds = \int_0^T f(s)\, ds.$$ (1.6)

The result (1.6) is intuitively obvious since it states that the integral of a periodic function of period T over an interval of length T is independent of where the interval is located. A simple analytical proof follows. Let

$$I(t) = \int_t^{t+T} f(s)\ \mathrm{d}s.$$

Then $I'(t) = f(t+T) - f(t) = 0$ by virtue of (1.5) and, hence, $I(t)$ is a constant equal to $I(0)$.

The result (1.3) shows that in general the fundamental matrix $X(t)$ for (1.1) is not periodic and, hence, that the general solution of (1.1) is not periodic, in general. To proceed we shall need to examine the structure of the relation (1.3) in more detail. Note that, since (1.3) is true for all t, the constant matrix B can be expressed in terms of the fundamental matrix by putting $t = 0$:

$$B = X^{-1}(0) X(T). \qquad (1.7)$$

It is often useful to choose $X(t)$ to be the principal fundamental matrix, so that $X(0) = E$, and then $B = X(T)$.

Definition: Let the eigenvalues of B (defined by (1.3) or (1.7)) be $\rho_1, \ldots, \rho_n$, called the **characteristic multipliers** for (1.1). The **characteristic exponents** $\mu_1, \ldots, \mu_n$ are defined by

$$\rho_1 = e^{\mu_1 T}, \qquad \ldots, \qquad \rho_n = e^{\mu_n T}. \qquad (1.8)$$

The characteristic exponents are sometimes also called the Floquet exponents, and the theory being developed here is called Floquet theory. The characteristic exponents $\mu_1, \ldots, \mu_n$ are not unique as we can replace μ_i by $\mu_i + 2\pi i k / T$ $(i = 1, \ldots, n)$ for any integer $k = \pm 1, \pm 2, \ldots$ without altering the definition (1.8). However, we shall see that this non-uniqueness is not a matter for concern. The characteristic multipliers and, hence, the characteristic exponents, do not depend on the particular choice of fundamental matrix $X(t)$ and are intrinsic properties of the equation (1.1). For suppose $\hat{X}(t)$ is another fundamental matrix, leading to a constant matrix $\hat{B}$ given by

$$\hat{X}(t+T) = \hat{X}(t) \hat{B}.$$

But there exists a non-singular constant matrix C such that

$$\hat{X}(t) = X(t) C.$$

Using (1.3) and these two relations, it is easily shown that

$$\hat{B} = C^{-1} B C.$$

Hence $\hat{B}$ and B have the same eigenvalues. If we now choose $X(t)$ to be the principal fundamental matrix, so that $B = X(T)$, the characteristic multipliers are the eigenvalues of $X(T)$, where $X(0) = E$.

Theorem 3.2: Let ρ be a characteristic multiplier for (1.1) and let μ be the corresponding characteristic exponent so that $\rho = e^{\mu T}$. Then there exists a solution $x(t)$ of (1.1) such that

$$x(t+T) = \rho x(t) \quad \text{(for all } t\text{)}. \tag{1.9}$$

Further, there exists a periodic function $p(t)$ (i.e. $p(t+T) = p(t)$ for all t) such that

$$x(t) = e^{\mu t}p(t) \quad \text{(for all } t\text{)}. \tag{1.10}$$

Proof: Let b be an eigenvector of B corresponding to the eigenvalue ρ, so that

$$Bb = \rho b.$$

Then put

$$x(t) = X(t)b,$$

and so $x(t)$ is a solution of (3.1). But now

$$x(t+T) = X(t+T)b = X(t)Bb = X(t)\rho b = \rho x(t),$$

where the first step uses (1.3). Next put

$$p(t) = x(t)e^{-\mu t},$$

so that

$$p(t+T) = x(t+T)e^{-\mu t}e^{-\mu T} = \rho x(t)e^{-\mu t}e^{-\mu T} = p(t),$$

where we have used (1.8) and (1.9). Note here that if μ is replaced by $\mu + 2\pi ik/T$ for any integer k $(k = \pm 1, \pm 2, \ldots)$, then (1.10) is replaced by

$$x(t) = e^{\mu t}\left\{p(t)\exp\frac{2\pi ikt}{T}\right\}.$$

The quantity in the braces, $\{\cdot\}$, is again a periodic function of t with period T. Hence the structure of (1.10) is not changed and the non-uniqueness of μ does not affect the theory being developed here.

We are now in a position to determine the structure of the general solution of (1.1). Suppose first that $b^1, \ldots, b^n$ are n linearly independent eigenvectors of B corresponding to the eigenvalues $\rho_1, \ldots, \rho_n$. This is the generic case and occurs, in particular, when the eigenvalues are distinct. It follows from theorem 3.2 that there then exist n linearly independent solutions of (1.1) given by

$$x^i(t) = e^{\mu_i t}p^i(t) \quad (i = 1, \ldots, n), \tag{1.11}$$

where each $p^i(t)$ is a periodic function with period T. The key components here are the factors $e^{\mu_i t}$, since, modulo a periodic function, it is these factors which determine the behaviour of the solutions. There is an obvious analogy with the solutions of linear equations of constant coefficients (see, for instance, (5.5) of section 2.5), which can be exploited in the following way. Let

$$P_0(t) = [p^1(t), \ldots, p^n(t)]. \tag{1.12}$$

Then $P_0(t)$ is an $n \times n$ matrix function of t which is non-singular and is a periodic function of t, so that $P_0(t+T) = P_0(t)$ for all t. Next we form the fundamental matrix $X_0(t)$ for (1.1) from the n linearly independent solutions (1.11), so that

$$X_0(t) = [x^1(t), \ldots, x^n(t)]$$

$$= P_0(t) Y_0(t), \tag{1.13}$$

where

$$Y_0(t) = \mathrm{diag}[e^{\mu_1 t}, \ldots, e^{\mu_n t}].$$

Note that $Y_0(t)$ satisfies the equation

$$Y_0' = D_0 Y_0, \quad \text{where } D_0 = \mathrm{diag}[\mu_1, \ldots, \mu_n], \tag{1.14}$$

which is a matrix differential equation with constant coefficients. The general form for a fundamental matrix for (1.1) is then $X_0(t) C$, where C is an arbitrary constant matrix. We can now see that (1.1) will have a periodic solution of period T if and only if there exists a characteristic exponent $\mu = 0$ (modulo $2\pi i/T$) or, correspondingly, a characteristic multiplier $\rho = 1$ (see (1.9) and (1.10)). Of course, there may exist periodic solutions of period mT ($m = 2, 3, \ldots$) for $\mu = 2\pi i/mT$ (modulo $2\pi i/T$) or, correspondingly, $\rho = \exp 2\pi i/m$, so that $\rho^m = 1$. However, such solutions are generally exceptional, and usually the solutions of (1.1) will exhibit a time-scale which is not commensurate with T.

As an illustration, consider the following example:

$$x_1' = \left(1 + \frac{\cos t}{2 + \sin t}\right) x_1, \qquad x_2' = x_1 - x_2.$$

Here the period $T = 2\pi$. Of course, this example is very special as the first equation involves x_1 alone and the general solution can be obtained by elementary means. Thus, we find that

$$x_1 = c_1 e^t (2 + \sin t),$$

where c_1 is an arbitrary constant. With x_1 known, the second equation can be integrated to give

$$x_2 = c_2 e^t + c_1 e^t (2 + \tfrac{1}{2} \sin t - \tfrac{1}{2} \cos t),$$

where c_2 is the second arbitrary constant. A fundamental matrix is

$$X(t) = \begin{bmatrix} e^t(2 + \sin t) & 0 \\ e^t(2 + \tfrac{1}{2} \sin t - \tfrac{1}{2} \cos t) & e^{-t} \end{bmatrix}.$$

Clearly the characteristic exponents are ± 1 but, for illustrative purposes, we shall follow the procedure described above. Thus B is found from (1.7):

$$B = \begin{bmatrix} e^{2\pi} & 0 \\ 0 & e^{-2\pi} \end{bmatrix}.$$

The eigenvalues $\rho_{1,2}$ are $e^{2\pi}$ and $e^{-2\pi}$ and $\mu_{1,2} = \pm 1$. The remaining details are left as an exercise. In particular, it may be shown that

$$P_0(t) = \begin{bmatrix} 2 + \sin t & 0 \\ 2 + \frac{1}{2}\sin t - \frac{1}{2}\cos t & 1 \end{bmatrix}, \qquad Y_0(t) = \mathrm{diag}[e^t, e^{-t}],$$

while here $X(t)$ is just $X_0(t)$ (1.13). In general, while the result (1.13) is of considerable theoretical value in giving insight into the structure of the solutions of (1.1), it is rarely of immediate practical value. Useful methods of obtaining the characteristic exponents $\mu_1, \ldots, \mu_n$ and the periodic matrix $P_0(t)$ will be described for particular cases in sections 3.2, 3.3 and 3.4.

To this point we have tacitly assumed that each characteristic multiplier ρ, and each corresponding characteristic exponent μ, is real-valued. Of course, the preceding analysis remains valid when ρ, and hence μ, are complex-valued, and then $X_0(t)$ is also complex-valued. Nevertheless, it is useful to see how the procedure may be modified to produce a real-valued fundamental matrix. First we note that when ρ is real-valued and positive ($\rho > 0$) then μ is also real-valued and $-\infty < \mu < \infty$. However, if ρ is real-valued and negative ($\rho < 0$) then μ is complex-valued. But in this case we may put

$$\mu = \frac{i\pi}{T} + \nu, \quad \text{where } e^{\nu T} = -\rho. \tag{1.15}$$

Thus ν is real-valued and (1.10) is replaced by

$$x(t) = e^{\nu t} q(t), \quad \text{where } q(t) = \exp\left(\frac{i\pi t}{T}\right) p(t). \tag{1.16}$$

It is readily seen that $q(t)$ is periodic, of period $2T$ (i.e. $q(t+2T) = q(t)$ for all t), since $\exp(i\pi t/T)$ has period $2T$ and $p(t)$ has period T and, hence, also period $2T$. Also we may clearly choose $q(t)$ to be real-valued. The fundamental matrix is now modified in an obvious way. For each characteristic multiplier $\rho_i < 0$, we define ν_i by (1.15) and $q^i(t)$ by (1.16). Then the column $p^i(t)$ in $P_0(t)$ (1.12) is replaced by $q^i(t)$, the corresponding entry in $Y_0(t)$ is replaced by $e^{\nu_i t}$ and the entry μ_i in D_0 is replaced by ν_i.

Next, when ρ is complex-valued, ρ and $\bar{\rho}$ occur as a complex-conjugate pair of eigenvalues of B and, correspondingly, μ and $\bar{\mu}$ are both characteristic exponents. The modifications necessary to produce real-valued solutions of (1.1) are analogous to those described in section 2.5 for linear equations with constant coefficients. To illustrate this, consider the case $n = 2$, so that B is a 2×2 matrix with eigenvalues ρ and $\bar{\rho}$, while the corresponding characteristic exponents are μ and $\bar{\mu}$. We let $\mu = \nu + i\sigma$, where $|\rho| = e^{\nu T}$ and $\arg\rho = \sigma T$. The procedure described in theorem 3.2 remains valid and produces a complex-conjugate pair of solutions given by (1.10). Let $p(t) = q(t) + ir(t)$, where $q(t)$ and $r(t)$ are real-valued periodic functions of t whith period T. Then the real and imaginary parts of (1.10) yield a pair of real-valued linearly independent solutions of (1.1) given by

$$\text{Re}\{e^{\mu t}p(t)\} = e^{\nu t}\{\cos \sigma t\, q(t) - \sin \sigma t\, r(t)\},$$

$$\text{Im}\{e^{\mu t}p(t)\} = e^{\nu t}\{\sin \sigma t\, q(t) + \cos \sigma t\, r(t)\}. \tag{1.17}$$

The matrix $P_0(t)$ (1.12) is now replaced by the periodic matrix

$$Q_0(t) = [q(t), r(t)] \tag{1.18}$$

and we form the fundamental matrix $X_0(t)$ for (1.1) from the two linearly independent solutions defined by (1.17). Thus (1.13) is replaced by

$$X_0(t) = Q_0(t)\, Y_0(t), \quad \text{where } Y_0(t) = e^{\nu t}\begin{bmatrix} \cos \sigma t & \sin \sigma t \\ -\sin \sigma t & \cos \sigma t \end{bmatrix}. \tag{1.19}$$

Note that $Y_0(t)$ satisfies the equation

$$Y_0' = F_0 Y_0, \quad \text{where } F_0 = \begin{bmatrix} \nu & \sigma \\ -\sigma & \nu \end{bmatrix}. \tag{1.20}$$

To summarize, in the general case when A is an $n \times n$ matrix, the canonical fundamental matrix $X_0(t)$ is given by (1.13), where the columns of $P_0(t)$ are periodic functions of t with period T (or period $2T$ if the corresponding characteristic multiplier ρ is real-valued and negative), while $Y_0(t)$ is a matrix whose only non-zero elements are diagonal entries of the form $e^{\mu t}$, corresponding to real-valued characteristic multipliers, and 2×2 sub-matrices of the form (1.19) centred on the main diagonal, corresponding to each complex-conjugate pair of characteristic multipliers. Correspondingly, the canocical matrix D_0 (1.14) is a matrix whose only non-zero elements are either diagonal entries μ or 2×2 sub-matrices of the form F_0 (1.20) centred on the main diagonal.

The discussion so far has assumed that $b^1, \ldots, b^n$ are linearly independent eigenvectors of B. This is the generic case, which occurs most often in practice. Nevertheless, we shall discuss briefly the case when this assumption fails. Not surprisingly, the discussion parallels the corresponding case for linear equations with constant coefficients (see section 2.5). Indeed, it is useful to exploit here the concept of the exponential of a matrix (see (5.24) of section 2.5) in order to reduce the study of (1.1) to the study of a corresponding equation with constant coefficients. Since B is non-singular (see (1.7)), it can be shown that there exists a matrix D such that

$$B = e^{DT}. \tag{1.21}$$

We shall not give the proof in the general case here (but see, for instance, Coddington and Levinson, 1955, pp. 65–66). It may also be shown that B has the eigenvalues $\rho_1, \ldots, \rho_n$ if and only if D has the eigenvalues $\mu_1, \ldots, \mu_n$, where

$\rho_i = e^{\mu_i T}$ $(i = 1,\ldots,n)$ (see (1.8)). Next, we put

$$X(t) = P(t)\,Y(t), \quad \text{where } Y(t) = e^{Dt}. \tag{1.22}$$

Then, using (1.3) and (1.22), respectively, it follows that

$$X(t+T) = P(t)e^{Dt}B = P(t)e^{Dt}e^{DT}$$

and

$$X(t+T) = P(t+T)e^{Dt+DT} = P(t+T)e^{Dt}e^{DT}.$$

Here the last line is a consequence of the elementary fact that e^{Dt} and e^{DT} commute, which can readily be established from the definition of an exponential of a matrix (see (5.24) of section 2.5). Hence

$$P(t+T) = P(t) \tag{1.23}$$

and the $n \times n$ matrix $P(t)$ is a periodic function of t with period T. Note that $Y(t)$ satisfies the equation

$$Y' = DY. \tag{1.24}$$

In effect the transformation (1.22) has reduced (1.1) to the study of (1.24), which is an equation with constant coefficients.

Similarly, we may put

$$x(t) = P(t)\,y(t), \quad \text{where } y' = Dy. \tag{1.25}$$

The general theory for linear equations with constant coefficients (see section 2.5) may now be applied. In particular, for every eigenvalue μ of D, (1.25) has solutions for y of the form $e^{\mu t}u$, where u is a constant vector. Thus we have recovered the result (1.10). Further, if the algebraic multiplicity of μ is m and $m > p$, where p is the geometric multiplicity, then the full complement of solutions of (1.25) for y, corresponding to μ, takes the form $e^{\mu t}u(t)$, where $u(t)$ is a linear combination of m linearly independent vector-valued polynomials in t of degree $s-1$, where $1 \leqslant s \leqslant m-p+1$. It follows that, corresponding to each characteristic exponent μ of multiplicity m, there exist m linearly independent solutions of (1.1) of the form

$$x^1 = e^{\mu t}p^1(t), \quad \ldots, \quad x^m = e^{\mu t}p^m(t), \tag{1.26}$$

where $p^1(t),\ldots,p^m(t)$ are m linearly independent vector-valued polynomials in t of degree $s-1$, whose coefficients are themselves periodic functions of t with period T. The construction of these functions is achieved either by finding the matrix D from (1.21) and then solving (1.25) for y by the method described in section 2.5, or directly from the matrix B using a method analogous to that described in section 2.5 for linear equations with constant coefficients. This latter approach implies that if ρ is an eigenvalue of B with multiplicity m, we first seek solutions of the form $x(t) = X(t)b$, where (as in the proof of theorem

3.2) b is an eigenvector of B. If there are p linearly independent eigenvectors, this leads to p linearly independent solutions of (1.1) of the form (1.10). Next we suppose that c is a generalized eigenvector of B such that (cf. (5.17) of section 2.5)

$$Bc = \rho c + b \quad \text{and} \quad Bb = \rho b.$$

We then put

$$x(t) = X(t)c,$$

and it can then readily be shown that

$$x(t+T) = \rho x(t) + e^{\mu t}p(t),$$

where we recall that $p(t)$ is a periodic function of t with period T. This relation implies that

$$x(t) = e^{\mu t}q(t), \quad \text{where} \quad q(t+T) = q(t) + \rho^{-1}p(t).$$

Finally, this functional equation for $q(t)$ can be solved to give

$$q(t) = p_1(t) + (\rho T)^{-1}tp(t),$$

where $p_1(t)$ is a periodic function of t with period T. If there are q linearly independent generalized eigenvectors c this leads to a further q linearly independent solution of (1.1). If $p+q = m$ there is no need to proceed further but, if $p+q < m$, we must continue the process and eventually obtain m linearly independent solutions of the form (1.26).

It is also of interest to re-examine the generic case when $b^1,\ldots,b^n$ are linearly independent from the point of view of the penultimate paragraph. For this case we may define the non-singular matrix

$$S = [b^1,\ldots,b^n],$$

where

$$B = SB_0S^{-1} \quad \text{and} \quad B_0 = \text{diag}[\rho_1,\ldots,\rho_n]. \tag{1.27}$$

Now, using the definition (1.8), it is easily seen that

$$B_0 = e^{D_0 T} = \text{diag}[e^{\mu_1 T},\ldots,e^{\mu_n T}], \tag{1.28}$$

where D_0 is the diagonal matrix (1.14). But since S is non-singular, we may write

$$B = SB_0S^{-1} = Se^{D_0 T}S^{-1} = e^{DT},$$

where

$$D = SD_0S^{-1}. \tag{1.29}$$

The fact that B has the eigenvalues $\rho_1,\ldots,\rho_n$ if and only if D has the eigenvalues $\mu_1, \ldots, \mu_n$ is an immediate consequence of (1.28), since (1.27) shows that B and B_0 have the same eigenvalues, while (1.29) shows that D and D_0 likewise have the same eigenvalues. Now the proof of theorem 3.2 shows that $X_0(t) = X(t)S$, where we recall that $X_0(t)$ is defined by (1.13). We define

$$P(t) = P_0(t)S^{-1} \quad \text{and} \quad Y(t) = SY_0(t)S^{-1}, \tag{1.30}$$

where $P_0(t)$ and $Y_0(t)$ are also defined by (1.13). Hence $X(t) = P(t)Y(t)$, where $Y(t) = e^{Dt}$ and satisfies (1.24). This last statement is a simple consequence of (1.14) and (1.29), and we see that (1.14) is the canonical form of (1.24). Thus we have established (1.22) and shown that $X_0(t)$ and $P_0(t)$ are effectively the canonical forms of $X(t)$ and $P(t)$, respectively.

We conclude this section with a brief discussion of the inhomogeneous equation

$$x' = A(t)x + g(t), \tag{1.31}$$

where $A(t)$ is periodic (1.2), but $g(t)$ may or may not be periodic. The general solution is given by (3.2) of section 2.3, or

$$x = X(t)c + X(t) \int_a^t X^{-1}(s)g(s)\,ds. \tag{1.32}$$

Just as for the case of linear equations with constant coefficients (section 2.5) the form of the particular solution in (1.32) can be simplified. Since $Y(t)$ in (1.22) satisfies the equation (1.24) which has constant coefficients, it follows from (5.28) of section 2.5 that

$$Y(t)Y^{-1}(s) = Y(t-s), \tag{1.33}$$

where we note that here $Y(0) = E$. This result can, of course, also be directly obtained from (1.22). Hence, since $X(t) = P(t)Y(t)$ (see (1.22)) we see that (1.32) becomes

$$x = P(t)\left(Y(t)c + \int_a^t Y(t-s)P^{-1}(s)g(s)\,ds \right). \tag{1.34}$$

3.2 Parametric resonance

Here, and in the next section, we shall describe how the theory of the preceding section can be utilised to study the single second-order equation for $u(t)$,

$$u'' + a(t)u = 0, \quad \text{where } a(t+T) = a(t) \text{ (for all } t). \tag{2.1}$$

An equation of this form is sometimes called a Hill's equation. An example of a physical system which can be described by an equation of the form (2.1) is a simple pendulum whose point of support is made to move along a vertical line with period T (see figure 3.1). Thus if θ is the angle between the pendulum and

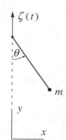

Fig. 3.1 A simple pendulum whose point of support is moving along a vertical line.

the vertical, where the pendulum is constrained to move in a vertical plane, then the equation of motion is, when $|\theta| \ll 1$,

$$\theta'' + (\omega^2 + \zeta'')\theta = 0. \qquad (2.2)$$

Here $\omega^2 = g/l$, where l is the length of the pendulum, g is the acceleration due to gravity and $l\zeta(t)$ is the vertical displacement of the point of support. The derivation of (2.2) is left as an exercise, but it is useful to note that $g + l\zeta''$ is the effective acceleration due to gravity in the accelerated frame of reference attached to the point of support. When $\zeta(t)$ is periodic with period T, (2.2) is an equation of the form (2.1). Note that if $\zeta = 0$ (i.e. the point of support is fixed) then (2.2) reduces to the equation for a simple harmonic oscillator of frequency ω and natural period $T_0 = 2\pi/\omega$. More generally (2.2) may be regarded as a model of a mechanical system which can be described by a simple harmonic oscillator of natural frequency ω which is subjected to an internally imposed forcing of period T. Our interest in these systems is due to the instabilities which can occur when there is a resonance between the imposed period T and the natural period T_0. We shall show later in this section that the resonance condition is that T, or $2T$, is an integer multiple of T_0. It is called parametric resonance as the imposed period T is contained in the coefficients of the equation, and can be contrasted with the external resonance which arises when the imposed period T is contained in an external forcing term (see chapter 8). For example, a simple harmonic oscillator subjected to external forcing is described by an equation of the form

$$\theta'' + \omega^2\theta = f(t),$$

where $f(t)$ is a specified periodic function of t, of period T. External resonance occurs here if $T = T_0$.

To utilize the theory of section 3.1 for the study of (2.1), we first form the equivalent first-order system, using the procedure of section 2.4. Thus we put (see (4.2) of section 2.4)

$$x_1 = u, \qquad x_2 = u', \qquad (2.3)$$

and (2.1) is equivalent to the first-order system

$$x_1' = x_2, \qquad x_2' = -a(t)x_1. \qquad (2.4)$$

The matrix $A(t)$ in (1.1) is here the 2×2 matrix

$$A(t) = \begin{bmatrix} 0 & 1 \\ -a(t) & 0 \end{bmatrix}.$$

Note that here $\operatorname{tr} A(t) = 0$. Next form the fundamental matrix $X(t)$ for (2.4) such that $X(0) = E$. Hence (see (4.8) of section 2.4)

$$X = \begin{bmatrix} u^1 & u^2 \\ u^{1\prime} & u^{2\prime} \end{bmatrix}, \tag{2.5}$$

where u^1 and u^2 are linearly independent solutions of (2.1) such that

$$\begin{aligned} u^1(0) &= 1, & u^2(0) &= 0, \\ u^{1\prime}(0) &= 0, & u^{2\prime}(0) &= 1. \end{aligned} \tag{2.6}$$

The matrix B is given by (1.7), where here $X(0) = E$, so that

$$B = \begin{bmatrix} u^1(T) & u^2(T) \\ u^{1\prime}(T) & u^{2\prime}(T) \end{bmatrix}. \tag{2.7}$$

We note that, since here $\operatorname{tr} A(t) = 0$, it is a consequence of (1.4) that

$$\det B = 1. \tag{2.8}$$

Indeed, more is true, since $\operatorname{tr} A(t) = 0$ implies that the Wronskian of (2.5) is constant (see (2.9) or (4.10) of chapter 2) and, hence (see (14.9) of chapter 2),

$$u^1 u^{2\prime} - u^2 u^{1\prime} = 1 \quad \text{(for all } t),$$

where we have used (2.6) to evaluate the constant term on the right-hand side. The characteristic multipliers ρ are the eigenvalues of B and, hence, are given by

$$\rho^2 - 2\phi\rho + 1 = 0, \quad \text{where } \phi = \tfrac{1}{2}\{u^1(T) + u^{2\prime}(T)\}. \tag{2.9}$$

Thus the eigenvalues $\rho_{1,2}$ are functions of the single parameter ϕ and are given by

$$\rho_{1,2} = \phi \pm \sqrt{\phi^2 - 1}. \tag{2.10}$$

Note the useful relations

$$\rho_1 \rho_2 = 1, \qquad \rho_1 + \rho_2 = 2\phi, \tag{2.11}$$

where the first expression is a consequence of (2.8). The characteristic exponents are $\mu_{1,2}$, where $\rho_{1,2} = e^{\mu_{1,2} T}$, and, consequent to (2.11),

$$\mu_1 + \mu_2 = 0, \qquad \cosh \mu_1 T = \phi. \tag{2.12}$$

Although ϕ (2.9) is not known explicitly, it is useful to characterize the properties of $\rho_{1,2}$ (or $\mu_{1,2}$) in terms of ϕ. Note that the constraint (2.8) is significant in this respect and is a consequence of the fact that (2.1) contains no dissipative term (i.e. no term in u').

(i) $\phi > 1$: Here $\rho_{1,2}$ are both real and positive (see (2.10)) and $\rho_1 > 1 > \rho_2 > 0$. Consequently μ_1 (2.12) is real and positive, while μ_2 ($= -\mu_1$) is real and negative. From the general theory of section 3.1 (see, for instance, (1.13)) we can deduce that the general solution of (2.1) is

$$u = c_1 e^{\mu_1 t} p_1(t) + c_2 e^{-\mu_1 t} p_2(t), \tag{2.13}$$

where

$$p_{1,2}(t+T) = p_{1,2}(t) \quad \text{(for all } t\text{)}.$$

There are no periodic solutions and, in general, $|u| \to \infty$ as $t \to \infty$. We shall say that (2.1) describes unstable behaviour, although the precise definitions of stable and unstable solutions will be deferred until chapter 4.

(ii) $\phi < -1$: Here $\rho_{1,2}$ are both real and negative (see (2.10)) and $\rho_2 < -1 < \rho_1 < 0$. Consequently we put (see (1.15), but note the change of sign)

$$\mu_1 = \frac{i\pi}{T} - \gamma, \qquad \cosh \gamma T = -\phi. \tag{2.14}$$

The general solution of (2.1) is now (see (1.16))

$$u = c_1 e^{-\gamma_1 t} q_1(t) + c_2 e^{\gamma_1 t} q_2(t), \tag{2.15}$$

where

$$q_{1,2}(t+2T) = q_{1,2}(t) \quad \text{(for all } t\text{)}.$$

Note that now, in contrast to case (i), the underlying period is $2T$. Again there are no periodic solutions and, in general, $|u| \to \infty$ as $t \to \infty$, so that (2.1) describes unstable behaviour.

(iii) $-1 < \phi < 1$: Here $\rho_{1,2}$ are both complex-valued, with unit magnitude (i.e. $|\rho_{1,2}| = 1$). Indeed,

$$\rho_{1,2} = \exp(\pm i\sigma T), \qquad \mu_1 = i\sigma \tag{2.16}$$

where

$$\cos \sigma T = \phi \quad (0 < \sigma T < \pi).$$

The general solution of (2.1) is now (see(1.17))

$$u = c_1 \operatorname{Re}\{e^{i\sigma t} p(t)\} + c_2 \operatorname{Im}\{e^{i\sigma t} p(t)\}, \tag{2.17}$$

where

$$p(t+T) = p(t) \quad \text{(for all } t\text{)}.$$

Here, of course, $p(t)$ is a complex-valued periodic function. The solutions are bounded and oscillatory, and we say that (2.1) describes stable behaviour.

There are no periodic solutions of period T, or $2T$, but there are, exceptionally, periodic solutions of period mT whenever $\sigma T = 2\pi/m$ for $m = 3, 4, \ldots$.

(iv) $\phi = 1$: This case is the boundary between cases (i) and (iii). There is just a single characteristic multiplier, $\rho_1 = 1$, and a single characteristic exponent $\mu_1 = 0$. It can be regarded as the limit $\mu_1 \to 0$ in case (i), or $\sigma \to 0$ in case (iii). The general solution is

$$u = c_1 p_1(t) + c_2\{ktp_1(t) + p_2(t)\}, \tag{2.18}$$

where

$$p_{1,2}(t + T) = p_{1,2}(t) \quad \text{(for all } t\text{)}.$$

Here k is a constant, which may equal zero. This case is one of marginal stability and, significantly, we note that, choosing $c_2 = 0$, there exists a solution of period T.

(v) $\phi = -1$: This case is the boundary between cases (ii) and (iii). There is again just a single characteristic multiplier, $\rho_1 = -1$, and a single characteristic exponent, $\mu_1 = i\pi/T$. It can be regarded as the limit $\gamma \to 0$ in case (ii), or $\sigma \to \pi/T$ in case (iii). The general solution is

$$u = c_1 q_1(t) + c_2\{ktq_1(t) + q_2(t)\}, \tag{2.19}$$

where

$$q_{1,2}(t + 2T) = q_{1,2}(t) \quad \text{(for all } t\text{)}.$$

Again k is a constant, which may equal zero. This case is also one of marginal stability, and we note that, choosing $c_2 = 0$, there exists a solution of period $2T$.

The interesting and significant result from this analysis is that the boundary between stable and unstable behaviour (cases (iv) and (v)) is characterized by the existence of periodic solutions of (2.1) with period T or $2T$. This suggests that the mechanism of instability is the occurrence of a resonance between the imposed period T and some natural period. To explore this concept further, we consider the case when (2.1) has the form (2.2), so that

$$u'' + \{\omega^2 + \epsilon b(t)\} u = 0, \tag{2.20}$$

where

$$b(t + T) = b(t) \quad \text{(for all } t\text{)}.$$

For $\epsilon = 0$ this equation describes a simple harmonic oscillation of frequency ω and period $T_0 = 2\pi/\omega$. For $\epsilon \neq 0$, the term $\epsilon b(t)$ represents an internally imposed forcing of period T. Equation (2.20) contains two parameters, ω^2 and ϵ, and consequently $\phi = \phi(\omega^2, \epsilon)$. The stability boundaries $\phi = 1$ and $\phi = -1$

are now curves in the ω^2-ϵ plane, and the main aim in the analysis of (2.20) is the determination of these curves.

Since (2.20) is analytic in ω^2 and ϵ, it follows that the solutions u, and hence also ϕ, are analytic in ω^2 and ϵ. We shall exploit this by considering the limit $\epsilon \to 0$, when the general solution of (2.20) is

$$u = c_1 \cos \omega t + c_2 \sin \omega t \quad \text{(for } \epsilon = 0\text{)}$$

and is periodic with a period $T_0 = 2\pi/\omega$. Also, from (2.6), we find that, for $\epsilon = 0$,

$$u^1 = \cos \omega t, \qquad u^2 = \frac{1}{\omega} \sin \omega t.$$

Hence, from (2.9),

$$\phi = \cos \omega T \quad \text{(for } \epsilon = 0\text{)}.$$

Of course, with $\epsilon = 0$ (2.20) reduces to the simple harmonic oscillator equation, whose solutions are well understood. However, because $\phi(\omega^2, \epsilon)$ is analytic in ω^2 and ϵ, we can now use its value at $\epsilon = 0$ to obtain information about the stability boundaries when ϵ is small, but finite. Thus, for $\epsilon = 0$, $-1 \leqslant \phi \leqslant 1$ for all values of ω, and there is stability unless $\phi = \pm 1$, or, equivalently, ωT is an integral multiple of π. The stability boundaries for $\epsilon = 0$ are given by

$$\begin{aligned} \phi = 1, \qquad \omega T = 2k\pi \qquad &\text{or} \qquad T = kT_0; \\ \phi = -1, \qquad \omega T = (2k+1)\pi \quad &\text{or} \quad 2T = (2k+1)\,T_0, \end{aligned} \qquad (2.21)$$

where $k = 0, 1, 2, \ldots$. These are the conditions for parametric resonance. Note that when $\phi = 1$, or $T = kT_0$, there is a solution u with period T, while, when $\phi = -1$, or $2T = (2k+1)T_0$, there is a solution u with period $2T$. The fundamental resonance in the first case is $k = 1$, or $T = T_0$, and is a direct resonance. In the second case the fundamental resonance is $k = 0$, or $2T = T_0$, and is a subharmonic resonance (i.e. the response period T_0 is twice the forcing period).

For $\epsilon = 0$ and $\omega T \neq 2k\pi$ or $(2k+1)\pi$ we have shown that $-1 < \phi < 1$. It follows by continuity that for $\omega T \neq 2k\pi$ or $(2k+1)\pi$ and sufficiently small $|\epsilon|$, it will remain true that $-1 < \phi < 1$ and, hence, the solutions of (2.20) will be bounded and oscillatory and represent stable behaviour (i.e. case (iii) of section 3.2 applies). Hence, if an instability occurs for small $|\epsilon|$, it must do so in the vicinity of the points defined by (2.21), where $\epsilon = 0$ and $\omega T = 2k\pi$ or $(2k+1)\pi$. In general, we can expect the stability boundaries in the ω^2-ϵ plane to cross the $\epsilon = 0$ axis at these points. For sufficiently small $|\epsilon|$ the structure of these boundaries can be calculated by expanding the solutions of (2.20) as a power series in ϵ. The details of this approach are described in section 3.3, for the special case of Mathieu's equation when $b(t) = \cos 2t$. For larger values of ϵ it is generally necessary to employ a numerical procedure, and one such approach will also be described briefly in the next section.

3.3 Perturbation methods for the Mathieu equation

The Mathieu equation is the special case of (2.20) when $b(t) = \cos 2t$, so that

$$u'' + (\delta + \epsilon \cos 2t)\, u = 0. \tag{3.1}$$

Here we have put $\omega^2 = \delta$, and we will find it useful to allow δ to take all real values, $-\infty < \delta < \infty$. The period $T = \pi$, and the conditions (2.21) for parametric resonance become

$$\delta = (2k)^2 \quad \text{or} \quad \delta = (2k+1)^2, \tag{3.2}$$

where $k = 0, 1, 2, \dots$. Note that when $\epsilon = 0$ and $\delta = (2k)^2$ there is a solution of (3.1) with period π, but when $\epsilon = 0$ and $\delta = (2k+1)^2$ there is a solution of (3.1) with period 2π (see (2.21) and the following discussion). Based on the discussion in section 3.2 we can conclude that the instability boundaries in the δ-ϵ plane will cross the $\epsilon = 0$ axis at the points defined by (3.2).

However, in section 3.2 we assumed that $\delta = \omega^2$ is non-negative. Hence it remains here to consider the case when δ is negative. In the limit $\epsilon \to 0$, the general solution of (3.1) is

$$u = c_1 \cosh \sigma t + c_2 \sinh \sigma t \quad \text{(for } \epsilon = 0),$$

where

$$\sigma^2 = -\delta > 0.$$

From (2.6) we find that, for $\epsilon = 0$,

$$u^1 = \cosh \sigma t, \qquad u^2 = \frac{1}{\sigma} \sinh \sigma t$$

and, hence, from (2.9),

$$\phi = \cosh \pi \sigma \quad \text{(for } \epsilon = 0).$$

Thus $\phi > 1$ for all $\sigma > 0$, and there is instability with $\mu = \pm \sigma$ (see (2.12)). By continuity, it follows that, for $\delta < 0$ and sufficiently small $|\epsilon|$, it will remain true that $\phi > 1$ and, hence, the solutions of (3.1) will describe unstable behaviour (see case (i) of section 3.2).

Based on the discussion in section 3.2 we can anticipate that, when $\delta \approx n^2$, where $n = 0, 1, 2, \dots$, and ϵ is small, the solutions of (3.1) will take the form

$$u(t) = e^{\mu t} q(t), \tag{3.3}$$

where $q(t)$ is a periodic function of period π when n is an even integer (i.e. $n = 2k$) and has period 2π when n is an odd integer (i.e. $n = 2k+1$). Further $\mu \to 0$ as $\epsilon \to 0$. Note that when n is an even integer, μ is the characteristic exponent, but when n is an odd integer, μ differs from the characteristic exponent as previously defined by i (see (2.14)). In both cases stability is determined by the sign of $\text{Re}\,\mu$. Away from the stability boundaries (i.e. $\phi \neq \pm 1$), all solutions of (3.1) will have the form (3.3), but on the stability boundary (i.e. $\phi = \pm 1$) only one solution can generally be expected to have the form (3.3). On substituting (3.3) into (3.1), we find that

$$q'' + 2\mu q' + \mu^2 q + (\delta + \epsilon \cos 2t) q = 0. \tag{3.4}$$

The condition that the solutions of (3.4) have the required period π, or 2π, determines μ as a function of δ and ϵ, and, hence, the stability boundaries.

Because the analyticity of the solutions of (3.4) with respect to the parameter ϵ can be assumed, we are allowed to seek solutions of (3.4) in the form

$$q = q_0(t) + \epsilon q_1(t) + \epsilon^2 q_2(t) + \ldots, \quad \mu = \epsilon \mu_1 + \epsilon^2 \mu_2 + \ldots, \quad \delta = n^2 + \epsilon \delta_1 + \epsilon^2 \delta_2 + \ldots. \tag{3.5}$$

Note that the leading order term for μ (i.e. μ_0) is zero. When the expansions (3.5) are substituted into the left-hand side of (3.4), the result will be a power series in ϵ and, clearly, since this power series must vanish, the coefficient of each power of ϵ must be zero. Applying this process to the coefficients of ϵ^m ($m = 0, 1, 2, \ldots$), we find that

$$\begin{aligned} q_0'' + n^2 q_0 &= 0, \\ q_1'' + n^2 q_1 &= f_1, \\ q_2'' + n^2 q_2 &= f_2, \\ &\vdots \\ q_m'' + n^2 q_m &= f_m, \end{aligned} \tag{3.6}$$

where the terms on the right-hand side are given by

$$\begin{aligned} f_1 &= -2\mu_1 q_0' - (\delta_1 + \cos 2t) q_0, \\ f_2 &= -2\mu_1 q_1' - (\delta_1 + \cos 2t) q_1 - 2\mu_2 q_0' - \mu_1^2 q_0 - \delta_2 q_0. \end{aligned} \tag{3.7}$$

We see that f_1 depends only on q_0, f_2 depends only on q_0 and q_1, and it can be shown that f_m depends only on $q_0, q_1, \ldots, q_{m-1}$. Hence each term $q_0, q_1, \ldots, q_m$ can be determined in succession. Further, the condition that each term $q_0, q_1, q_2, \ldots, q_m$ is periodic, with period π or 2π according as $n = 2k$ or $2k + 1$ for $k = 0, 1, 2, \ldots$, will determine successively the coefficients $\mu_1, \mu_2, \ldots, \mu_m$ in terms of $\delta_1, \delta_2, \ldots, \delta_m$ and, hence, enable the construction of μ as a function of $\delta - n^2$ and ϵ, with an error of $O(\epsilon^{m+1})$. In practice the algebraic manipulation involved increases rapidly with m. However, in most cases, it is rarely necessary to go beyond the terms $m = 1$ or 2 to obtain significant information about the stability boundaries, where $\operatorname{Re} \mu = 0$. We shall illustrate the procedure by considering the cases $n = 0, 1, 2$. Because the case $n = 0$ proceeds rather differently from all the other cases, we shall defer its consideration and begin our discussion with the case $n = 1$.

(i) $n = 1$ ($\delta \approx 1$). Since n is here an odd integer, $q(t)$ and, hence, each term $q_0(t)$, $q_1(t)$, $q_2(t)$, $\ldots$, must be periodic functions of t with period 2π. From

(3.6) we see that

$$q_0'' + q_0 = 0,$$

so that

$$q_0 = A_0 \cos t + B_0 \sin t.$$

Thus q_0 is periodic with period 2π for all values of the constants A_0 and B_0. Next we see from (3.6) that

$$q_1'' + q_1 = f_1,$$

where f_1 is given by (3.7). Substituting the above expression for q_0,

$$f_1 = (-\delta_1 A_0 - \tfrac{1}{2}A_0 - 2\mu_1 B_0)\cos t + (-\delta_1 B_0 + \tfrac{1}{2}B_0 + 2\mu_1 A_0)\sin t$$
$$- \tfrac{1}{2}A_0 \cos 3t - \tfrac{1}{2}B_0 \sin 3t.$$

Here we have used the trigonometric identities

$$2\cos 2t \cos t = \cos 3t + \cos t, \qquad 2\cos 2t \sin t = \sin 3t - \sin t.$$

The general solution for q_1 is

$$q_1 = A_1 \cos t + B_1 \sin t + q_{1P},$$

where

$$q_{1P} = \tfrac{1}{2}(-\delta_1 A_0 - \tfrac{1}{2}A_0 - 2\mu_1 B_0)t \sin t - \tfrac{1}{2}(-\delta_1 B_0 + \tfrac{1}{2}B_0 + 2\mu_1 A_0)t \cos t$$
$$+ \tfrac{1}{16}A_0 \cos 3t + \tfrac{1}{16}B_0 \sin 3t.$$

In this solution for q_1 the terms involving the constants A_1 and B_1 represent the solution of the homogeneous equation, while q_{1P} is the particular solution corresponding to the 'forcing' function f_1. Now, for q_1 to be periodic with period 2π we must set to zero the coefficients of the terms $\tfrac{1}{2}t \sin t$ and $-\tfrac{1}{2}t \cos t$ in q_{1P}, since these are the only non-periodic terms. Note that these terms are, respectively, the particular solutions corresponding to the terms $\cos t$ and $\sin t$ in the 'forcing' function f_1. Consequently we call these the 'resonant' terms, as they are precisely the terms which arise in the solution of the homogeneous equation for q_1. In contrast the particular solutions corresponding to the terms $\cos 3t$ and $\sin 3t$ in the 'forcing' function f_1 are $-\tfrac{1}{8}\cos 3t$ and $-\tfrac{1}{8}\sin 3t$, respectively, and these are periodic. Note that in identifying the 'resonant' terms it is essential that we first express f_1 as a Fourier series in terms of the functions $\{1, \cos rt, \sin rt; r = 1, 2, 3, \ldots\}$. Then the 'resonant' terms correspond to $r = 1$.

Removing the coefficients of the non-periodic terms in q_{1P}, we get

$$-(\delta_1 + \tfrac{1}{2})A_0 + 2\mu_1 B_0 = 0,$$
$$2\mu_1 A_0 + (\delta_1 - \tfrac{1}{2})B_0 = 0. \tag{3.8}$$

For a non-trivial solution

$$4\mu_1^2 = \tfrac{1}{4} - \delta_1^2.$$

Recalling the definitions (3.5), this corresponds to

$$4\mu^2 \approx \tfrac{1}{4}\epsilon^2 - (\delta - 1)^2, \tag{3.9}$$

where the error term is $O(\epsilon^3)$. Thus for $(\delta-1)^2 < \tfrac{1}{4}\epsilon^2$ the two solutions for μ are both real-valued with $\mu_1 > 0$ and $\mu_2 = -\mu_1 < 0$; this corresponds to unstable behaviour. On the other hand, for $(\delta-1)^2 > \tfrac{1}{4}\epsilon^2$ the two solutions for μ are both pure imaginary and correspond to stable behaviour. The stability boundary here corresponds to $\mu = 0$, and is given by

$$\delta - 1 \approx \pm\tfrac{1}{2}\epsilon, \tag{3.10}$$

where the error term is $O(\epsilon^2)$. The situation is sketched in figure 3.2, where we see that there is a zone of instability bounded, for sufficiently small ϵ, by the straight lines (3.10). For $\delta - 1 \approx \pm\tfrac{1}{2}\epsilon$, we see that either $A_0 = 0$ or $B_0 = 0$, respectively, and, as expected, we see that on the stability boundaries there is a solution of (3.1) which is periodic with period 2π.

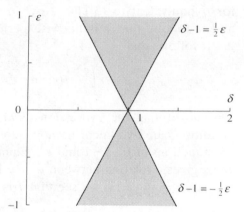

Fig. 3.2 The zone of instability for the Mathieu equation (3.1). The zones are symmetric in ϵ, so only the region $\epsilon \geqslant 0$ is shown. The shaded regions correspond to unstable zones.

With q_1 now determined we may proceed to consider q_2. The procedure is to express the forcing function f_2 (3.7) as a Fourier series in terms of the functions $\{1, \cos rt, \sin rt; r = 1, 2, 3, \ldots\}$, where the 'resonant' terms correspond to $r = 1$, viz. $\cos t$ and $\sin t$. The coefficients of the 'resonant' terms must be removed in order for q_2 to be periodic. This will then determine μ_2 and hence the $O(\epsilon^2)$ correction term to the stability boundaries (3.9). The details are left to the exercises. The process can, in principle, be continued indefinitely.

Clearly the derivation of the conditions (3.8) is the key result of the analysis and, before proceeding to consider the other cases $n = 2$ and $n = 0$, it is instructive to describe an alternative but equivalent procedure for obtaining (3.8). We write (3.4) in the form

$$q'' + q = f, \tag{3.11}$$

where

$$f = -2\mu q' - \mu^2 q - \{(\delta - 1) + \epsilon \cos 2t\} q.$$

Further, let us impose the initial conditions

$$q(0) = A, \qquad q'(0) = B. \tag{3.12}$$

It then follows from theorem 2.1 (section 2.1) that there exists a unique solution $q = q(t; A, B, \mu, \delta - 1, \epsilon)$ which is an analytic function of the parameters A, B, μ, $\delta - 1$ and ϵ. Further, since (3.11) is a linear equation, q will be a linear function of the parameters A and B.

Next, from (4.14) of section 2.4 it follows that

$$q = A \cos t + B \sin t + \int_0^t f(s) \sin(t - s) \, ds. \tag{3.13}$$

The derivation of this result in section 2.4 assumed that $f(t)$ was a known function, but it clearly remains valid when, as in this case, f is not known but depends on q and q'. Equation (3.13) is thus an integro-differential equation for q equivalent to (3.11). Now, q is a periodic function of t with period 2π and it follows that f is likewise periodic. Hence, from theorem 2.4 of section 2.4 it follows that

$$-\int_0^{2\pi} f(s) \sin s \, ds = \int_0^{2\pi} f(s) \cos s \, ds = 0. \tag{3.14}$$

On substituting the expression (3.11) for f into (3.14), we see that these two conditions are two linear homogeneous equations for A and B whose coefficients are functions of μ, $\delta - 1$ and ϵ. Equating to zero the determinant of these equations gives a relation between μ, $\delta - 1$ and ϵ.

Now, from (3.11) we see that f is $O(\mu, \delta - 1, \epsilon)$ and, hence, (3.13) shows that

$$q = A \cos t + B \sin t + O(\mu, \delta - 1, \epsilon),$$

and hence

$$f \approx -2\mu(-A \sin t + B \cos t) - \{(\delta - 1) + \epsilon \cos 2t\}(A \cos t + B \sin t),$$

where the error term in the expression for f is quadratic in the small parameters μ, $\delta - 1$ and ϵ. Note that the leading order term in q is essentially just q_0, since we expect $A, B \to A_0, B_0$ as $\epsilon \to 0$, and it follows that the leading-order term in f is ϵf_1. Substituting this approximate expression for f into (3.14) gives

$$-2\mu A + \{(\delta - 1) - \tfrac{1}{2}\epsilon\} B \approx 0,$$
$$-\{(\delta - 1) + \tfrac{1}{2}\epsilon\} A - \qquad 2\mu B \approx 0.$$

These are clearly equivalent to (3.8), and the criterion for a non-trivial solution is again (3.9).

Higher-order terms can now be obtained by substituting the leading-order expression for f into (3.13), and thus obtaining a higher-order approximation to q. This in turn can be substituted into the expression (3.11) for f to obtain a higher-order approximation, which may then be used in the conditions (3.14) to obtain ultimately a correction term for (3.9). Clearly this iterative process is equivalent to the expansion procedure described earlier. In practice the expansion procedure is to be preferred as its implementation is more straight-forward. However, the importance of the second method is that it demonstrates that the rigorous conditions (3.14) are equivalent to the removal of the 'resonant' terms in the first method. Indeed, if f is expanded as a Fourier series in terms of the functions $\{1, \cos rt, \sin rt; r = 1, 2, 3, \ldots\}$, it is immediately clear from (3.14) that the 'resonant' terms correspond to $r = 1$.

(ii) $n = 2$ ($\delta \approx 4$). Since n is an even integer, $q(t)$, and each term $q_0(t)$, $q_1(t)$, $q_2(t)$, $\ldots$, must be periodic functions of t with period π. From (3.6) we see that

$$q_0'' + 4q_0 = 0,$$

so that

$$q_0 = A_0 \cos 2t + B_0 \sin 2t.$$

Thus q_0 is periodic with period π for all values of the constants A_0 and B_0. Next, from (3.6),

$$q_1'' + 4q_1 = f_1,$$

where f_1 is given by (3.7). Substituting the above expression for q_0,

$$f_1 = (-\delta_1 A_0 - 4\mu_1 B_0)\cos 2t + (-\delta_1 B_0 + 4\mu_1 A_0)\sin 2t$$

$$- \tfrac{1}{2}A_0(1 + \cos 4t) - \tfrac{1}{2}B_0 \sin 4t.$$

Here we have used trigonometric identities to express f_1 as a Fourier series in terms of the functions $\{1, \cos rt, \sin rt; r = 1, 2, 3, \ldots\}$. In this case the 'resonant' terms correspond to $r = 2$ and are $\cos 2t$ and $\sin 2t$, since these are just the terms which arise in the solution of the homogeneous equation for q_1 and, hence, would produce non-periodic terms in the general solution for q_1.

Removing the coefficients of the 'resonant' terms in f_1, we find that

$$\begin{aligned} -\delta_1 A_0 - 4\mu_1 B_0 &= 0, \\ -4\mu_1 A_0 + \delta_1 B_0 &= 0. \end{aligned} \tag{3.15}$$

For a non-trivial solution

$$16\mu_1^2 = -\delta_1^2.$$

Recalling the definitions (3.5), this corresponds to

$$16\mu^2 \approx -(\delta - 4)^2, \tag{3.16}$$

where the error term is $O(\epsilon^3)$. For $\delta \neq 4$ the two solutions for μ are both pure imaginary and correspond to stable behaviour. The stability boundary here corresponds to $\mu = 0$ and is given by $\delta = 4$. To the order considered this unstable zone has not been resolved, as its width is $O(\epsilon^2)$.

To resolve the unstable zone, we must proceed to the next order and, for this purpose, we may put $\mu_1 = \delta_1 = 0$. In this case the general solution for q_1 is

$$q_1 = A_1 \cos 2t + B_1 \sin 2t - \tfrac{1}{8} A_0 + \tfrac{1}{24} A_0 \cos 4t + \tfrac{1}{24} B_0 \sin 4t.$$

Then, from (3.6),

$$q_2'' + 4q_2 = f_2,$$

where f_2 is given by (3.7). Substituting the above expressions for q_0 and q_1 we find that

$$f_2 = (-\delta_2 A_0 + \tfrac{5}{48} A_0 - 4\mu_2 B_0)\cos 2t + (-\delta_2 B_0 - \tfrac{1}{48} B_0 + 4\mu_2 A_0)\sin 2t$$
$$- \tfrac{1}{48} A_0 \cos 6t - \tfrac{1}{48} B_0 \sin 6t - \tfrac{1}{2} A_1 (1 + \cos 4t) - \tfrac{1}{2} B_1 \sin 4t.$$

Here we have again used standard trigonometric identities to express f_2 as a Fourier series. Removing the coefficients of the 'resonant' terms, $\cos 2t$ and $\sin 2t$, we get

$$\begin{aligned} -(\delta_2 - \tfrac{5}{48})A_0 - 4\mu_2 B_0 &= 0, \\ -4\mu_2 A_0 + (\delta_2 + \tfrac{1}{48}) B_0 &= 0. \end{aligned} \tag{3.17}$$

For a non-trivial solution

$$16\mu_2^2 = -(\delta_2 - \tfrac{5}{48})(\delta_2 + \tfrac{1}{48}).$$

Recalling the definitions (3.5), this corresponds to

$$16\mu^2 \approx -\{(\delta - 4) - \tfrac{5}{48}\epsilon^2\}\{(\delta - 4) + \tfrac{1}{48}\epsilon^2\}, \tag{3.18}$$

where the error term is $O(\epsilon^5)$. Thus for

$$\delta - 4 > \tfrac{5}{48}\epsilon^2 \quad \text{or} \quad \delta - 4 < -\tfrac{1}{48}\epsilon^2,$$

the two solutions for μ are both pure imaginary and correspond to stable behaviour. On the other hand, for

$$-\tfrac{1}{48}\epsilon^2 < \delta - 4 < \tfrac{5}{48}\epsilon^2,$$

the two solutions for μ are real-valued with $\mu_1 > 0$ and $\mu_2 = -\mu_1 < 0$; this corresponds to unstable behaviour. The stability boundary here corresponds to $\mu = 0$ and is given by

$$\delta - 4 \approx -\tfrac{1}{48}\epsilon^2 \quad \text{or} \quad \tfrac{5}{48}\epsilon^2, \tag{3.19}$$

where the error term is $O(\epsilon^3)$. The situation is sketched in figure 3.3, where we see that there is a zone of instability bounded, for sufficiently small ϵ, by the

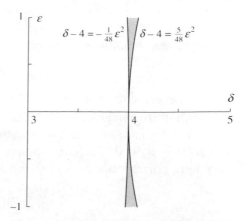

Fig. 3.3 As for figure 3.2, but now $\delta \approx 4$.

parabolas (3.19). On each boundary, either $A_0 = 0$ or $B_0 = 0$, respectively, and, as expected, there is a solution of (3.1) which is periodic with period π.

As for case (i), the expansion may now be continued to higher order. However, there is not much to be gained in doing so as the main result being sought is the stability boundary, and this has already been obtained to leading order in (3.19). Further calculation will only provide small correction terms to this result.

(iii) $n = 0$ ($\delta \approx 0$). As in case (ii), $q(t)$, and each term $q_0(t)$, $q_1(t)$, $q_2(t)$, ..., must be periodic functions of t with periodic π. From (3.6)

$$q_0'' = 0,$$

so that

$$q_0'' = A_0 + B_0 t.$$

For q_0 to be periodic we must put $B_0 = 0$. Next from (3.6) and (3.7),

$$q_1'' = f_1, \quad \text{where} \quad f_1 = -(\delta_1 + \cos 2t)A_0,$$

so that

$$q_1 = A_1 + B_1 t - \tfrac{1}{2}\delta_1 A_0 t^2 + \tfrac{1}{4}A_0 \cos 2t.$$

For q_1 to be periodic we must put $\delta_1 = 0$ and $B_1 = 0$. At this stage μ_1 has not been determined, so we must proceed to the next order. We find that

$$q_2'' = f_2,$$

where

$$f_2 = -(\mu_1^2 + \delta_2 + \tfrac{1}{8})A_0 + \mu_1 A_0 \sin 2t - \tfrac{1}{8}A_0 \cos 4t - A_1 \cos 2t.$$

In this case the 'resonant' term is the constant term and, removing its coefficient, we get

$$\mu_1^2 + \delta_2 + \tfrac{1}{8} = 0.$$

Recalling the definitions (3.5), this corresponds to

$$\mu^2 + \delta + \tfrac{1}{8}\epsilon^2 \approx 0, \tag{3.20}$$

where the error term is $O(\epsilon^3)$. Thus, for $\delta + \tfrac{1}{8}\epsilon^2 > 0$, the two solutions for μ are both pure imaginary and correspond to stable behaviour. On the other hand, for $\delta + \tfrac{1}{8}\epsilon^2 < 0$, the two solutions for μ are both real-valued with $\mu_1 > 0$ and $\mu_2 = -\mu_1 < 0$; this corresponds to unstable behaviour. The stability boundary here corresponds to $\mu = 0$ and is given by

$$\delta \approx -\tfrac{1}{8}\epsilon^2, \tag{3.21}$$

where the error term is $O(\epsilon^3)$. The situation is sketched in figure 3.4, where we see that there is a zone of stability separated from a zone of instability by (3.21). Interestingly note that, whereas $\delta < 0$ is always unstable for $\epsilon = 0$, there are regions of stable behaviour with $\delta < 0$ for ϵ small but non-zero.

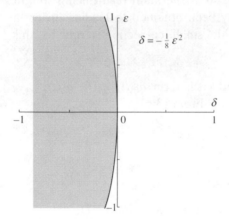

Fig. 3.4 As for figure 3.2, but now $\delta \approx 0$.

A similar procedure can be used for the other values of n, $n \geq 3$. However, we shall not give details as the complexity of the algebra increases rapidly with n. Indeed, it is necessary to proceed to the nth order in the expansion to determine the stability boundary. Interestingly, it can be shown that the unstable zone has a bandwidth of $O(\epsilon^n/(2n)^{n-1})$ as $n \to \infty$ and, hence, for sufficiently small ϵ, the bandwidth decreases rapidly as n increases. This result, and more complete expansions for the stability boundaries than those given here, can be found in Abramowitz and Stegun, 1965, pp. 722–730.

To this point it has been assumed that ϵ is sufficiently small for the perturbation procedure to be applicable. When this is not the case, an alternative method is needed, and we shall briefly discuss one such method here. The discussion in section 3.2 shows that the stability boundaries in the δ-ϵ plane are characterized by the presence of solutions which are periodic with period π or 2π. Hence, we can attempt to determine these boundaries directly by seeking a solution of (3.1) in the form

$$u = \tfrac{1}{2}a_0 + \sum_{n=1}^{\infty} (a_n \cos nt + b_n \sin nt). \tag{3.22}$$

Although this is a Fourier series for u with period 2π, it can also be used to obtain the solutions with period π by setting to zero the Fourier coefficients corresponding to n being an odd integer. On substituting this Fourier series into (3.1) and using the trigonometric identities,

$$2\cos nt \cos 2t = \cos(n+2)t + \cos(n-2)t,$$

$$2\sin nt \cos 2t = \sin(n+2)t + \sin(n-2)t,$$

we find that

$$\tfrac{1}{2}(\delta a_0 + \epsilon a_2) + \sum_{n=1}^{\infty} \{(\delta - n^2)a_n + \tfrac{1}{2}\epsilon(a_{n+2} + a_{n-2})\}\cos nt$$

$$+ \{(\delta - n^2)b_n + \tfrac{1}{2}\epsilon(b_{n+2} + b_{n-2})\}\sin nt = 0,$$

where we have temporarily adopted the convention that $b_0 = 0$, $a_{-1} = a_1$ and $b_{-1} = -b_1$. Since this Fourier series must vanish for all t, every coefficient must vanish and, hence, we have obtained the following infinite sequence of equations.

$$\delta a_0 + \epsilon a_2 = 0, \qquad\qquad (\delta - 1)b_1 + \tfrac{1}{2}\epsilon(b_3 - b_1) = 0,$$

$$(\delta - 1)a_1 + \tfrac{1}{2}\epsilon(a_3 + a_1) = 0, \qquad\qquad (\delta - 4)b_2 + \tfrac{1}{2}\epsilon b_4 = 0,$$

$$(\delta - 4)a_2 + \tfrac{1}{2}\epsilon(a_4 + a_0) = 0, \qquad\qquad (\delta - 9)b_3 + \tfrac{1}{2}\epsilon(b_5 + b_1) = 0,$$

$$\vdots \qquad\qquad\qquad\qquad \vdots$$

$$(\delta - n^2)a_n + \tfrac{1}{2}\epsilon(a_{n+2} + a_{n-2}) = 0, \qquad (\delta - n^2)b_n + \tfrac{1}{2}\epsilon(b_{n+2} + b_{n-2}) = 0. \tag{3.23}$$

In fact, it can be seen that there are four infinite subsequences of equations. Two of these are for $a_0, a_2, \ldots, a_{2m}$ and $b_2, b_4, \ldots, b_{2m}$, and represent solutions with period π. For a solution to exist, the corresponding infinite determinant of the coefficients must vanish, thus determining the stability boundaries in the δ-ϵ plane. For this case these boundaries are those which reduce to $\delta = (2k)^2$ ($k = 0, 1, 2, \ldots$) when $\epsilon \to 0$. The remaining two subsequences are for $a_1, a_3, \ldots, a_{2m+1}$ and $b_1, b_3, \ldots, b_{2m+1}$, and correspond to those stability boundaries which reduce to $\delta = (2k+1)^2$ ($k = 0, 1, 2, \ldots$) when $\epsilon \to 0$. When the coefficients $\{a_0, a_n, b_n; n = 1, 2, \ldots\}$ satisfy the system of equations (3.23), the corresponding solutions (3.22) of the Mathieu equation (3.1) are called Mathieu functions.

To find numerical values for the stability boundaries, the infinite sequence of equations (3.23) must be truncated after the mth term, where m is a suitably large integer. The corresponding finite determinants can then be evaluated numerically for any specified values of δ and ϵ. The zeros, corresponding to the stability boundaries, are found by specifying ϵ and then searching for a δ which gives a zero (or vice versa). For instance, the determinant corresponding to the subsequence $a_0, a_2, \ldots, a_{2m}$ is

$$
D_n = \det \begin{bmatrix}
1 & V_0 & 0 & 0 & \cdots & 0 & 0 & 0 \\
V_2 & 1 & V_2 & 0 & \cdots & 0 & 0 & 0 \\
0 & V_4 & 1 & V_4 & \cdots & 0 & 0 & 0 \\
0 & 0 & V_6 & 1 & \cdots & 0 & 0 & 0 \\
\vdots & \vdots & \vdots & \vdots & \ddots & \vdots & \vdots & \vdots \\
0 & 0 & 0 & 0 & \cdots & 1 & V_{n-4} & 1 \\
0 & 0 & 0 & 0 & \cdots & V_{n-2} & 1 & V_{n-2} \\
0 & 0 & 0 & 0 & \cdots & 0 & V_n & 1
\end{bmatrix}, \quad (3.24)
$$

where

$$
V_0 = \epsilon \delta^{-1}, \qquad V_r = \epsilon \{2(\delta - r^2)\}^{-1} \quad (r = 1, 2, \ldots)
$$

Here, of course, $n = 2m$ and is an even integer. This sequence of determinants is best evaluated from the following recurrence relation:

$$
D_n = D_{n-2} - V_n V_{n-2} D_{n-4} \quad (n \geq 4), \tag{3.25}
$$

which is readily established from (3.24). To start the iterative process we observe that

$$
D_0 = 1, \qquad D_2 = 1 - V_0 V_2.
$$

A similar procedure can be followed for the other subsequences. For instance, the determinant corresponding to the subsequence $b_2, b_4, \ldots, b_{2m}$ is just D_n with the first row and column deleted. The recurrence relation is again (3.25) for $n \geq 6$, but the starting values are now given by

$$
D_2 = 1, \qquad D_4 = 1 - V_2 V_4.
$$

For the subsequences $a_1, a_3, \ldots, a_{2m+1}$ and $b_1, b_3, \ldots, b_{2m+1}$ the determinant is

$$
D_n = \det \begin{bmatrix}
1 \pm V_1 & V_1 & 0 & 0 & \cdots & 0 & 0 & 0 \\
V_3 & 1 & V_3 & 0 & \cdots & 0 & 0 & 0 \\
0 & V_5 & 1 & V_5 & \cdots & 0 & 0 & 0 \\
0 & 0 & V_7 & 1 & \cdots & 0 & 0 & 0 \\
\vdots & \vdots & \vdots & \vdots & \ddots & \vdots & \vdots & \vdots \\
0 & 0 & 0 & 0 & \cdots & 1 & V_{n-4} & 1 \\
0 & 0 & 0 & 0 & \cdots & V_{n-2} & 1 & V_{n-2} \\
0 & 0 & 0 & 0 & \cdots & 0 & V_n & 1
\end{bmatrix}. \quad (3.26)
$$

Here $n = 2m+1$ and is an odd integer. The recurrence relation is again (3.25), now for $n \geqslant 5$, and the starting values are now

$$D_1 = 1 \pm V_1, \qquad D_3 = 1 \pm V_1 - V_1 V_3.$$

Here the alternate signs refer to the sequences $a_1, a_3, \dots, a_{2m+1}$ and $b_1, b_3, \dots, b_{2m+1}$, respectively.

The result of this procedure for calculating the stability boundaries is shown in figure 3.5. Note that the boundaries are symmetric with respect to ϵ, and so only the region $\epsilon \geqslant 0$ is shown. It is apparent that, as $|\epsilon|$ increases, the unstable regions increase in width and, eventually, the stable regions are restricted to bands of extremely small width.

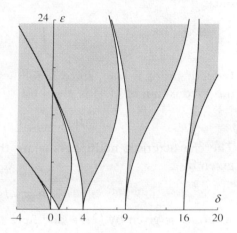

Fig. 3.5 Zones of instability for the Mathieu equation (3.1). The zones are symmetric in ϵ, so only the region $\epsilon \geqslant 0$ is shown. The shaded regions correspond to unstable zones.

3.4 The Mathieu equation with damping

It is instructive to consider the effect of including a linear damping term in the Mathieu equation (3.1). Thus we consider

$$u'' + \nu u' + (\delta + \epsilon \cos 2t) u = 0. \tag{4.1}$$

Our main interest is in the case when ν is small and positive, but initially we shall reconsider the general theory of section 3.2 for parametric resonance, without any restriction on ν.

First form the equivalent first-order system, using the procedure of section 2.4 of chapter 2. Thus we put (see (4.2) of section 2.4)

$$x_1 = u, \qquad x_2 = u', \tag{4.2}$$

and (4.1) is equivalent to the first-order system

$$x_1' = x_2, \qquad x_2' = -(\delta + \epsilon \cos 2t) x_1 - \nu x_2. \tag{4.3}$$

The matrix $A(t)$ in (1.1) is here

$$A(t) = \begin{bmatrix} 0 & 1 \\ -(\delta + \epsilon \cos 2t) & -\nu \end{bmatrix},$$

and we note that now $\operatorname{tr} A(t) = -\nu$. Next we form the fundamental matrix $X(t)$ of (4.3) such that $X(0) = E$. It is again given by (2.5), where u^1 and u^2 are linearly independent solutions of (4.1) which satisfy the initial conditions (2.6). The matrix B (see (1.7)) is again given by (2.7), but we note that now (see (1.4))

$$\det B = e^{-\pi\nu},$$

since

$$\int_0^\pi \operatorname{tr} A(t) \, dt = -\pi\nu. \tag{4.4}$$

Indeed, more is true, since it follows from (4.9) and (4.10) of section 2.4 that the Wronskian of (4.1) is given by

$$u^1 u^{2\prime} - u^2 u^{1\prime} = e^{-\nu t} \quad \text{(for all } t\text{)}.$$

The characteristic multipliers ρ are the eigenvalues of B. Using (4.4), they are given by

$$\rho^2 - 2\phi\rho + e^{-\pi\nu} = 0, \tag{4.5}$$

where ϕ is given by (2.9), and here the period $T = \pi$. Hence the characteristic multipliers $\rho_{1,2}$ are given by

$$\rho_{1,2} = \phi \pm \sqrt{\phi^2 - e^{-\pi\nu}}. \tag{4.6}$$

Note the useful relations

$$\rho_1 \rho_2 = e^{-\pi\nu}, \qquad \rho_1 + \rho_2 = \phi. \tag{4.7}$$

The characteristic exponents are $\mu_{1,2}$, where $\rho_{1,2} = e^{\pi\mu_{1,2}}$, and, consequent to (4.7),

$$\mu_1 + \mu_2 = -\nu, \qquad \cosh(\mu_1\pi + \tfrac{1}{2}\nu\pi) = \phi e^{\frac{1}{2}\pi\nu}. \tag{4.8}$$

First we note from either (4.7) or (4.8) that, in the case of negative damping when $\nu < 0$, at least one solution will always describe unstable behaviour. Indeed it is an immediate consequence of (4.7) that

$$|\rho_1||\rho_2| = e^{-\pi\nu} > 1 \quad \text{(if } \nu < 0\text{)}.$$

Thus at least one of $|\rho_1|$ and $|\rho_2|$ is greater than one, and the corresponding solution clearly describes unstable behaviour. Equivalently, (4.8) shows that

$$\mathrm{Re}\,\mu_1 + \mathrm{Re}\,\mu_2 = -\nu > 0 \quad (\text{if } \nu < 0)$$

and at least one of $\mathrm{Re}\,\mu_1$ and $\mathrm{Re}\,\mu_2$ is positive. Of course, the fact that negative damping will cause instability regardless of the values of δ and ϵ is intuitively obvious, but it is reassuring that it can be established analytically.

Next when $\nu > 0$, we must proceed, as in section 3.2, to discuss several possibilities, depending on the value of ϕ.

(i) $\phi > \mathrm{e}^{-\frac{1}{2}\pi\nu}$: Here $\rho_{1,2}$ are both real and positive. If $\mathrm{e}^{-\frac{1}{2}\pi\nu} < \phi < \frac{1}{2}(1 + \mathrm{e}^{-\pi\nu})$ then $0 < \rho_2 < \rho_1 < 1$, which corresponds to stable behaviour. Equivalently, $-\nu < \mu_2 < -\frac{1}{2}\nu < \mu_1 < 0$. However, if $\phi > \frac{1}{2}(1 + \mathrm{e}^{-\pi\nu})$ then $0 < \rho_2 < 1 < \rho_1$, which corresponds to unstable behaviour. Equivalently, $\mu_1 > 0$ and $\mu_2 < -\nu$.

(ii) $\phi < -\mathrm{e}^{-\frac{1}{2}\pi\nu}$: Here $\rho_{1,2}$ are both real and negative. If $\mathrm{e}^{-\frac{1}{2}\pi\nu} < -\phi < \frac{1}{2}(1 + \mathrm{e}^{-\pi\nu})$ then $0 > \rho_2 > \rho_1 > -1$, which corresponds to stable behaviour. In this case we must put $\mu_1 = \mathrm{i} - \gamma_1$ and $\mu_2 = -\mathrm{i} - \gamma_2$ (see (2.14)) and, equivalently, $0 < \gamma_2 < \frac{1}{2}\nu < \gamma_1 < \nu$. However, if $-\phi > \frac{1}{2}(1 + \mathrm{e}^{-\pi\nu})$ then $0 > \rho_1 > -1 > \rho_2$ which corresponds to unstable behaviour. Equivalently, $\gamma_1 > \nu$ and $\gamma_2 < 0$.

(iii) $-\mathrm{e}^{-\frac{1}{2}\pi\nu} < \phi < \mathrm{e}^{-\frac{1}{2}\pi\nu}$: Here $\rho_{1,2}$ are both complex-valued and $|\rho_1| = |\rho_2| = \mathrm{e}^{-\pi\nu}$, which corresponds to stable behaviour. Equivalently, $\mathrm{Re}\,\mu_1 = \mathrm{Re}\,\mu_2 = -\nu$.

The stability boundaries are thus defined by

$$\phi = \pm\tfrac{1}{2}(1 + \mathrm{e}^{-\pi\nu}).$$

In the first case $\rho_1 = 1$ and $\mu_1 = 0$ and there is a solution of (4.1) which is periodic with period π. Also $\rho_2 = \mathrm{e}^{-\pi\nu}$ and $\mu_2 = -\nu$, which corresponds to a solution of the form $\mathrm{e}^{-\nu t}p(t)$, where $p(t)$ is periodic with period π. In the second case $\rho_2 = -1$ and $\mu_2 = -\mathrm{i}$ and there is a solution of (4.1) which is periodic with period 2π. Also $\rho_1 = -\mathrm{e}^{-\pi\nu}$ and $\mu_1 = \mathrm{i} - \nu$, which corresponds to a solution of the form $\mathrm{e}^{-\nu t}q(t)$, where $q(t)$ is periodic with period 2π.

Next we consider the limit as $\epsilon \to 0$, with δ and ν fixed. The general solution of (4.1) is

$$u = \mathrm{e}^{\frac{1}{2}\nu t}(c_1 \cos \omega t + c_2 \sin \omega t) \quad (\text{for } \epsilon = 0),$$

where

$$\omega^2 = \delta - \tfrac{1}{4}\nu^2,$$

and, initially, we assume that $\delta > \frac{1}{4}\nu^2$. The characteristic exponents are

$-\frac{1}{2}\nu\pm i\omega$ and the solutions are always stable. Next, if $\delta < \frac{1}{4}\nu^2$, the general solution of (4.1) is

$$u = e^{-\frac{1}{2}\nu t}(c_1 \cosh \sigma t + c_2 \sinh \sigma t) \quad (\text{for } \epsilon = 0),$$

where

$$\sigma^2 = \frac{1}{4}\nu^2 - \delta.$$

The characteristic exponents are now $-\frac{1}{2}\nu\pm\sigma$ and the solutions are stable if $\delta > 0$, but one solution is unstable if $\delta < 0$. We can conclude that, for $\nu > 0$, the stability boundaries do not cross the axis $\epsilon = 0$, except at $\delta = 0$.

To discuss parametric resonance, it is necessary to suppose that $\nu, \epsilon \to 0$ simultaneously. In this joint limit the stability boundaries are given by $\phi = \pm 1$, just as in section 3.2, and it follows that the conditions for parametric resonance are again given by (3.2). To analyse the stability boundaries when $\delta \approx n^2$, where $n = 0, 1, 2, \ldots$, and ν and ϵ are both small, we seek a solution of (4.1) of the form (cf. (3.3))

$$u = e^{\mu t}q(t), \tag{4.9}$$

where $q(t)$ is periodic of period π when n is an even integer, and has period 2π when n is an odd integer. Substituting (4.9) into (4.1), we get

$$q'' + (2\mu + \nu)q' + (\mu^2 + \nu\mu)q + (\delta + \epsilon \cos 2t)q = 0. \tag{4.10}$$

The condition that the solutions of (4.10) have the required periodicity determines μ as a function of ν, δ and ϵ.

As in section 3.3, we seek solutions of (4.10) as power series in ϵ. Thus we assume the expansions (3.5) for q, μ and δ, and add to them the corresponding expansion of ν:

$$\nu = \epsilon\nu_1 + \epsilon^2\nu_2 + \ldots. \tag{4.11}$$

Substituting these expansions into (4.10) leads to a series of equations of the form (3.6), and the requirement that each $q_m(t)$ ($m = 0, 1, 2, \ldots$) is periodic will determine $\mu_1, \mu_2, \ldots$ in terms of $\delta_1, \nu_1, \delta_2 \nu_2, \ldots$. The procedure is analogous to that used in section 3.3, but we shall not give details here as in this instance an alternative and simpler method is available. Let

$$\hat{\mu} = \mu + \frac{1}{2}\nu, \qquad \hat{\delta} = \delta - \frac{1}{4}\nu^2. \tag{4.12}$$

Then (4.10) becomes

$$q'' + 2\hat{\mu}q' + \hat{\mu}^2 q + (\hat{\delta} + \epsilon \cos 2t)q = 0, \tag{4.13}$$

which is identical in form to the corresponding equation (3.4) for the undamped Mathieu equation, provided we replace $\hat{\mu}$ with μ and $\hat{\delta}$ with δ. The transformation (4.12) is a special case of a more general transformation which is described in the exercises.

Since (4.13) has the same form as (3.4) we can utilize the results of section 3.3 to obtain the corresponding results for (4.13) and, hence, for (4.10).

(i) $n = 1$ ($\delta \approx 1$). From (3.9) we find that

$$4\hat{\mu}^2 \approx \tfrac{1}{4}\epsilon^2 - (\hat{\delta} - 1)^2,$$

with an error of $O(\epsilon^3)$. Since ν is $O(\epsilon)$, it follows from (4.12) that

$$(2\mu + \nu)^2 \approx \tfrac{1}{4}\epsilon^2 - (\delta - 1)^2. \qquad (4.14)$$

Thus, for $(\delta - 1)^2 > \tfrac{1}{4}\epsilon^2$, the two solutions are complex-valued, but $\mathrm{Re}\,\mu_1 = \mathrm{Re}\,\mu_2 = -\tfrac{1}{2}\nu$, corresponding to stable behaviour. For $(\delta - 1)^2 < \tfrac{1}{4}\epsilon^2$ the two solutions are real-valued, but $\mathrm{Re}\,\mu_2 < \mathrm{Re}\,\mu_1 < 0$ when $(\delta - 1)^2 > \tfrac{1}{4}\epsilon^2 - \nu^2$, corresponding again to stable behaviour. Otherwise, $\mathrm{Re}\,\mu_1 > 0 > \mathrm{Re}\,\mu_2$ and there is unstable behaviour. The stability boundary here corresponds to $\mu = 0$ and is given by

$$\nu^2 + (\delta - 1)^2 \approx \tfrac{1}{4}\epsilon^2, \qquad (4.15)$$

where the error term is $O(\epsilon^3)$. For each fixed $\nu > 0$, this is a hyperbola in the δ-ϵ plane, which is asymptotic to the straight lines $\delta - 1 \approx \pm \tfrac{1}{2}\epsilon$. Note that these lines are stability boundaries when $\nu = 0$ (see (3.10)). The situation is sketched in figure 3.6. Note that the unstable zone is at least a distance 2ν from the axis $\epsilon = 0$.

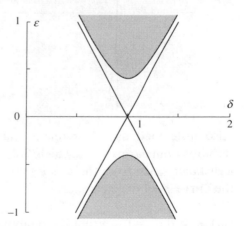

Fig. 3.6 The zones of instability for the damped Mathieu equation (4.1) when $\delta \approx 1$ and $|\epsilon|$ and ν are small. The shaded regions correspond to unstable zones. The case shown is for $\nu = 0.2$.

(ii) $n = 2$ ($\delta \approx 4$). From (3.16) we find that

$$16\hat{\mu}^2 \approx -(\hat{\delta} - 4)^2,$$

with an error of $O(\epsilon^3)$. Hence, from (4.12),

$$(4\mu + 2\nu)^2 \approx -(\delta - 4)^2, \qquad (4.16)$$

where the error term is again $O(\epsilon^3)$. For $\delta \neq 4$, the two solutions for μ are complex-valued with $\operatorname{Re}\mu = -\frac{1}{2}\nu$ and, hence, there is stable behaviour for all $\nu > 0$.

To resolve the unstable zone we must proceed to the next order. We put $\mu_1 = \delta_1 = \nu_1 = 0$ so that μ, $\delta - 4$ and ν are all $O(\epsilon^2)$. From (3.18),

$$16\hat{\mu}^2 \approx -\{(\hat{\delta}-4) - \tfrac{5}{48}\epsilon^2\}\{(\hat{\delta}-4) + \tfrac{1}{48}\epsilon^2\},$$

and so, using (4.12),

$$(4\mu + 2\nu)^2 \approx -\{(\delta-4) - \tfrac{5}{48}\epsilon^2\}\{(\delta-4) + \tfrac{1}{48}\epsilon^2\}, \qquad (4.17)$$

where the error term is $O(\epsilon^5)$. The stability boundary again corresponds to $\mu = 0$, or

$$4\nu^2 + \{(\delta-4) - \tfrac{5}{48}\epsilon^2\}\{(\delta-4) + \tfrac{1}{48}\epsilon^2\} \approx 0. \qquad (4.18)$$

The situation is sketched in figure 3.7, where we see that the unstable zones lie above (for $\epsilon > 0$) and below (for $\epsilon < 0$) the curves given by (4.18).

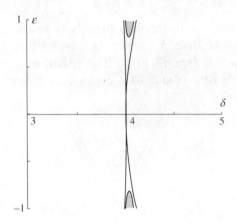

Fig. 3.7 As for figure 3.6, but now $\delta \approx 4$ and $\nu = 0.02$.

Also note that, as $|\epsilon|$ increases, these curves asymptote to the corresponding stability boundaries (3.9) which hold when $\nu = 0$. The unstable zone is at least a distance $\sqrt{32\nu}$ from the axis $\epsilon = 0$, and the point of closest approach occurs on the curve $\delta - 4 \approx \tfrac{1}{24}\epsilon^2$.

(iii) **$n = 0$** ($\delta \approx 0$). From the analysis in section 3.3 $\hat{\delta}_1 = 0$, which implies that $\hat{\delta}$ and, hence, δ (see (4.12)), is $O(\epsilon^2)$. Then, from (3.20), we get

$$\hat{\mu}^2 + \hat{\delta} + \tfrac{1}{8}\epsilon^2 \approx 0,$$

with an error of $O(\epsilon^3)$. Hence, from (4.12),

$$\mu^2 + \nu\mu + \delta + \tfrac{1}{8}\epsilon^2 \approx 0, \qquad (4.19)$$

where the error term is again $O(\epsilon^3)$. The stability boundary again corresponds to $\mu = 0$, or

$$\delta \approx -\tfrac{1}{8}\epsilon^2. \tag{4.20}$$

To this order, the presence of positive damping has not altered the position of the stability boundary (see (3.21)), although the characteristic exponents are affected.

Problems

1 For each of the following linear systems with periodic coefficients of period T, identify the period, find a fundamental matrix $X(t)$, the matrix B such that $X(T) = X(0)B$, the characteristic multipliers and the characteristic exponents. Also show that $X(t) = P(t)Y(t)$, where $P(t)$ is periodic, $Y' = DY$ and D is a constant matrix.

(a)
$$x_1' = -x_1 + x_2, \qquad x_2' = \left(1 + \cos t - \frac{\sin t}{2 + \cos t}\right)x_2.$$

(b) $x_1' = (1 + 2\cos 2t)x_1 + (1 - 2\sin 2t)x_2, \qquad x_2' = -(1 + 2\sin 2t)x_1 + (1 - 2\cos 2t)x_2.$

[*Hint*: Try the transformation $x = Q(t)\hat{x}$, where

$$Q = \begin{bmatrix} \cos t & \sin t \\ -\sin t & \cos t \end{bmatrix}.]$$

(c)
$$x_1' = \left(1 + \frac{\cos t}{2 + \sin t}\right)x_1, \qquad x_2' = x_2 + 2x_1.$$

2 For the Mathieu equation

$$u'' + (\delta + \epsilon \cos 2t)u = 0,$$

find the characteristic exponents when ϵ is small and $(\delta - n^2)$ is $O(\epsilon)$ for $n = 0, 1, 2, \ldots$. Hence, or otherwise, show that the boundaries which separate stable from unstable behaviour are given by

$$\delta = -\tfrac{1}{8}\epsilon^2 + \tfrac{7}{2048}\epsilon^4 + \ldots, \qquad \text{(for } n = 0\text{)};$$

$$\delta - 1 = \pm\tfrac{1}{2}\epsilon - \tfrac{1}{32}\epsilon^2 + \ldots, \qquad \text{(for } n = 1\text{)};$$

$$\delta - 4 = \tfrac{5}{48}\epsilon^2 \text{ or } -\tfrac{1}{48}\epsilon^2 + \ldots, \qquad \text{(for } n = 2\text{)};$$

$$\delta - n^2 = \frac{\epsilon^2}{8(n^2 - 1)} + \ldots, \qquad \text{(for } n \geqslant 3\text{)}.$$

3 For the equation

$$u'' + (\delta + \epsilon \cos t + \epsilon \sin 2t)u = 0,$$

show that the conditions for parametric resonance are $\epsilon = 0$ and $\delta = k^2$ or $(k + \tfrac{1}{2})^2$ $(k = 0, 1, 2, \ldots)$. Use a perturbation method to find the boundaries which separate stable from unstable behaviour for the cases $k = 0, 1, 2$.

4 Consider Hill's equation

$$u'' + \{\delta + \epsilon b(t)\}u = 0,$$

where $b(t + \pi) = b(t)$ for all t. Show that the conditions for parametric resonance are $\epsilon = 0$ and $\delta = k^2$ $(k = 0, 1, 2, \ldots)$. Suppose that $b(t)$ is given by the Fourier series

$$b(t) = \sum_{n=1}^{\infty} (c_n \cos 2nt + d_n \sin 2nt.)$$

Use a perturbation method to show that the boundaries which separate stable from unstable behaviour are given by

$$\delta - k^2 \approx \pm \tfrac{1}{2} \epsilon \sqrt{c_k^2 + d_k^2} \quad (\text{for } k = 1, 2, 3, \dots),$$

where the error term is $O(\epsilon^2)$ and

$$\delta \approx -\tfrac{1}{8} \epsilon^2 \sum_{n=1}^{\infty} \frac{c_n^2 + d_n^2}{n^2} \quad (\text{for } k = 0),$$

where the error term is $O(\epsilon^3)$. Note that here

$$\int_0^{\pi} b(t) \, dt = 0,$$

and comment on how these results would change if this condition does not hold.

5 The stability boundaries for the Mathieu equation (3.1) can be obtained by finding the zeros of the determinants D_n, given by (3.24), (3.24) modified by omission of the first row and column, or (3.26). The integer n must be chosen large enough to ensure that some specified requirement for accuracy is satisfied. Verify the recurrence relation (3.25), and use the sequence of determinants to obtain the stability boundaries correct to $O(\epsilon^2)$. Compare your answers with those obtained in problem 2.

6 For Meissner's equation

$$u'' + (\delta + \epsilon b(t)) u = 0,$$

where $b(t + \pi) = b(t)$ for all t and

$$b(t) = \begin{cases} 1 & (0 < t < \tfrac{1}{2}\pi), \\ -1 & (\tfrac{1}{2}\pi < t < \pi), \end{cases}$$

find an expression for the Floquet exponents and hence find an equation for the curves in the δ-ϵ plane which are stability boundaries. Solve this for sufficiently small $|\epsilon|$ in the vicinity of the points $\epsilon = 0$, $\delta = n^2$ $(n = 0, 1, 2, \dots)$. Compare these latter results with those which would be obtained using the approach of problem 4. [*Hint*: Assume that u and u' are continuous at $t = \tfrac{1}{2}\pi$ and construct u^1 and u^2 such that

$$u^1(0) = u^{2\prime}(0) = 1, \qquad u^{1\prime}(0) = u^2(0) = 0.]$$

7 For the differential equation

$$u'' + 2\nu \sin^2 t \, u' + (\delta + \epsilon \sin 2t) u = 0,$$

show that the characteristic multipliers are determined by the roots of the quadratic equation

$$\rho^2 - 2\phi\rho + e^{-\pi\nu} = 0,$$

where

$$\phi = \tfrac{1}{2} \{ u^1(\pi) + u^{2\prime}(\pi) \}.$$

Here u^1 and u^2 are solutions such that

$$u^1(0) = u^{2\prime}(0) = 1, \qquad u^{1\prime}(0) = u^2(0) = 0.$$

Deduce that there is unstable behaviour when $\nu < 0$. For $\nu > 0$ use a perturbation method to find the stability boundaries when ν and $|\epsilon|$ are both small, in the vicinity of the points $\epsilon = 0$, $\delta = n^2$ $(n = 0, 1, 2)$.

8 (a) For the differential equation

$$u'' + a_1(t)u' + a_2(t)u = 0,$$

where $a_i(t + T) = a_i(t)$ for all t $(i = 1, 2)$, show that the characteristic multipliers ρ_1 and ρ_2 satisfy the relation

$$\rho_1\rho_2 = \exp\left\{ -\int_0^T a_1(t) \, dt \right\}.$$

(b) Use the transformation

$$u = v \exp\left\{ -\tfrac{1}{2}\int_0^t a_1(s) \, ds \right\}$$

to show that

$$v'' + a(t)v = 0, \qquad a(t) = a_2(t) - \tfrac{1}{2}a_1' - \tfrac{1}{4}a_1^2.$$

If the characteristic multipliers for this equation are σ_1 and σ_2, show that

$$\rho_{1,2} = \sigma_{1,2}\exp\left\{ -\tfrac{1}{2}\int_0^T a_1(s) \, ds \right\}$$

and, hence, confirm that $\sigma_1\sigma_2 = 1$.

(c) For the case when

$$a_1 = 2\nu \sin^2 t, \qquad a_2 = \delta + \epsilon \sin 2t,$$

use a perturbation method applied to the equation for v to determine the characteristic exponents when ν and $|\epsilon|$ are both small, in the vicinity of the points $\epsilon = 0$, $\delta = n^2$ $(n = 0, 1, 2)$. Hence determine the stability boundaries for the equation for u, and compare your results with those obtained in problem 7.

9 (a) Consider the Hill's equation

$$u'' + a(t)u = 0,$$

where $a(t + T) = a(t)$ for all t. Show that, if $a(t) < 0$ for all t, then the solution satisfying the initial condition

$$u(0) = u'(0) = 1$$

is unbounded as $t \to \infty$. Hence deduce that this solution describes unstable behaviour. [*Hint*: Use the expression

$$u' = 1 - \int_0^t a(s)u(s) \, ds$$

to deduce that $u'(t) \geqslant 1$ for all $t \geqslant 0$ and, hence, conclude that $u(t) \to \infty$ as $t \to \infty$.]

(b) Next suppose that $a(t) > 0$ for all t and

$$\int_0^T a(t) \, dt < \frac{4}{T}.$$

It may then be shown that all solutions are bounded as $t \to \infty$ (see, for instance, Cesari, 1962, p. 60). Use this result and that of part (a) to estimate the stable and unstable zones in the δ-ϵ plane for the Mathieu equation (3.1) and Meissner's equation (problem 6).

10 A two-species population model is described by the following linear system

$$x_1' = -\nu x_1 + a(t) x_1,$$
$$x_2' = -\nu x_2 + ka(t) x_1,$$

where $a(t + T) = a(t)$ for all t and ν and k are positive constants. Find the characteristic exponents and show that there is a periodic solution of period T if and only if

$$\int_0^T a(t) \, dt = \nu T.$$

CHAPTER FOUR
STABILITY

4.1 Preliminary definitions

Consider the first-order system

$$x' = f(x, t), \qquad (1.1)$$

where, to avoid unnecessary complications, we shall suppose that $f(x, t)$ is defined and continuous for all x and all $t \geq t_0$, and satisfies a Lipschitz condition in x in any bounded domain. Then, for the initial-value problem

$$x(t_0) = x_0, \qquad (1.2)$$

the uniqueness and existence theorems of section 1.3 (theorems 1.3 and 1.4) show that there is a unique solution

$$x = x(t; x_0, t_0) \quad (t_0 \leq t < T),$$

where, by theorem 1.7 (section 1.4), either $|x| \to \infty$ as $t \to T$ or $T = \infty$. We shall suppose here that the latter alternative holds, and so $x(t; x_0, t_0)$ is defined for all $t \geq t_0$. This will not be a significant restriction as in the applications to follow $x(t)$ will either be a constant or a periodic function of t.

Stability is concerned with the question as to whether solutions which are in some sense close to $x(t)$ at some instant will remain close for all subsequent times. Clearly, stability is a desirable property as dynamical processes, modelled here by the system of equations (1.1), are often subject to small, unpredictable disturbances. Unstable solutions are thus extremely difficult to realize either experimentally or numerically, as an arbitrarily small disturbance will eventually cause large deviation from the unstable solution. It is important to distinguish here between stability and the Hadamard notion that the initial-value problem be well-set. We have already shown in theorem 1.2 (section 1.3) and theorem 1.5 (section 1.4) that the initial-value problem for (4.1) is well-set in that the solutions, *inter alia*, are continuous functions of the initial value x_0. Thus solutions which are close to $x(t)$ at $t = t_0$, will remain close for some interval $t_0 \leq t \leq t_0 + \delta$. The important point here is that, in general, these results hold only for some **finite** interval of time t (i.e. δ is finite), whereas stability is concerned with the behaviour of neighbouring solutions to $x(t)$ for **all** times $t \geq t_0$.

To study the stability of $x(t)$, consider the neighbouring solution $y = y(t; y_0, t_0)$, where

$$y' = f(y,t), \qquad y(t_0) = y_0.$$

We are concerned here with the difference between x and y and, hence, we put

$$z = y - x(t). \tag{1.3}$$

It follows that

$$z' = F(z,t), \tag{1.4}$$

where

$$F(z,t) = f(x(t)+z,t) - f(x(t),t).$$

Note that the t-dependence of $F(z,t)$ includes that due to the solution $x(t)$. Here $F(0,t) = 0$ for all $t \geqslant t_0$, and so the function $z = 0$ for all $t \geqslant t_0$ is a solution of (1.4). Thus the stability of $x(t)$ as a solution of (1.1) is reduced to the stability of the zero solution of (1.4). Note that, although we can assume that $x(t)$ exists for all $t \geqslant t_0$, we cannot necessarily assume that $z(t)$ (or $y(t)$) likewise exist for all $t \geqslant t_0$. Three notions of stability will be found useful and their definitions follow.

Definition: The solution $z = 0$ of (1.4) is said to be:
   (i) **stable**, if for all $\epsilon > 0$ and any $t_1 \geqslant t_0$, there exists a $\delta(\epsilon, t_1)$ such that $|z(t_1)| < \delta$ implies that $|z(t)| < \epsilon$ for all $t \geqslant t_1$;
   (ii) **uniformly stable**, if stable and $\delta = \delta(\epsilon)$ is independent of t_1;
   (iii) **asymptotically stable**, if stable and $|z(t_1)| < \delta$ implies that $|z(t)| \to 0$ as $t \to \infty$.

Definition (i) is also sometimes called **Liapunov** stability. If (i) does not hold, the solution $z = 0$ is said to be **unstable**.

Before proceeding with the general theory, we shall illustrate these definitions with the following examples. In the first three cases $n = 1$ and we temporarily write z for the scalar component z_1. Consider

$$z' = -z,$$

so that

$$z = z_0 \exp\{-(t-t_0)\}.$$

Clearly the zero solution is uniformly and asymptotically stable. Indeed, here we may put $\delta = \epsilon$, since $|z(t_1)| < \epsilon$ implies that $|z| < \epsilon$ for all $t \geqslant t_1$, and $|z| \to 0$ as $t \to \infty$. Next consider

$$z' = z,$$

so that

$$z = z_0 \exp\{t-t_0\}.$$

Here the zero solution is unstable since $|z| \to \infty$ as $t \to \infty$ for all $z_0 \neq 0$. For an example where the zero solution is stable but not uniformly stable, consider the function

$$z = \frac{C}{1+t} \exp\{(C^2 - 1)t\},$$

where C is an arbitrary constant. Elimination of C between this expression and the analogous expression for z' will yield a first-order differential equation for z, although, in this instance, it cannot be written down explicitly. The zero solution corresponds to $C = 0$ and is asymptotically stable. Indeed, for any $t_1 \geqslant 0$, we can choose δ to be the minimum of ϵ and $(1+t_1)^{-1}$. Also $|z| \to 0$ as $t \to \infty$ whenever $|z(t_1)| < \delta$. However, δ now depends on t_1 and reference to figure 4.1 shows that this dependence cannot be removed. Note that $z(0) = C$ and that, for $0 \leqslant |C| \leqslant 1$, $|z| \to 0$ as $t \to \infty$, but, for $|C| > 1$, all the solution curves diverge as $t \to \infty$. But, as $|C| \to 1+$, the diverging solution curves can be made to approach the axis $z = 0$ arbitrarily closely.

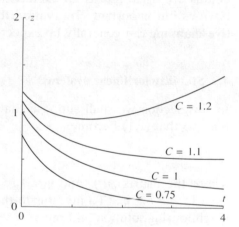

Fig. 4.1 An example of a situation when the zero solution is stable, but not uniformly stable. A plot of $z = C(1+t)^{-1} \exp\{(C^2 - 1)t\}$.

The next example is the equation for a simple harmonic oscillator,

$$u'' + \omega^2 u = 0,$$

which, on putting $u = z_1$ and $u' = \omega z_2$, is equivalent to the first-order system

$$z_1' = \omega z_2, \qquad z_2' = -\omega z_1.$$

The general solution is

$$z_1 = c_1 \cos \omega t + c_2 \sin \omega t, \qquad z_2 = -c_1 \sin \omega t + c_2 \cos \omega t.$$

It may now be shown that

$$|z| \leqslant \sqrt{2}|z(t_1)|$$

and, hence, the zero solution is uniformly stable. The details are left as an exercise. However, since z oscillates, but does not tend to zero as $t \to \infty$, the zero solution cannot be asymptotically stable. The final example is the equation for the damped simple harmonic oscillator,

$$u'' + \nu u' + \omega^2 u = 0,$$

which, on putting $u = z_1$ and $u' = \omega z_2$, is equivalent to the first-order system

$$z_1' = \omega z_2, \qquad z_2' = -\omega z_1 - \nu z_2.$$

Now the general solution is

$$z_1 = e^{-\frac{1}{2}\nu t}\{c_1 \cos \sigma t + c_2 \sin \sigma t\}, \qquad z_2 = z_1' \omega^{-1},$$

where

$$\sigma^2 = \omega^2 - \tfrac{1}{4}\nu^2.$$

Clearly, the zero solution is now asymptotically stable whenever $\nu > 0$. The details are again left as an exercise. This last example illustrates the intuitively obvious but important observation that the presence of a small amount of positive damping can generally be expected to improve the stability.

4.2 Stability for linear systems

In this section we shall suppose that the first-order system (1.1) is a linear system, so that (1.1) becomes

$$x' = A(t)x + g(t), \tag{2.1}$$

where the matrix $A(t)$, and $g(t)$, are continuous for all $t \geq t_0$. Hence, solutions exist for all $t \geq t_0$. To investigate the stability of a solution $x(t)$, we let $y(t)$ be a neighbouring solution and put $z = y - x$ (1.3). Then (1.4) for z becomes

$$z' = A(t)z. \tag{2.2}$$

The significant point here is that equation (2.2) for z is just the homogeneous equation corresponding to (2.1) and, hence, does not depend on the specific solution $x(t)$ whose stability is being studied. Thus all solutions of (2.1) will have the same stability, which is just the stability of the zero solution of (2.2). In this sense we can refer to the stability of the system of equations (2.1). Note that this cannot be done for a genuinely nonlinear equation, since then the stability equation (1.4) will depend explicitly on the specific solution $x(t)$ whose stability is being studied and, in general, different solutions of the same equation will have different stability properties. The discussion in chapter 5 of critical points provides a simple illustration of this.

When the matrix $A(t)$ is a constant, A_0, equation (2.2) can be solved explicitly (see section 2.5), and the following theorem can be established.

Theorem 4.1: Suppose that the matrix $A(t)$ in (2.2) is a constant, A, with eigen-values $\lambda_1, \lambda_2, \ldots, \lambda_n$. Then the stability of the zero solution of (2.2) (and hence of any solution of (2.1)) is determined according to the following criteria.

  (i) If $\text{Re}\,\lambda_i < 0$ for all $i = 1, \ldots, n$, then there is uniform and asymptotic sta-bility.

 (ii) If $\text{Re}\,\lambda_i \leq 0$ for all $i = 1, \ldots, n$ and the algebraic multiplicity equals the geometric multiplicity whenever $\text{Re}\,\lambda_j = 0$ for any j, then there is uniform stability.

(iii) If $\text{Re}\,\lambda_j > 0$ for at least one j, or the algebraic multiplicity is greater than the geometric multiplicity should $\text{Re}\,\lambda_j = 0$, then there is instability.

Proof: From the discussion in section 2.5 (see, for instance, (5.5), (5.23) or (5.25) of section 2.5), it can be established that a fundamental matrix for (2.2) has the form

$$Z(t) = [e^{\lambda_1 t}p^1(t), \ldots, e^{\lambda_n(t)}p^n(t)], \tag{2.3}$$

where $p^1(t), \ldots, p^n(t)$ are n linearly independent vector-valued polynomials in t of degree $s_i - 1$, where $1 \leq s_i \leq m_i - p_i + 1$, and m_i and p_i are, respectively, the algebraic and geometric multiplicities of a given eigenvalue λ_i. If any eigen-value λ_i is complex-valued, then λ_i and $\bar{\lambda}_i$ are both eigenvalues and the corresponding columns of $Z(t)$ are replaced by their real and imaginary parts. The general solution of (2.2) is

$$z = Z(t)c,$$

where c is an arbitrary constant vector. There is no loss of generality in choos-ing $Z(t)$ to be the principal fundamental matrix, so that $Z(0) = E$. Next, since $Z(t - t_1)$ is also a fundamental matrix for any t_1, it follows that

$$z = Z(t - t_1)c_1,$$

where the constant vector c_1 is $z(t_1)$.

(i) In this case, since $\text{Re}\,\lambda_i < 0$ for all $i = 1, \ldots, n$, we can deduce from (2.3) that there exists an $\alpha > 0$ such that

$$|Z(t)| \leq Me^{-\alpha t} \quad \text{(for } t \geq 0), \tag{2.4}$$

where M is a positive constant. Indeed, if all the eigenvalues have the same algebraic and geometric multiplicities (i.e. $s_i = 1$), then $p^1, \ldots, p^n$ are just con-stant vectors and α is just the minimum value of $-\text{Re}\,\lambda_i$ ($i = 1, \ldots, n$). If $s_i > 1$, we observe that there exists a constant δ, which can be chosen arbitrarily small, such that

$$|p^i(t)| \leq Ke^{\delta t} \quad \text{(for } i = 1, \ldots, n).$$

Thus α is now the minimum value of $\delta - \text{Re}\,\lambda_i$ ($i = 1, \ldots, n$) and can be made positive by choosing δ sufficiently small. The result (2.4) implies that, since $c_1 = z(t_1)$,

$$|z| \leq M|z(t_1)|\exp\{-\alpha(t-t_1)\} \quad (\text{for } t \geq t_1).$$

Uniform and asymptotic stability now follows, since, for any $\epsilon > 0$, $|z(t_1)| < \epsilon/M$ implies that $|z| < \epsilon$ for all $t \geq t_1$, and $|z| \to 0$ as $t \to \infty$.

(ii) In this case,

$$|Z(t)| \leq M \quad (\text{for } t \geq 0),$$

where M is a positive constant. The argument is similar to that for case (i). We now have

$$|z| \leq M|z(t_1)| \quad (\text{for } t \geq t_1)$$

and uniform stability follows, since, for any $\epsilon > 0$, $|z(t_1)| < \epsilon/M$ implies that $|z| < \epsilon$ for all $t \geq t_1$.

(iii) In this case there exists a solution

$$z^j = kp^j(t-t_1)\exp\{\lambda_j(t-t_1)\},$$

where either $\text{Re}\,\lambda_j > 0$, or if $\text{Re}\,\lambda_j = 0$ then the algebraic multiplicity of λ_j is greater than the geometric multiplicity (i.e. $s_j > 1$). In this latter case $p^j(t)$ is non-trivially a polynomial of degree $s_j - 1$. Thus, although $|z^j(t_1)|$ can be made arbitrarily small by a suitable choice of the constant k, independently of t_1, $|z^j(t)| \to \infty$ as $t \to \infty$. Hence there is instability.

Since stability is primarily concerned only with the behaviour of solutions as $t \to \infty$, it seems natural to anticipate that the properties of the matrix $A(t)$ as $t \to \infty$ play a dominant role in determining the type of stability. The following theorem gives a useful result in this direction.

Theorem 4.2: Suppose that the constant matrix A_0 has eigenvalues $\lambda_1, \lambda_2, \ldots, \lambda_n$ such that $\text{Re}\,\lambda_i < 0$ for all $i = 1, \ldots, n$. Let

$$A(t) = A_0 + B(t), \tag{2.5}$$

where $B(t)$ is continuous for all $t \geq t_0$ and $|B(t)| \to 0$ as $t \to \infty$. Then the zero solution of (2.2) (and hence any solution of (2.1)) is uniformly and asymptotically stable.

Proof: Consider first the linear system

$$z' = A_0 z$$

and let $Z(t)$ be the principal fundamental matrix for this system. Then, using

(2.5), we can write (2.2) in the form

$$z' = A_0 z + g, \quad \text{where } g = B(t)z.$$

But then, using (5.29) of section 2.5, it follows that, for $t \geq t_1 \geq t_0$,

$$z = Z(t-t_1)c_1 + \int_{t_1}^t Z(t-s)B(s)z(s)\,\mathrm{d}s. \tag{2.6}$$

The derivation of this result in section 2.5 assumed that $g(t)$ was a known function, but it clearly remains valid when, as here, g depends on z as well. As in the proof of theorem 4.1, the constant vector c_1 is $z(t_1)$. It follows from (2.6) that, for $t \geq t_1$,

$$|z| \leq |Z(t-t_1)||c_1| + \int_{t_1}^t |Z(t-s)||B(s)||z(s)|\,\mathrm{d}s.$$

But the fundamental matrix $Z(t)$ has the form given in (2.3) and, as in the proof of theorem 4.1, satisfies the inequality (2.4) since $\operatorname{Re}\lambda_i < 0$ for all $i = 1,\ldots,n$. Hence

$$|z| \leq M|c_1|\exp\{-\alpha(t-t_1)\} + M\int_{t_1}^t |B(s)||z(s)|\exp\{-\alpha(t-s)\}\,\mathrm{d}s.$$

Let $r(t) = |z|\exp\{\alpha(t-t_1)\}$, so that this inequality becomes

$$0 \leq r \leq M|c_1| + M\int_{t_1}^t |B(s)|r(s)\,\mathrm{d}s.$$

Gronwall's lemma (see section 1.3 and problem 5 of chapter 1) now implies that, for $t \geq t_1$,

$$r \leq M|c_1|\exp\left\{M\int_{t_1}^t |B(s)|\,\mathrm{d}s\right\},$$

or

$$|z| \leq M|c_1|\exp\left\{-\alpha(t-t_1) + M\int_{t_1}^t |B(s)|\,\mathrm{d}s\right\}.$$

But, since $|B(t)| \to 0$ as $t \to \infty$, for any $\beta > 0$ there exists a $t_2(\beta)$ such that $t > t_2(\beta)$ implies that $|B(t)| < \beta$. Also, there exists a constant γ such that $|B(t)| \leq \gamma$ for all $t \geq t_0$, where γ depends only on t_0. Hence, for any $t_1 \geq t_0$,

$$|z| \leq M_1|c_1|\exp\{-(\alpha-\beta)(t-t_1)\} \quad \text{(for } t \geq t_1),$$

where

$$M_1 = M\exp\{\gamma|t_2-t_0|\}.$$

Now choose β so that $0 < \beta < \alpha$; for instance, we could choose $\beta = \frac{1}{2}\alpha$. This fixes $t_2(\beta)$, and we also note that the constant M_1 does not depend on t_1. Uniform and asymptotic stability now follows since, for any $\epsilon > 0$, $|z(t_1)| < \epsilon/M_1$ implies that $|z| < \epsilon$ for all $t \geq t_1$, and $|z| \to 0$ as $t \to \infty$.

Theorem 4.2 can be regarded as extending the scope of theorem 4.1 (i) from constant matrices to matrices $A(t)$ which become constant only in the limit $t \to \infty$. However, theorem 4.1 (ii) cannot be extended in an analogous way without some modifications, as the following example demonstrates. We put $n = 1$ and temporarily write z for the scalar component z_1. Consider

$$z' = 2tz(t^2+1)^{-1}.$$

In the terminology of theorem 4.2, $A_0 = 0$ and $B = 2t(t^2+1)^{-1}$; of course both are here 1×1 matrices. A_0 has a single eigenvalue, $\lambda_1 = 0$, and $|B(t)| \to 0$ as $t \to \infty$. The reduced system $z' = A_0 z$ is uniformly stable, since $\operatorname{Re} \lambda_1 = 0$, but is not asymptotically stable. By analogy with theorem 4.2 it might have been expected that the full system (i.e. (2.2) with $A(t)$ given by (2.5)) is also uniformly stable. However, this is not the case in this example, since the general solution is

$$z = c(t^2+1),$$

where c is an arbitrary constant. For $c \neq 0$, $|z| \to \infty$ as $t \to \infty$ and the zero solution is unstable. Fortunately the situation can be remedied by slightly strengthening the conditions on $B(t)$. Thus the following result holds.

Corollary 4.2: Suppose that the constant matrix A_0 has eigenvalues $\lambda_1, \lambda_2, \ldots, \lambda_n$ such that $\operatorname{Re} \lambda_i \leq 0$ for all $i = 1, \ldots, n$, and that the algebraic multiplicity equals the geometric multiplicity whenever $\operatorname{Re} \lambda_j = 0$ for any j. Let $A(t)$ in (2.2) be given by (2.5), where now $B(t)$ is continuous for all $t \geq t_0$, and

$$\int_{t_0}^{t} |B(s)| \, ds < K \quad \text{(for all } t \geq t_0),$$

where K is a constant. Then the zero solution of (2.2) (and hence any solution of (2.1)) is uniformly stable.

The proof is similar to that of theorem 4.2 and is left as an exercise.

Analogous results to theorems 4.1 and 4.2 hold when $A(t)$ is a periodic matrix, $A_0(t)$. The following theorem is the analogue of theorem 4.1.

Theorem 4.3: Suppose that the matrix $A(t)$ in (2.2) is a periodic matrix $A_0(t)$, so that $A_0(t+T) = A_0(t)$ for all t. Let the characteristic exponents of the system (2.2) be $\mu_1, \mu_2, \ldots, \mu_n$. Then the stability of the zero solution of (2.2) (and hence of any solution of (2.1)) is determined according to the following criteria.

(i) If $\mathrm{Re}\,\mu_i < 0$ for all $i = 1,\ldots,n$, then there is uniform and asymptotic stability.
(ii) If $\mathrm{Re}\,\mu_i \leqslant 0$ for all $i = 1,\ldots,n$ and the algebraic multiplicity equals the geometric multiplicity whenever $\mathrm{Re}\,\mu_j = 0$ for any j, then there is uniform stability.
(iii) If $\mathrm{Re}\,\mu_j > 0$ for at least one j, or the algebraic multiplicity is greater than the geometric multiplicity should $\mathrm{Re}\,\mu_j = 0$ for any j, then there is instability.

Proof: From the discussion in section 3.1 (see, for instance, (1.11) and (1.26)) it can be established that a fundamental matrix for (2.2) has the form:

$$Z(t) = [e^{\mu_1 t}p^1(t),\ldots,e^{\mu_n t}p^n(t)], \tag{2.7}$$

where $p^1(t),\ldots,p^n(t)$ are n linearly independent vector-valued polynomials in t of degree $s_i - 1$, whose coefficients are themselves periodic functions of t with period T. Here $1 \leqslant s_i \leqslant m_i - p_i + 1$, where m_i and p_i are, respectively, the algebraic and geometric multiplicities of a given characteristic exponent μ_i, which are defined to be the corresponding multiplicities of the characteristic multipliers as eigenvalues of the matrix B (see (1.7) and (1.8) of section 3.1). The proof now proceeds similarly to that for theorem 4.1. In particular, it is useful to note that in case (i) the fundamental matrix (2.7) will satsfy an estimate of the form (2.4).

Alternatively, we use the result of section 3.1 (see (1.22) or (1.25)) that there exists a periodic non-singular matrix $P(t)$ such that, if

$$z = P(t)y, \tag{2.8}$$

then (2.2), with $A = A_0(t)$, becomes

$$y' = Dy. \tag{2.9}$$

Here D is a constant matrix whose eigenvalues are the characteristic exponents $\mu_1, \mu_2, \ldots, \mu_n$. Theorem 4.1 applied to equation (2.9) for y can now be used to establish theorem 4.3.

Similarly, there is an analogue for theorem 4.2. Suppose that $A_0(t + T) = A_0(t)$ for all t and that the characteristic exponents of the system

$$z' = A_0(t)z$$

are $\mu_1, \mu_2, \ldots, \mu_n$, where $\mathrm{Re}\,\mu_i < 0$ for all $i = 1,\ldots,n$. Let $A(t)$ in (2.2) be given by

$$A(t) = A_0(t) + B(t),$$

where $B(t)$ is continuous for all $t \geqslant t_0$ and $|B(t)| \to 0$ as $t \to \infty$. Now apply the transformation (2.8), so that (2.2) becomes

$$y' = Dy + P(t)^{-1}B(t)P(t)y.$$

Theorem 4.2 can now be used to show that the zero solution of (2.2) (and hence any solution of (2.1)) is uniformly and asymptotically stable. Note that here, since $P(t)$ is a continuous, periodic matrix function of t, $|P(t)|$ is bounded for all t and, since $P(t)$ is also a non-singular matrix, $|P(t)^{-1}|$ is likewise a bounded function of t.

4.3 Principle of linearized stability

We now return to a consideration of the stability of a solution $x(t)$ of the non-linear equation (1.1). Here stability is determined by the behaviour of the solutions $z(t)$ of equation (1.4). Since the zero solution of (1.4) corresponds to the solution $x(t)$ of (1.1), we anticipate that it may be useful to study equation (1.4) when $|z|$ is small. Hence we replace (1.4) with

$$z' = A(t)z + H(z, t), \tag{3.1}$$

where $|H(z, t)|$ is $o(|z|)$ as $|z| \to 0$. In order to relate (3.1) to (1.4) more precisely, we shall assume that $f(x, t)$ is continuous and has continuous first-order partial derivatives with respect to x, for all x and t. These conditions are not very restrictive, and are usually satisfied in applications. Indeed, for most of the applications in this text $f(x, t)$ will be an analytic function of x and t. It follows that

$$A(t) = \frac{\partial f}{\partial x}(x(t), t). \tag{3.2}$$

Here $\partial f / \partial x$ is the $n \times n$ matrix $[\partial f_i / \partial x_j]$. Further, the mean-value theorem can be used to show that (1.4) may be written in the form (3.1), where $|H(z, t)|$ is $o(|z|)$ as $|z| \to 0$ uniformly for t in any finite interval. In practice, we shall strengthen this condition to hold uniformly in t for $t \geq t_0$, so that

$$\frac{|H(z, t)|}{|z|} \to 0 \quad (\text{as } |z| \to 0, \text{ uniformly for } t \geq t_0). \tag{3.3}$$

In particular, this will be the case when equation (1.1) is autonomous (i.e. $f = f(x)$ is independent of t), or when $f(x, t)$ is a periodic function of t (i.e. $f(x, t + T) = f(x, t)$ for all t).

If the nonlinear term $H(z, t)$ is omitted from (3.1), we obtain the linear equation

$$z' = A(t)z, \tag{3.4}$$

where $A(t)$ is given by (3.2). This equation is called the first variational equation for the solution $x(t)$ of (1.1). It is natural to ask under what conditions equation (3.4) can be used to determine the stability of $x(t)$. Later in this section we shall establish some theorems which establish such conditions, and these theorems are collectively called the principle of linearized stability. First, however, we observe that solutions of (3.4) can often be constructed explicitly from just a knowledge of $x(t)$. Suppose that

$$x = x(t; c_1, \dots, c_m)$$

is a solution of (1.1), where $c_1, \dots, c_m$ are m constants of integration. For instance, with $m = n$, the constants c_i may be the n components of the initial condition x_0 (1.2). Then, since $f(x, t)$ in (1.1) does not depend explicitly on c_i ($i = 1, \dots, m$), differentiation of (1.1) with respect to c_i shows that

$$\frac{\partial x'}{\partial c_i} = \frac{\partial f}{\partial x}(x(t), t)\frac{\partial x}{\partial c_i} \tag{3.5}$$

and so $\partial x / \partial c_i$, for each $i = 1, \dots, m$, is a solution of (3.4). Here, of course, we must assume that $f(x, t)$ is sufficiently smooth for these derivatives to exist. In particular, theorem 4.1 shows that, if $f(x, t)$ is analytic in x and t, and if the initial condition $x_0 = x_0(c_1, \dots, c_m)$ is an analytic function of $c_1, \dots, c_m$, then $x(t; c_1, \dots, c_m)$ is likewise an analytic function of t and $c_1, \dots, c_m$. An important special case occurs when the constants c_i are the n components of the initial condition x_0. Then

$$Z' = A(t)Z, \quad \text{where } Z = \frac{\partial x}{\partial x_0}. \tag{3.6}$$

If the components of x_0 can be varied independently, this is the principal fundamental matrix for (3.4), since $Z(t_0) = E$.

Next we return to a consideration of (3.1) and suppose that the matrix $A(t)$ is a constant, A_0. This will occur, for instance, if $f = f(x)$ is autonomous and x is just the constant solution for which $f(x) = 0$. The study of such special solutions, called critical points, will be taken up in chapter 5 for the planar case $n = 2$. When $A(t)$ is a constant, A_0, we may use theorem 4.1 to determine the stability properties of the first variational equation (3.4). The following two theorems show that parts (i) and (iii) of that theorem carry over to (3.1) as well, provided (3.3) is satisfied.

Theorem 4.4: Suppose that the matrix $A(t)$ in (3.1) is a constant, A_0, whose eigenvalues $\lambda_1, \lambda_2, \dots, \lambda_n$ are such that $\text{Re } \lambda_i < 0$ for all $i = i, \dots, n$. Let $H(z, t)$ be defined and continuous for all z and all $t \geq t_0$, and satisfy a Lipschitz condition in z in any bounded domain. Further, suppose that $H(z, t)$ satisfies (3.3). Then the zero solution of (3.1) is uniformly and asymptotically stable.

Proof: The continuity and Lipschitz conditions on $H(z, t)$ are to ensure the local existence of solutions of (3.1). They can be inferred, if necessary, from the corresponding conditions on $f(x, t)$. However, part of the proof is to establish the existence of solutions of (3.1) for all $t \geq t_1 \geq t_0$, provided that $|z(t_1)|$ is sufficiently small.

First, from the existence and uniqueness theorem, theorem 1.4, and the continuation theorem, theorem 1.7, we can conclude that the solutions of (3.1) which satisfy the initial condition

$$z(t_1) = c_1,$$

will exist for all $t \geqslant t_1$, provided $|z(t)|$ is bounded. Then let $Z(t)$ be the principal fundamental matrix for the first variational equation (3.4), so that

$$Z' = A_0 Z, \qquad Z(0) = E.$$

Now, (3.1) can be written in the form

$$z' = A_0 z + g, \quad \text{where } g = H(z,t).$$

But then, using (5.29) of section 2.5, it follows that, for $t \geqslant t_1 \geqslant t_0$,

$$z = Z(t-t_1)c_1 + \int_{t_1}^{t} Z(t-s)H(z(s),s) \, ds.$$

Hence

$$|z| \leqslant |Z(t-t_1)||c_1| + \int_{t_1}^{t} Z(t-s)H(z(s),s) \, ds.$$

But, just as in the proof of theorem 4.1, $Z(t)$ satisfies the inequality (2.4), since $\operatorname{Re} \lambda_i < 0$ for all $i = 1,\ldots,n$. It follows that

$$|z| \leqslant M|c_1|\exp\{-\alpha(t-t_1)\} + M \int_{t_1}^{t} |H(z(s),s)|\exp\{-\alpha(t-s)\} \, ds.$$

This inequality remains valid for all $t \geqslant t_1$ such that $|z(t)|$ is bounded.

Next, the condition (3.3) implies that, for any $\beta > 0$, there exists a $\delta(\beta)$ such that $|z| < \delta(\beta)$ implies that

$$|H(z,t)| < \beta|z|,$$

uniformly for all $t \geqslant t_0$. Choose $t_1 \geqslant t_0$ and $|z(t_1)| < \delta(\beta)$. Then, for all $t \geqslant t_1$ such that $|z(t)| < \delta(\beta)$, it follows that

$$|z| \leqslant M|c_1|\exp\{-\alpha(t-t_1)\} + M\beta \int_{t_1}^{t} |z(s)|\exp\{-\alpha(t-s)\} \, ds.$$

Let $r(t) = |z|\exp\{\alpha(t-t_1)\}$, so that this inequality becomes

$$0 \leqslant r \leqslant M|c_1| + M\beta \int_{t_1}^{t} r(s) \, ds.$$

Gronwall's lemma (see section 1.3) now implies that, for $t \geqslant t_1$,

$$r \leqslant M|c_1|\exp\{M\beta(t-t_1)\}$$

or

$$|z| \leq M|c_1|\exp\{-(\alpha - M\beta)(t - t_1)\}.$$

Now choose β so that $0 < M\beta < \alpha$; for instance, we could choose $M\beta = \frac{1}{2}\alpha$. This fixes $\delta(\beta)$. Hence, recalling that $z(t_1) = c_1$ if $|z(t_1)| < \delta/M$, it follows that $|z| < \delta$ for all $t \geq t_1$. This establishes both the existence of $z(t)$ for all $t \geq t_1$ and the validity of this inequality for $|z|$ for all $t \geq t_1$. Uniform and asymptotic stability now follows since, for any $\epsilon > 0$, $|z(t_1)| < \epsilon/M$ implies that $|z| < \epsilon$ for all $t \geq t_1$, and $|z| \to 0$ as $t \to \infty$.

There are a number of extensions of theorem 4.4 (see, for instance, Coddington and Levinson, 1955, ch. 13, or Cesari, 1962, ch. III, section 6). We shall give just one extension here which combines theorem 4.2 and theorem 4.4.

Corollary 4.4: Suppose that the constant matrix A_0 has eigenvalues $\lambda_1, \lambda_2, \ldots, \lambda_n$ such that $\mathrm{Re}\,\lambda_i < 0$ for all $i = 1, \ldots, n$. Let $A(t)$ in (3.1) be given by $A_0 + B(t)$, where $B(t)$ is continuous for all $t \geq t_0$, and $|B(t)| \to 0$ as $t \to \infty$. Let $H(z,t)$ be defined and continuous for all z and all $t \geq t_0$, satisfy a Lipschitz condition in z in any bounded domain and satisfy (3.3). Then the zero solution of (3.1) is uniformly and asymptotically stable.

Proof: The proof is similar to that of theorem 4.4, the modifications being analogous to those used in the proof of theorem 4.2.

The counterpart to theorem 4.4 follows, and gives conditions under which instability of the first variational equation (3.4) corresponds to instability of the zero solution of (3.1).

Theorem 4.5: Suppose that the matrix $A(t)$ in (3.1) is a constant A_0, with eigenvalues $\lambda_1, \lambda_2, \ldots, \lambda_n$, and $\mathrm{Re}\,\lambda_j > 0$ for at least one j. Let $H(z,t)$ be defined and continuous for all z and all $t \geq t_0$, satisfy a Lipschitz condition in z in any bounded domain, and satisfy (3.3). Then the zero solution of (3.1) is unstable.

Proof: First, introduce a linear transformation $z = Sw$, so that (3.1) becomes

$$w' = B_0 w + G(w, t), \tag{3.7}$$

where

$$B_0 = S^{-1}A_0 S \quad \text{and} \quad G(w, t) = S^{-1}H(Sw, t).$$

Suppose that A_0 has k eigenvalues λ_j $(j = 1, \ldots, k)$ such that $\mathrm{Re}\,\lambda_j > 0$, and that all the remaining eigenvalues λ_j, $(j = k+1, \ldots, n)$ are such that $\mathrm{Re}\,\lambda_j \leq 0$. Note that the hyphotheses of the theorem require that $1 \leq k \leq n$. Now choose S to be a non-singular matrix such that B_0 has the form

$$B_0 = \begin{bmatrix} B_+ & 0 \\ 0 & B_- \end{bmatrix},$$

where B_+ is a $k \times k$ matrix whose eigenvalues are λ_j $(j = 1, \ldots, k)$, where $\mathrm{Re}\,\lambda_j > 0$, and B_- is an $l \times l$ matrix $(l = n - k)$, whose eigenvalues are λ_j $(j = k+1, \ldots, n)$, where $\mathrm{Re}\,\lambda_j \le 0$. For instance, if A_0 has a full set of linearly independent eigenvectors, then S is chosen to be the matrix whose columns are just these eigenvectors (or their real and imaginary parts if λ_j is complex-valued), and then B_0 is a matrix whose only non-zero elements are diagonal entries, corresponding to real eigenvalues, and 2×2 sub-matrices of the form Ω (see (5.13) of section 2.5) centred on the main diagonal, corresponding to each complex-conjugate pair of eigenvalues. For the general case, S may be chosen to be the matrix whose columns are the generalized eigenvectors of A (see, for instance, Coddington and Levinson, 1955, pp. 62–67). Since S is non-singular, $G(w, t)$ will satisfy the same hypotheses with respect to w as $H(z, t)$ satisfies with respect to z. Further, it is sufficient to show that the zero solution of (3.7) is unstable, since this is clearly equivalent to the instability of the zero solution of (3.1).

Next, decompose w into a k-vector u and an l-vector v, so that

$$w = \begin{bmatrix} u \\ v \end{bmatrix}.$$

Similarly, let G be decomposed into a k-vector G_+ and an l-vector G_-. Then (3.6) becomes

$$u' = B_+ u + G_+(u, v, t), \qquad v' = B_- v + G_-(u, v, t).$$

It will be convenient here to measure the magnitude of w by $\|w\|$, where (see section 1.2)

$$\|w\|^2 = \sum_{i=1}^{n} w_i^2.$$

Similarly we put

$$\|u\|^2 = \sum_{j=1}^{k} u_j^2, \qquad \|v\|^2 = \sum_{j=1}^{l} v_j^2.$$

Note that $\|w\|^2 = \|u\|^2 + \|v\|^2$. Let

$$M = \|u\|, \qquad N = \|v\|,$$

so that

$$MM' = u^\top u', \qquad NN' = v^\top v'.$$

Here $u^\top$ and $v^\top$ are, respectively, the row vectors corresponding to the column vectors u and v. Thus

$$MM' = u^\top B_+ u + u^\top G_+(u, v, t), \qquad NN' = v^\top B_- v + v^\top G_-(u, v, t).$$

Now our hypotheses on the matrices B_+ and B_- imply that there exist constants α and σ, with $0 \leq \sigma < \alpha$, such that

$$u^\top B_+ u \geq \alpha u^\top u = \alpha M^2, \qquad v^\top B_- v \leq \sigma v^\top v = \sigma N^2.$$

This result is readily established when B_0 has the canonical form described in the preceding paragraph, in which case α is the minimum value of $\mathrm{Re}\,\lambda_j$ $(j = 1,\ldots,k)$ and $\sigma = 0$. The proof in the general case is left as an exercise.

The proof of the theorem is by contradiction. Suppose that the zero solution of (3.7) is stable, so that, for every $\epsilon > 0$, there exists a $\delta > 0$ such that $\|w(t_1)\| < \delta$ implies that $\|w(t)\| < \epsilon$ for all $t \geq t_1$. Now, condition (3.3) implies that, for every $\beta > 0$, there exists a $\gamma(\beta)$ such that, for $\|w\| < \gamma(\beta)$,

$$|G(w, t)| < \beta \|w\|,$$

uniformly for all $t \geq t_0$. Choose $t_1 \geq t_0$, $\epsilon < \gamma(\beta)$ and $\|w(t_1)\|$ less than the minimum of δ and $\gamma(\beta)$. It then follows from the expressions above for M' and N' that, for $t \geq t_1$,

$$MM' \geq \alpha M^2 - \beta M(M+N), \qquad NN' \leq \sigma N^2 + \beta N(M+N).$$

Hence,

$$(M-N)' \geq \alpha M - \sigma N - 2\beta(M+N).$$

Choose β so that $\alpha - 2\beta > \nu > \sigma + 2\beta > 0$; for instance, we could choose $\beta = \frac{1}{8}(\alpha - \sigma)$ and $\nu = \frac{1}{2}(\alpha + \sigma)$. Thus

$$(M-N)' \geq \nu(M-N),$$

or

$$M - N \geq (M-N)(t_1) \exp \nu(t - t_1).$$

Now choose $w(t_1)$ so that $M(t_1) > N(t_1)$; for instance, we could choose $N(t_1) = 0$. It follows that $M \to \infty$ as $t \to \infty$, which contradicts the hypothesis that $\|w(t)\| < \epsilon$.

Theorems 4.4 and 4.5 together constitute the principle of linearized stability when the matrix $A(t)$ in (3.1) is a constant matrix A_0. Significantly, neither theorem is applicable when the eigenvalues $\lambda_1, \lambda_2, \ldots, \lambda_n$ of A_0 are such that $\mathrm{Re}\,\lambda_i \leq 0$ $(i = 1,\ldots,n)$ and $\mathrm{Re}\,\lambda_j = 0$ for at least one j. Depending on the multiplicity of the eigenvalue λ_j, this corresponds to case (ii) or (iii) of theorem 4.1 for the stability of the first variational equation (4.1). However, the following examples demonstrate that in this case the first variational equation cannot, in general, be used to infer stability, or otherwise, for the full equation (3.1). For instance, putting $n = 1$ and temporarily writing z for the scalar component z_1, consider

$$z' = -\tfrac{1}{2} z^3.$$

In the terminology of (3.1), the matrix $A_0 = 0$ and the nonlinear term
$H(z,t) = -z^3$. Here A_0 is a 1×1 matrix with a zero eigenvalue. The general
solution is

$$z = c_1 [1 + c_1^2(t-t_1)]^{-\frac{1}{2}},$$

where $c_1 = z(t_1)$. Hence, the zero solution is uniformly and asymptotically
stable, since $|z(t_1)| < \epsilon$ implies that $|z| < \epsilon$ for all $t \geq t_1$, and $|z| \to 0$ as
$t \to \infty$. On the other hand, consider

$$z' = z^2.$$

Again $A_0 = 0$, but now $H(z,t) = z^2$. The general solution is

$$z = c_1 [1 - c_1(t-t_1)]^{-1},$$

where again $c_1 = z(t_1)$. For $c_1 > 0$, $z \to \infty$ as $t \to t_1 + c_1^{-1}$ and, hence, the zero
solution is unstable.

Just as in section 4.2, there are analogues to theorems 4.4 and 4.5 when $A(t)$
is a periodic matrix $A_0(t)$. Thus the following results hold.

Theorem 4.6: Suppose that the matrix $A(t)$ in (3.1) is a periodic matrix $A_0(t)$, so
that $A_0(t+T) = A_0(t)$ for all t. Let the characteristic exponents of the first variational equation (3.4) be $\mu_1, \mu_2, \ldots, \mu_n$. Let $H(z,t)$ satisfy the hypotheses of
theorem 4.4.
 (i) If $\operatorname{Re}\mu_i < 0$ for all $i = 1, \ldots, n$, then the zero solution of (3.1) is uniformly
     and asymptotically stable.
(ii) If $\operatorname{Re}\mu_j > 0$ for at least one j, then the zero solution of (3.1) is unstable.

Proof: As in the proof of theorem 4.3, we use the tranformation (2.8), so that
(3.1) becomes

$$y' = Dy + P(t)^{-1}H(P(t)y,t), \qquad (3.8)$$

where we recall that D is a constant matrix whose eigenvalues are $\mu_1, \mu_2, \ldots, \mu_n$
and $P(t)$ is a continuous, periodic and non-singular matrix function of t. Hence
$|P(t)|$ and $|P(t)^{-1}|$ are both bounded functions of t. Case (i) is now established
from theorem 4.4, and case (ii) from theorem 4.5.

4.4 Stability for autonomous systems

We recall that equation (1.1) is said to be autonomous when the function
$f = f(x)$ is explicitly independent of t, so that (1.1) becomes

$$x' = f(x). \qquad (4.1)$$

Here we shall assume that $f(x)$ is continuous for all x and satisfies a Lipschitz
condition in x in any bounded domain. An important observation about

autonomous systems is that, if $x(t)$ is a solution, so is $x(t-t_0)$ for any t_0. Hence, one of the n constants of integration is t_0 and corresponds to an arbitrary translation along the t-axis. While all the previous results remain valid for autonomous systems, the redundancy implied by the essential equivalence of $x(t)$ and $x(t-t_0)$ necessitates a reconsideration of our stability theory.

For autonomous systems it is useful to represent the solutions as **orbits** in **phase space**. Here the phase space is just the set of all points x and has n dimensions, while an orbit is the path, or trajectory, traced out in the phase space as t varies. In this representation t is just a parameter along the orbit, and attention is focused on the geometrical structure of the solution paths. Note that, if $x(t)$ corresponds to an orbit, then so does $x(t-t_0)$ for any t_0. As an illustration, consider the two-dimensional system (i.e. $n = 2$):

$$x_1' = \omega x_2, \qquad x_2' = -\omega x_1. \tag{4.2}$$

Here the phase space is the x_1-x_2 plane and is commonly called the phase plane. These equations are equivalent to those for a simple harmonic oscillator and the general solution is

$$x_1 = c_1 \cos \omega t + c_2 \sin \omega t, \qquad x_2 = -c_1 \sin \omega t + c_2 \cos \omega t.$$

The orbits are circles $r = r_0$, where $r_0 = \sqrt{c_1^2 + c_2^2}$ and

$$r = \sqrt{x_1^2 + x_2^2}. \tag{4.3}$$

They are sketched in figure 4.2; the arrows indicate the sense in which t is increasing. The solution point $(x_1(t), x_2(t))$ travels around a circle with a period of $2\pi/\omega$.

Next we show that, although each orbit corresponds to an infinite set of solutions, it is uniquely defined in phase space.

Theorem 4.7: There is a unique orbit through each point of phase space.

Proof: Suppose that $x^1(t)$ passes through a given point P at $t = t_1$, and $x^2(t)$ passes through the same point P at $t = t_2$. Let $t_0 = t_1 - t_2$. Then $x^1(t)$ and $x^2(t-t_0)$ both pass through point P at $t = t_1$. Hence, by the uniqueness result, theorem 1.3, they are the same solution, so that

$$x^1(t) = x^2(t-t_0)$$

for all t. Hence they define the same orbit.

The following example illustrates why it is necessary to reconsider the notions of stability defined in section 4.1 when the system being discussed is autonomous. Let

$$x_1' = r^2 x_2, \qquad x_2' = -r^2 x_1, \tag{4.4}$$

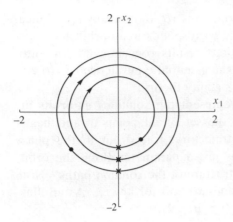

Fig. 4.2 A plot of orbits for equations (4.2) and (4.4). × denotes $t = \frac{1}{2}\pi$ for (4.2) with $\omega = 1$, while ● denotes $t = \frac{1}{2}\pi$ for (4.4). The plots are for $r_0 = 0.75$, 1 and 1.25.

where r is given by (4.3). The equations are solved by introducing the polar coordinates (r, θ), where

$$x_1 = r\cos\theta, \qquad x_2 = r\sin\theta. \tag{4.5}$$

Then,

$$rr' = x_1 x_1' + x_2 x_2', \qquad r^2\theta' = x_1 x_2' - x_2 x_1'. \tag{4.6}$$

Thus (4.4) becomes

$$r' = 0, \qquad \theta' = -r^2,$$

so that

$$r = r_0, \qquad \theta = \theta_0 - r_0^2 t.$$

The general solution for x_1, x_2 is

$$x_1 = r_0\cos(r_0^2 t - \theta_0), \qquad x_2 = -r_0\sin(r_0^2 t - \theta_0).$$

As for the simple harmonic oscillator equation (4.2), the orbits are $r = r_0$ (see figure 4.2), but the period is now $2\pi/r_0^2$. The zero solution, $r_0 = 0$, of (4.4) is uniformly stable, but any other solution, with $r_0 \neq 0$, is unstable. The instability occurs because any small perturbation δr_0 in r_0 will alter the period by an amount proportional to δr_0 (approximately $-4\pi\delta r_0/r_0^3$) and, hence, after a time proportional to $(\delta r_0)^{-1}$, the perturbation will have deviated significantly. Note, however, that, for the simple harmonic oscillator equation (4.2), all solutions are stable, since it is a linear equation and the zero solution is stable; or, more pertinently, the period is $2\pi/\omega$ and does not depend on the amplitude r_0. Returning to the nonlinear equation (4.4) we note that although the solutions are unstable in the strict Liapunov sense (i.e. according to the definition in section 4.1), it is stable in an orbital sense, since a small perturbation in the orbit remains small.

This notion of orbital, or Poincaré, stability is made more precise in the following definition. For simplicity we shall consider stability only for periodic solutions of (4.1), for which the orbits are closed curves in phase space.

Definition: Let $x(t)$ be a periodic solution of (4.1) (i.e. $x(t+T) = x(t)$ for all t) defining an orbit C in phase space.

(i) Suppose that $y(t)$ is a solution of (4.1) such that, for all $\epsilon > 0$ and any t_1, there is a $\delta(\epsilon) > 0$ such that $d(y(t_1), C) < \delta$ implies that $d(y(t), C) < \epsilon$ for all $t \geq t_1$. Then C is said to be **orbitally**, or **Poincaré**, stable.

(ii) If C is orbitally stable, and $d(y(t), C) \to 0$ as $t \to \infty$, then C is said to be **asymptotically orbitally**, or **Poincaré**, stable.

Here

$$d(y(t), C) = \min_{x \in C}|x - y(t)|$$

is the distance of $y(t)$ from the orbit C. It can be shown that Liapunov stability (see section 4.1) implies orbital stability, but the example (4.4) discussed above shows that the converse is not generally true.

For the autonomous equation (4.1) the first variational equation (3.4) becomes, with $z = y - x$,

$$z = A(t)z, \quad \text{where } A(t) = \frac{\partial f}{\partial x}(x(t)), \tag{4.7}$$

and $A(t)$ is a periodic matrix function of t. But, since $x(t)$ is a solution of (4.1), differentiation of (4.1) with respect to t shows that

$$z = x'(t)$$

is a solution of (4.7). Alternatively, since $x(t - t_0)$ is a solution of (4.1) for any t_0, it follows from (3.5) that $\partial x / \partial t_0$ is a solution of (4.7) and the result follows since $x' = -\partial x / \partial t_0$. Since $x'(t)$ is periodic with period T, the associated characteristic exponent is zero. Thus the principle of linearized stability in the form of theorem 4.6 cannot be used to infer stability. Instead, the following theorem holds.

Theorem 4.8: Let $x(t)$ be a periodic solution of (4.1), where $f(x)$ is continuous and has continuous first-order partial derivatives for all x. Suppose that the first variational equation (4.7) has characteristic exponents $0, \mu_2, \ldots, \mu_n$, where $\text{Re}\,\mu_i < 0$ for all $i = 2, \ldots, n$. Then the orbit C, which represents $x(t)$ in phase space, is asymptotically orbitally stable.

Proof: The proof is beyond the scope of this text (but see, for instance, Coddington and Levinson, 1955, pp. 321–327). Note that the zero exponent corresponds to the solution $x'(t)$ of (4.7), and the theorem shows that, in effect, this can be put aside when considering stability.

4.5 Liapunov functions

As its name suggests, the principle of linearized stability reduces the determination of stability to a consideration of a linear equation, namely, the first variational equation (3.4). While this is the main technique used in this text, there are occasions when it fails to give useful information, or when it is inconvenient or difficult to solve. An alternative approach is to consider the fully nonlinear equation (1.4), using Liapunov's direct method. In essence, this method seeks a scalar function of z, which can be regarded as a measure of the 'energy' of the system (1.4), and then seeks to demonstrate that either this 'energy' decreases as $t \to \infty$, indicating stability, or it increases, indicating instability.

For simplicity, we shall consider only autonomous systems so that (1.4) becomes

$$z' = F(z), \tag{5.1}$$

where $F(z)$ is defined and continuous for all z, satisfies a Lipschitz condition in z in any bounded domain and is such that $F(0) = 0$. Let $V(z)$ be a scalar function of z, defined and continuous, with continuous partial derivatives, for $|z| \leqslant k$ $(k > 0)$, and such that $V(0) = 0$.

Definition:
(i) $V(z)$ is **positive (negative) definite** for $|z| \leqslant k$ if $V > 0$ $(V < 0)$ for all $z \neq 0$, $|z| \leqslant k$.
(ii) $V(z)$ is **positive (negative) semidefinite** for $|z| \leqslant k$ if $V \geqslant 0$ $(V \leqslant 0)$ for all z, $|z| \geqslant k$.

For example, the function

$$V = z_1^2 + z_2^2 + \ldots + z_n^2,$$

is positive definite, whereas

$$V = (z_1 + z_2 + \ldots + z_n)^2$$

is positive semidefinite.

Next let $z(t)$ be a solution of (5.1) and consider the function $V(t) = V(z(t))$. Then the derivative of V along the orbit $z(t)$ is

$$V' = \frac{d}{dt} V(z(t)) = (\nabla V)^\top z' = (\nabla V)^\top F(z). \tag{5.2}$$

Here ∇V is the gradient of $V(z)$ and is the n-vector whose components are $\partial V / \partial z_i$ $(i = 1, \ldots, n)$. Note that the last term in (5.2) allows us to regard V' as a function of z alone.

Theorem 4.9:
(i)   Let $V(z)$ be positive definite and V' be negative semidefinite for $|z| \leqslant k$. Then the zero solution of (5.1) is uniformly stable.

(ii) Let $V(z)$ be positive definite and V' be negative definite for $|z| \leqslant k$. Then the zero solution of (5.1) is uniformly and asymptotically stable.

Proof: When $V(z)$ satisfies the condition of the theorem it is said to be a Liapunov function. In essence, the proof uses the notion that $V(z)$ can be used as a measure of $|z|$.

(i) Let $z(t)$ be a solution of (5.1) such that $0 < |z(t_1)| < k$. Then $z(t)$ exists for all $t \geqslant t_1$, provided $|z(t)|$ is bounded. Let $t_2 > t_1$ be such that $|z(t)| < k$ for $t_1 \leqslant t < t_2$. Note that either $t_2 = \infty$ or, if $t_2 < \infty$, we may suppose that t_2 is the first point where $|z(t_2)| = k$. Then, for all t with $t_1 \leqslant t < t_2$, it follows that

$$V(z(t)) - V(z(t_1)) = \int_{t_1}^{t} V'(z(s)) \, dx \leqslant 0,$$

where we have used the information that V' is negative semidefinite. Since $z(t) \neq 0$ for any t (a consequence of the fact that $|z(t_1)| > 0$), it follows that, for $t_1 \leqslant t < t_2$,

$$0 < V(z(t)) \leqslant V(z(t_1)).$$

Next, given any $\epsilon > 0$ $(0 < \epsilon < k)$, consider the set S of all z such that $\epsilon \leqslant |z| \leqslant k$ and let $m(\epsilon)$ be the minimum of $V(z)$ in S. Since $z = 0$ is not in S and S is compact, $m > 0$. Now choose $\delta(\epsilon)$ such that $0 < \delta(\epsilon) < \epsilon$ and $V(z) < m$ for all $|z| < \delta$. Let $|z(t_1)| < \delta$. Hence, for $t_1 \leqslant t < t_2$,

$$0 < V(z(t)) \leqslant V(z(t_1)) < m.$$

It now follows that $|z(t)| < \epsilon$ for $t_1 \leqslant t < t_2$. Indeed, this is the case for $t = t_1$ and, if there is any t such that $|z(t)| = \epsilon$, then $V(z(t)) \geqslant m$, which is a contradiction. In particular, it follows that, since $\epsilon < k$, $t_2 = \infty$ and, hence, $|z(t)| < \epsilon$ for all $t \geqslant t_1$.

(ii) The proof of part (i) shows that the zero solution is uniformly stable, and that, for any $\epsilon > 0$ $(0 < \epsilon < k)$, there exists a $\delta(\epsilon)$ such that, if $|z(t_1)| < \delta$, then $|z(t)| < \epsilon$ for all $t \geqslant t_1$. Suppose now that it is possible to find such a solution $z(t)$, so that, for some $b > 0$, $V(z(t)) \geqslant b$ for all $t \geqslant t_1$. Choose h $(0 < h < k)$ so that $V(z) < b$ for all $|z| < h$. It follows that $|z(t)| \geqslant h$ for $t \geqslant t_1$. Next, since V' is negative definite, there exists a constant $m' > 0$ such that $-V' \geqslant m'$ for all $|z|$ $(h \leqslant |z| \leqslant k)$. But then, for all $t \geqslant t_1$,

$$V(z(t)) - V(z(t_1)) = \int_{t_1}^{t} V'(z(s)) \, ds \leqslant -m'(t - t_1)$$

and, hence, $V(z(t)) \to -\infty$ as $t \to \infty$. This is a contradiction, since V is positive definite. Hence, we have shown that, as t increases, eventually $V(z(t)) < b$, for any $b > 0$. Since $V' < 0$, it follows that $V(z(t)) \to 0$ as $t \to \infty$.

Finally we can conclude that $|z(t)| \to 0$ as $t \to \infty$. Indeed, as in the proof of part (i), given any $\epsilon > 0$, let $m(\epsilon)$ (> 0) be the minimum of $V(z)$ in the region $\epsilon \leqslant |z| \leqslant k$. Then, since $V(z(t)) \to 0$ as $t \to \infty$, there exists $T \geqslant t_1$ such that, for all $t \geqslant T$, $|V(z(t))| < m(\epsilon)$. Hence $|z(t)| < \epsilon$ for $t \geqslant T$.

The counterpart to theorem 4.9 follows, and gives conditions for which Liapunov's direct method may be used to determine the instability of the zero solution of (5.1).

Theorem 4.10: Let V' be positive definite for $|z| \leqslant k$, and suppose that in every neighbourhood of $z = 0$ there is at least one point z such that $V(z) > 0$. Then the zero solution of (5.1) is unstable.

Proof: As in theorem 4.9, it is assumed that $V(z)$ is defined and continuous, with continuous partial derivatives, for $|z| \leqslant k$, and is such that $V(0) = 0$. Now, given any $b > 0$ $(0 < b < k)$, there exists a z_b with $|z_b| < b$ and such that $V(z_b) > 0$. Let $z(t)$ be that solution of (5.1) such that $z(t_1) = z_b$. Then $z(t)$ exists for all $t \geqslant t_1$, provided $z(t)$ is bounded. Let $t_2 > t_1$ be such that $|z(t)| < k$ for $t_1 \leqslant t < t_2$. Then, for all t with $t_1 \leqslant t < t_2$, it follows that

$$V(z(t)) - V(z(t_1)) = \int_{t_1}^{t} V'(z(s)) \, \mathrm{d}s > 0.$$

Here we have used the information that V' is positive definite and that $z(t) \neq 0$ for any t. It follows that

$$V(z(t)) > V(z(t_1)) > 0.$$

We can now conclude that $z(t)$ is bounded away from the origin, and so, there exists $h > 0$ $(0 < h \leqslant b)$ such that $|z(t)| \geqslant h$ for all $t \geqslant t_1$. For if not, $z(t)$ may become arbitrarily small, and then, from the continuity of $V(z)$ and the fact that $V(0) = 0$, it would follow that $V(z(t))$ would become arbitrarily small, which is a contradiction.

Next, since V' is positive definite, there exists a constant $m' > 0$ such that $V' \geqslant m'$ for all $|z|$ $(h \leqslant |z| \leqslant k)$. But then, for all $t \geqslant t_1$,

$$V(z(t)) - V(z(t_1)) = \int_{t_1}^{t} V'(z(s)) \, \mathrm{d}s \geqslant m'(t - t_1).$$

Since $V(z)$ is continuous, it is bounded for $|z| \leqslant k$. But the right-hand side of the above expression is unbounded as $t \to \infty$. Hence, it follows that $|z(t_2)| = k$. Thus, given any $c > 0$ $(0 < c < k)$, and for any $b > 0$ $(0 < b < k)$, there is at least one solution such that $|z(t_1)| < b$ but $|z(t)| > c$ for some t $(t_1 \leqslant t < t_2)$. Therefore the zero solution of (5.1) is unstable.

To illustrate the use of Liapunov functions, consider the following examples. First, for the two-dimensional system (i.e. $n = 2$) let

$$z_1' = -2z_1 z_2^2 - z_1^3, \qquad z_2' = -z_2 + z_1^2 z_2.$$

Try the function $V = az_1^2 + z_2^2$, with a to be determined. Then, from (5.2),

$$V' = -2z_2^2 - 2az_1^4 - 2z_1^2 z_2^2(2a - 1).$$

Choose $a = 1$, so that

$$V' = -2z_2^2 - 2z_1^2(z_1^2 + z_2^2)$$

is negative definite, while $V = z_1^2 + z_2^2$ is positive definite. By theorem 4.9 (ii), the zero solution is uniformly and asymptotically stable. Note that any choice of $a \geq \frac{1}{2}$ would be just as useful here. Also, for this particular example, the first variational equation (3.4) has a matrix $A(t)$ given by

$$\begin{bmatrix} 0 & 0 \\ 0 & -1 \end{bmatrix}$$

whose eigenvalues are 0 and -1. Hence, the principle of linearized stability (theorem 4.4) cannot be used here.

Next, consider the two-dimensional system

$$z_1' = z_2, \qquad z_2' = -\omega^2 z_1 - \nu z_1^2 z_2.$$

Putting $u = z_1$, this is equivalent to the equation

$$u'' + \nu u^2 u' + \omega^2 u = 0,$$

which can be recognised as the equation for a simple harmonic oscillator of frequency ω, with a nonlinear damping term. Here, we try $V = \frac{1}{2}(z_2^2 + \omega^2 z_1^2)$, which may be interpreted as the energy of the undamped oscillator. Then, from (5.2),

$$V' = -\nu z_1^2 z_2^2.$$

For $\nu > 0$, V' is negative semidefinite, showing that, as expected, the energy is damped. Theorem 4.9 (ii) shows that the zero solution is uniformly stable. However, on physical grounds one would expect asymptotic stability in this case, but theorem 4.9 (ii) cannot be used here as V' is only negative semidefinite. Indeed, asymptotic stability does hold here, but the proof requires a more sophisticated version of theorem 4.9 (ii) (see, for instance, Hagedorn, 1988, section 2.3). Also note here that, for $\nu < 0$, V' is positive semidefinite, but this is not sufficient to enable us to use theorem 4.20, even though, on physical grounds, one would expect instability in this case.

This last example shows that, in seeking Liapunov functions, it is sometimes useful to identify V with the energy of the physical system which the equations describe. Thus, in general, a frictionless autonomous mechanical system can be described by the function $H(p,q)$, in terms of which the equations of motion are

$$p' = -\frac{\partial H}{\partial q}, \qquad q' = \frac{\partial H}{\partial p}. \tag{5.3}$$

Here p and q are m-vectors, and altogether these equations form a first-order system of order n, where $n = 2m$. The theory of these systems will be developed in chapter 11. Here we just note that $H(p,q)$ is called the Hamiltonian and that

$$H' = \frac{\partial H}{\partial p}p' + \frac{\partial H}{\partial q}q',$$

$$= \frac{\partial H}{\partial p}\left(-\frac{\partial H}{\partial q}\right) + \frac{\partial H}{\partial q}\frac{\partial H}{\partial p} = 0,$$

so that H = constant is an integral of the equations of motion. In many applications the Hamiltonian takes the form

$$H(p,q) = \tfrac{1}{2}p^\top T(q)p + U(q),$$

where $T(q)$ is an $m \times m$ positive-definite matrix function of q. The first term then represents the kinetic energy of the motion, while $U(q)$ is the potential energy. Suppose now that $U(q)$ has a stationary value at $q = 0$ (i.e. $\partial U/\partial q = 0$ at $q = 0$). Then the point $p = q = 0$ is a stationary solution of the equations of motion. Further, if $U(q)$ is a minimum at $q = 0$, then the zero solution $p = q = 0$ is uniformly stable. Indeed, without loss of generality we may assume that $U(0) = 0$, and so $H(p,q)$ is positive definite in some neighbourhood of $p = q = 0$. Then we may set the Liapunov function $V = H(p,q)$ and theorem 4.9(i) establishes the required result.

On the other hand, if $U(q)$ is a maximum at $q = 0$, then the zero solution is unstable. We shall not give the proof in the general case (but see Hagedorn, 1988, section 2.2), but illustrate the situation by the following example in which $m = 1$ (so that $n = 2$) and

$$H(p,q) = \tfrac{1}{2}p^2 - \tfrac{1}{4}q^4.$$

Hence the equations of motion (5.3) are

$$p' = q^3, \qquad q' = p.$$

For a Liapunov function we choose $V = pq$, so that, from (5.2),

$$V' = p^2 + q^4,$$

and is positive definite. Since $V > 0$ in $p, q > 0$, theorem 4.10 shows that the zero solution is unstable. Note here that the first variational equation (3.4) has a matrix $A(t)$ given by

$$\begin{bmatrix} 0 & 0 \\ 1 & 0 \end{bmatrix},$$

whose eigenvalues are both zero. Hence the principle of linearized stability (theorem 4.5) cannot be used here.

In general, the difficult practical aspect of Liapunov's direct method is the determination of a suitable Liapunov function $V(z)$. As the above examples illustrate, the simplest procedure is often trial and error, using physical considerations to guess a suitable form for $V(z)$. However, sometimes the linearized part of the right-hand side of (5.1) can be used to provide a suitable function. Thus, let us now suppose that (5.1) has the form

$$z' = Az + H(z), \tag{5.4}$$

where $H(z)$ is $O(|z|^2)$ (i.e. $F(z)$ in (5.1) has been expanded in a Taylor series), and A is a constant $n \times n$ matrix. Now seek a Liapunov function $V(z)$ in the form

$$V(z) = z^\top B z, \tag{5.5}$$

where B is a constant $n \times n$ symmetric matrix, yet to be determined. Then from (5.2)

$$V' = z^\top C z + 2 z^\top B H(z), \tag{5.6}$$

where

$$C = A^\top B + BA.$$

The aim now is to choose the matrix B so that the matrix C is either negative definite, or positive definite, and, hence, either theorem 4.9 (ii) or theorem 4.10 can be used.

We shall illustrate this for the case $n = 2$ (in the general case, see, for instance, Jordan and Smith, 1977, section 10.6). For simplicity, we shall suppose that the matrix A is in canonical form. Thus, if A has two real eigenvalues λ_1 and λ_2 (with linearly independent eigenvectors), then we suppose that

$$A = \begin{bmatrix} \lambda_1 & 0 \\ 0 & \lambda_2 \end{bmatrix}.$$

Assuming that neither eigenvalue is zero, we let

$$B = \begin{bmatrix} \lambda_1^{-1} & 0 \\ 0 & \lambda_2^{-1} \end{bmatrix} \operatorname{sign} \lambda_1,$$

and so

$$C = \begin{bmatrix} 2 & 0 \\ 0 & 2 \end{bmatrix} \operatorname{sign} \lambda_1.$$

Thus if $\lambda_1, \lambda_2 < 0$, $V(z)$ (5.5) is positive definite and C is negative definite. Then, from (5.6), V' will be negative definite in a neighbourhood of $z = 0$ and theorem 4.9 (ii) shows that the zero solution of (5.4) is uniformly and asymptotically stable. But if $\lambda_1 > 0$, $V(z)$ (5.5) can take positive values in every neighbourhood of $z = 0$ (for instance, let $z_2 = 0$ and $z_1 \neq 0$), while C is positive definite. Then, from (5.6), V' will be positive definite in a neighbourhood of $z = 0$ and theorem 4.10 shows that the zero solution of (5.4) is unstable.

Next, suppose that A has the complex-valued eigenvalues $\alpha \pm i\beta$ and the canonical form

$$A = \begin{bmatrix} \alpha & \beta \\ -\beta & \alpha \end{bmatrix}.$$

In this case, let $B = E$, the unit matrix, so that $V(z)$ (5.5) is positive definite. From (5.6),

$$C = \begin{bmatrix} 2\alpha & 0 \\ 0 & 2\alpha \end{bmatrix}.$$

Then, if $\alpha < 0$, C is negative definite and, from (5.6), V' is negative definite in a neighbourhood of $z = 0$. Then theorem 4.9 (ii) shows that the zero solution is uniformly and asymptotically stable. On the other hand, if $\alpha > 0$, C is positive definite and V' is positive definite in a neighbourhood of $z = 0$, and theorem 4.10 shows that the zero solution is unstable.

Of course, these results are just special cases of the principle of linearized stability (see theorems 4.4 and 4.5), and no particular advantage has been gained here by using Liapunov's direct method. Nevertheless, it is useful to have some general guidance how to choose a suitable Liapunov function. Indeed, in practical applications it is often useful to let V be a quadratic form in z whose coefficients are to be determined by inspection. This was the procedure followed in the first example discussed in this section. Note that this procedure is useful even when the eigenvalues of the matrix A are not known *a priori*.

Problems

1 Determine the stability of the following linear systems:

(a) $z_1' = z_1 + 6z_2,$ $z_2' = -z_1 - 4z_2;$

(b) $z_1' = -7z_1 + 10z_2,$ $z_2' = -4z_1 + 5z_2;$

(c) $z_1' = -5z_1 + 6z_2,$ $z_2' = -3z_1 + 4z_2.$

2 Use theorem 4.2 and problem 1 to determine the stability of the following linear systems:

(a) $z_1' = z_1 + (6 + e^{-t})z_2,$ $z_2' = -z_1 - 4 \tanh t \, z_2;$

(b) $z_1' = -7z_1 + 10t^2(1 + t^2)^{-1}z_2,$ $z_2' = -(4 + t^{-1})z_1 + 5z_2.$

3 Use the principle of linearized stability and problem 1 to determine the stability of the zero solution of the following systems:

(a) $z_1' = z_1 + 6z_2 - z_1^2,$ $z_2' = -z_1 - 4z_2 + z_1 z_2;$

(b) $z_1' = -7z_1 + 10z_2 + z_2 \sin z_1,$ $z_2' = -4z_1 + 5z_2 \cos z_1;$

(c) $z_1' = -5z_1 + 6z_2 + z_2^2,$ $z_2' = -3z_1 + 4z_2 - z_2^4.$

4 The logistic equation for growth of a single population is

$$x' = x(1-x),$$

where here x is a scalar variable (i.e. $n = 1$). Use the principle of linearized stability to show that the solution $x = 0$ is unstable, but the solution $x = 1$ is uniformly and asymptotically stable. Also show that the general solution is

$$x = (1 + ce^{-t})^{-1},$$

and sketch the solutions satisfying the initial condition $x(0) = x_0$ for the cases $x_0 > 1$, $0 < x_0 < 1$ and $x_0 < 0$.

5 The motion of a simple pendulum with linear damping is governed by the equation

$$u'' + vu' + g \sin u = 0,$$

where g and v are positive constants. Use the principle of linearized stability to show that the solution $u = 0$ is uniformly and asymptotically stable, but that the solution $u = \pi$ is unstable.

6 Verify that the equation

$$u'' + vu' + (1 - b^2 \cos^2 t)u + u^3 = -vb \sin t$$

has the solution $u = b \cos t$. Assuming that $v > 0$, use the principle of linearized stability to show that this solution is uniformly and asymptotically stable, provided b^2 is sufficiently small.

7 For the equations

$$z_1' = z_2 + kr^2 z_1, \qquad z_2' = -z_1 + kr^2 z_2,$$

where $r^2 = z_1^2 + z_2^2$, use the transformation (see (4.5) and (4.6))

$$z_1 = r \cos \theta, \qquad z_2 = r \sin \theta$$

to show that

$$r' = kr^3, \qquad \theta' = -1.$$

Deduce that the zero solution is uniformly and asymptotically stable if $k < 0$, but is unstable if $k > 0$. Note that in this example the first variational equation cannot be used to determine stability.

8 For the equations

$$z_1' = z_2 - tz_2^3, \qquad z_2' = -\tfrac{1}{2}z_2^3,$$

show that the zero solution is uniformly stable. Here the first variational equation (3.4) is such that the matrix $A(t)$ is

$$\begin{bmatrix} 0 & 1 \\ 0 & 0 \end{bmatrix}.$$

This has two zero eigenvalues, but the geometric multiplicity is only one, and so the first variational equation is unstable (theorem 4.1).

9 (a) For the equations

$$x_1' = x_2 + x_1(1-r), \qquad x_2' = -x_1 + x_2(1-r),$$

where $r^2 = x_1^2 + x_2^2$, use the transformation (4.5) to show that

$$r' = r(1-r), \qquad \theta' = -1.$$

Deduce that the periodic solution $r = 1$ is asymptotically orbitally stable. Also show that the zero solution is unstable.

(b) For the equations

$$x_1' = x_2 + x_1(1-r)(2-r), \qquad x_2' = -x_1 + x_2(1-r)(2-r),$$

use the transformation (4.5) as in part (a) to show that $r = 1$ and $r = 2$ are both periodic solutions, but that, while $r = 1$ is asymptotically orbitally stable, $r = 2$ is unstable. Also show that the zero solution is unstable.

10 For each of the following equations:

(a) $z_1' = z_2 + kr^2 z_1, \cdot \qquad z_2' = -z_1 + kr^2 z_2;$

(b) $z_1' = z_2 + z_1(1-r), \qquad z_2' = -z_1 + z_2(1-r),$

use the Liapunov function $V = \frac{1}{2}r^2$, where $r^2 = z_1^2 + z_2^2$, to determine the stability of the zero solution. Compare with the solution of problems 7 and 9 (i).

11 For each of the following equations:

(a) $z_1' = -2z_1 + z_2^3, \qquad\qquad z_2' = z_1 - z_2 + z_1 z_2^2;$

(b) $z_1' = z_1 - z_2 + z_1^2 \sin z_2, \qquad z_2' = -2z_2 + z_1^3;$

(c) $z_1' = 3z_1 + 2z_2 + z_2^3, \qquad\qquad z_2' = -10z_1 - 5z_2 - z_1^2 z_2,$

use a suitable Liapunov function $V(z_1, z_2)$ to determine the stability of the zero solution. In each case try

$$V = az_1^2 + 2bz_1 z_2 + cz_2^2$$

and determine the coefficients a, b and c to give V and V' suitable properties. Compare with the results obtained using the principle of linearized stability. [*Hint*: Choose a, b and c so that V' is approximately $\pm 2(z_1^2 + z_2^2)$, using the upper sign when the zero solution is unstable and the lower sign otherwise. See (5.6).]

12 The n-dimensional system

$$z' = \nabla U(z)$$

has the solution $z = 0$ (for all t) whenever $z = 0$ is a stationary point of $U(z)$ (i.e. $\nabla U(0) = 0$). Show that this zero solution is asymptotically stable whenever $U(z)$ has a maximum at $z = 0$, and is unstable whenever $U(z)$ has a minimum at $z = 0$. [*Hint*: Choose the Liapunov function $V(z) = \mp U(z)$, using the upper (lower) sign accordingly as $U(z)$ has a maximum (minimum) at $z = 0$. Without loss of generality we may assume that $U(0) = 0$.]

13 A single particle of mass m and position vector x (here $n = 3$) moves under the influence of a potential field $U(x)$, so that the equations of motion are

$$mx'' = -\nabla U(x).$$

Let $q = x$, $p = mx'$ and define a Hamiltonian

$$H(p,q) = \frac{p^2}{2m} + U(q).$$

Show that the equations of motion are equivalent to (5.3), i.e.

$$p' = -\frac{\partial H}{\partial q}, \qquad q' = \frac{\partial H}{\partial p}.$$

Also show that if $U(q)$ has a minimum at $q = 0$, then the stationary solution $p = q = 0$ is uniformly stable. [*Hint*: Choose the Liapunov function to be $V = H(p,q)$.]

CHAPTER FIVE
PLANE AUTONOMOUS SYSTEMS

5.1 Critical points

In this chapter we consider the autonomous system of equations

$$x' = f(x),\tag{1.1}$$

where $f(x)$ is continuous for all x and satisfies a Lipschitz condition in x in any bounded domain. In the following subsections of this chapter we shall confine attention to the planar case where $n = 2$, where we recall that n is the dimension of the vector x. However, we shall not make this restriction for the moment.

At the beginning of section 4.4 we introduced the notion of **phase space** in relation to the autonomous equation (1.1). For convenience we summarize that discussion here. First, recall that, if $x(t)$ is a solution of (1.1), so is $x(t-t_0)$ for any t_0. Thus, of the n arbitrary constants of integration one is t_0, and there are, in effect, only $n-1$ degrees of freedom for (1.1). This redundancy can be overcome by representing the solutions as **orbits** in the phase space, which is just the set of all points x. An orbit (or path, or trajectory) is the set of points traced out by the solution $x(t)$ as t varies. Note that $x(t)$ and $x(t-t_0)$ describe the same orbit for any t_0. Next, we recall theorem 4.7, which shows that there is a unique orbit through any point of phase space.

Let us now turn our attention to a special class of solutions of (1.1), namely, critical points.

Definition: A point c is a **critical** point of (1.1) if $f(c) = 0$.

Note that $x = c$ for all t ($-\infty < t < \infty$) is then a solution of (1.1) and, hence, a critical point is a stationary solution, and can be thought of as an equilibrium state. By the uniqueness theorem, theorem 4.7, it is the only solution such that $x(t_0) = c$, for any finite t_0. Thus, if a solution $x(t)$ of (1.1) approaches a critical point c (i.e. $x(t) \to c$), then it must do so either as $t \to \infty$ or as $t \to -\infty$. Further, if $x(t) \to c$ (and $x'(t) \to 0$) as $t \to \infty$, or as $t \to -\infty$, then c is a critical point; this result follows on substitution into (1.1).

These last two properties show that the importance of critical points lies in their role as sinks (if $x(t) \to c$ as $t \to \infty$) or sources (if $x(t) \to c$ as $t \to -\infty$), of nearby orbits. To analyze this further, we must consider the stability of a critical point. Let

$$z = x - c, \quad \text{where } f(c) = 0,$$

so that

$$z' = f(c + z). \tag{1.2}$$

To proceed further, we shall suppose that $f(x)$ is continuous, with continuous partial derivatives for all x. Then (1.2) becomes (see (3.1) of section 4.3)

$$z' = Az + H(z), \tag{1.3}$$

where

$$A = \frac{\partial f}{\partial x}(c) \quad \text{and} \quad \frac{|H(z)|}{|z|} \to 0 \quad \text{as } |z| \to 0.$$

Here we recall that $\partial f / \partial x$ is the $n \times n$ matrix $[\partial f_i / \partial x_j]$. Since A in (1.3) is a constant matrix, the principle of linearized stability (theorems 4.4 and 4.5) may be used, and it follows that the critical point is uniformly and asymptotically stable if all the eigenvalues of A have negative real parts, but is unstable if there is at least one eigenvalue of A with a positive real part.

To illustrate the dominant role that critical points can play, we consider the following planar example.

$$x_1' = lx_1 - px_1x_2, \qquad x_2' = -mx_2 + qx_1x_2. \tag{1.4}$$

Here $n = 2$ and x is a two-vector with components (x_1, x_2); l, m, p and q are positive constants. Equations (1.4) form a Lotka–Volterra predator-prey system in which x_1 is the prey and x_2 is the predator. Here $f(x)$ is a two-vector with components (f_1, f_2), where

$$f_1(x_1, x_2) = lx_1 - px_1x_2, \qquad f_2(x_1, x_2) = -mx_2 + qx_1x_2.$$

The matrix $A = \partial f / \partial x$ (see (1.3)) is thus the 2×2 matrix given by

$$A = \frac{\partial f}{\partial x} = \begin{bmatrix} l - px_2 & -px_1 \\ qx_2 & -m + qx_1 \end{bmatrix}.$$

There are two critical points, given by

$$c^{(1)} = (0, 0), \qquad c^{(2)} = \left(\frac{m}{q}, \frac{l}{p} \right)$$

To analyze (1.4), we first consider the first variational equation for each critical point. We recall from section 4.3 that this is obtained by omitting $H(z)$ in (1.3). Thus, for the critical point $c^{(1)}$,

$$A(c^{(1)}) = \begin{bmatrix} l & 0 \\ 0 & -m \end{bmatrix}$$

and the first variational equation is

$$z_1' = lz_1, \qquad z_2' = -mz_2. \tag{1.5}$$

Here we recall from (1.2) that $z = x - c^{(1)}$ and these equations are just the linearization of (1.4) about $c^{(1)}$. The solution of (1.5) is

$$z_1 = \alpha_0 e^{lt}, \qquad z_2 = \beta_0 e^{-mt}.$$

Here α_0 and β_0 are arbitrary constants. The orbits are sketched in figure 5.1, where the arrows are in the direction of t increasing. Clearly $c^{(1)}$ is unstable and, indeed, the eigenvalues of A are l and $-m$. In section 5.2 we shall classify all critical points for planar systems and, anticpating this classification, the orbits in figure 5.1 are those for a saddle point.

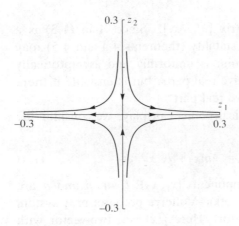

Fig. 5.1 A plot of the orbits for equation (1.5), for the case $l = m = 1$.

Next, for the critical point $c^{(2)}$,

$$A(c^{(2)}) = \begin{bmatrix} 0 & -\dfrac{pm}{q} \\ \dfrac{ql}{p} & 0 \end{bmatrix}$$

and the first variational equation is

$$z_1' = -\frac{pm}{q} z_2, \qquad z_2' = \frac{ql}{p} z_1. \tag{1.6}$$

Here $z = x - c^{(2)}$, and these equations are just the linearization of (1.4) about $c^{(2)}$. The solution of (1.6) is

$$z_1 = pmr_0 \cos(\omega t + \phi_0), \qquad z_2 = \omega q r_0 \sin(\omega t + \phi_0),$$

where

$$\omega = \sqrt{lm}.$$

Here r_0 and ϕ_0 are arbitrary constants and the orbits are sketched in figure 5.2. They are ellipses and are the orbits for a centre. In this case the eigenvalues of A are $\pm i\omega$, and the principle of linearized stability cannot be used to determine the stability. In fact, we show below that $c^{(2)}$ is uniformly stable.

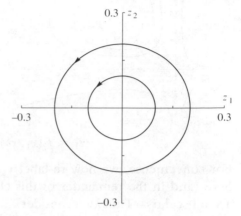

Fig. 5.2 A plot of the orbits for equation (1.6), for the case $l = m = p = q = 1$.

The information obtained from these local analyses near the two critical points is a useful guide to the structure of the orbits in the whole phase plane. For this particular example this can be readily demonstrated since (1.4) can be integrated exactly. First, however, we note from (1.4) that

$$x_1' \gtrless 0, \quad \text{according as } x_1(l - px_2) \gtrless 0,$$

$$x_2' \gtrless 0, \quad \text{according as } x_2(-m + qx_1) \gtrless 0.$$

The qualitative information gained from these inequalities is a useful guide to sketching the orbits and, indeed, this can be a useful approach even when exact solutions are not available. To integrate (1.4), we first put these equations into the form

$$\frac{\mathrm{d}x_2}{\mathrm{d}x_1} = \frac{x_2(-m + qx_1)}{x_1(l - px_2)}.$$

Then separation of variables leads to the solution

$$l \ln|x_2| - px_2 = -m \ln|x_1| + qx_1 + k_0,$$

where k_0 is an arbitrary constant. The orbits are sketched in figure 5.3. In particular, note the comparison with the local orbits near each critical point, sketched in figures 5.1 and 5.2.

5.2 Linear plane, autonomous systems

In the remainder of this chapter (and in the next chapter), we consider plane autonomous systems, so that in (1.1) the dimension $n = 2$. Thus (1.1) is, in components,

Chapter 5

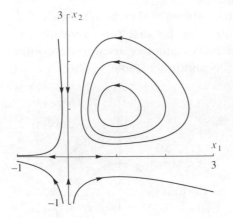

$$x_1' = f_1(x_1, x_2), \qquad x_2' = f_2(x_1, x_2).$$

For convenience, we now re-label x_1 by x, x_2 by y, f_1 by f and f_2 by g, where here (and in the remainder of this chapter, and the next chapter) x, y, f and g are all scalars. Thus, we consider

$$x' = f(x, y), \qquad y' = g(x, y). \tag{2.1}$$

The phase space is now the x-y plane, which we shall also call the **phase plane**. In this section we wish to consider the case when f and g are linear functions of x and y, so that (2.1) becomes

$$\begin{aligned} x' &= ax + by, \\ y' &= cx + dy, \end{aligned} \tag{2.2}$$

where a, b, c and d are constants. Of course, our interest in linear systems is that they occur as the first variational equations in the study of the stability of critical points (i.e. by the omission of the nonlinear term $H(z)$ in (1.3)).

The system (2.2) is a linear equation with constant coefficients. Methods for solving such equations have been described in section 2.5, where it was shown that the structure of the solutions depends on the eigenvalues of the matrix A_0, where

$$A_0 = \begin{bmatrix} a & b \\ c & d \end{bmatrix}. \tag{2.3}$$

Here, our interest is in determining the topology of the orbits in the phase plane and, for this purpose, it is useful to transform (2.2) so that the matrix A_0 is in canonical form. Thus, we put

$$\begin{bmatrix} x \\ y \end{bmatrix} = S \begin{bmatrix} u \\ v \end{bmatrix}, \tag{2.4}$$

where S is a non-singular 2×2 matrix. Then (2.2) becomes

$$\begin{bmatrix} u' \\ v' \end{bmatrix} = \Lambda \begin{bmatrix} u \\ v \end{bmatrix},$$

where

$$A_0 S = S\Lambda. \tag{2.5}$$

We shall choose S so that Λ adopts a simple, convenient form. Note that the transformation from (x, y) to (u, v) corresponds to choosing new axes (not necessarily at right angles) and a new measure of distance along each of these axes, but preserves the topology of the orbits. Also note that the transformation preserves the eigenvalues (i.e. A_0 and Λ have the same eigenvalues) and, hence, a classification of the topology of the orbits based on the eigenvalues is not affected by the transformation.

Recalling the discussion in section 2.5, we identify three canonical forms for the matrix A_0. First, if A_0 has eigenvalues λ_1 and λ_2 which are real and distinct, then the canonical form is

$$\Lambda = \begin{bmatrix} \lambda_1 & 0 \\ 0 & \lambda_2 \end{bmatrix}. \tag{2.6}$$

Second, if A_0 has a single real eigenvalue λ_1, of algebraic multiplicity two, then the canonical form is

$$\Lambda = \begin{bmatrix} \lambda_1 & \gamma \\ 0 & \lambda_1 \end{bmatrix}. \tag{2.7}$$

Here, if the eigenvalue has geometric multiplicity two then $\gamma = 0$, but if the eigenvalue only has geometric multiplicity one then $\gamma \neq 0$. Third, if A_0 has a complex-conjugate pair of eigenvalues, $\lambda_{1,2} = \alpha \pm i\beta$ ($\beta \neq 0$), then the canonical form is

$$\Lambda = \begin{bmatrix} \alpha & \beta \\ -\beta & \alpha \end{bmatrix}. \tag{2.8}$$

For a 2×2 matrix this exhausts the possibilities. Here we shall suppose also that the determinant of A_0 (i.e. $ad - bc$) is not zero, which is equivalent to requiring that no eigenvalue is zero. This condition is needed to ensure that the origin, which is a critical point for (2.2) (or (2.5)), is an isolated critical point, that is, there are no other critical points in a neighbourhood of the origin. Based on these canonical forms, we now proceed to classify the orbits of (2.5) (and, hence, those of (2.2)).

(i) **Nodes**: Here the eigenvalues λ_1 and λ_2 are real and have the same sign, that is, $\lambda_1\lambda_2 > 0$. Suppose first that they are also distinct. Then the canonical form is (2.6) and equations (2.5) become

$$u' = \lambda_1 u, \qquad v' = \lambda_2 v. \tag{2.9}$$

The solution is

$$u = u_0 e^{\lambda_1 t}, \qquad v = v_0 e^{\lambda_2 t}, \tag{2.10}$$

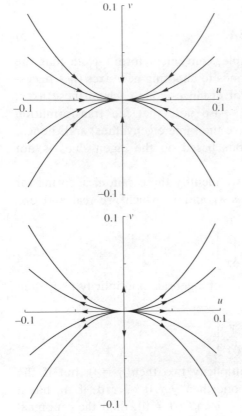

Fig. 5.4 The orbits for a stable node, for the case $\lambda_1 = -1$ and $\lambda_2 = -2$ in (2.9).

Fig. 5.5 The orbits for an unstable node, for the case $\lambda_1 = 1$ and $\lambda_2 = 2$ in (2.9).

where u_0 and v_0 are arbitrary constants. The orbits are sketched in figure 5.4 for $\lambda_2 < \lambda_1 < 0$, which describes a **stable node**, and in figure 5.5 for $\lambda_2 > \lambda_1 > 0$, which describes an **unstable node**. A node is characterized by a dominant orbit, here the u-axis, since $|\lambda_2| > |\lambda_1|$, which gathers to it all other orbits entering, or leaving, the origin, except one single orbit, here the v-axis. Note that the unstable case is analogous to the stable case, but the sense of time, t, is reversed. In the stable case, all the orbits approach the origin as $t \rightarrow \infty$, whereas in the unstable case all the orbits emanate from the origin as $t \rightarrow -\infty$.

Next suppose that the eigenvalues are equal, so that the canonical form is (2.7) and equations (2.5) become

$$u' = \lambda_1 u + \gamma v, \qquad v' = \lambda_1 v. \qquad (2.11)$$

The solution is

$$u = (u_0 + \gamma t v_0)e^{\lambda_1 t}, \qquad v = v_0 e^{\lambda_1 t}, \qquad (2.12)$$

where u_0 and v_0 are arbitrary constants. The orbits are sketched in figure 5.6 for $\gamma > 0$ and in figure 5.7 for $\gamma = 0$ (in both cases, (a) is $\lambda_1 < 0$ and (b) is $\lambda_1 > 0$). The origin is again a stable ($\lambda_1 < 0$) or unstable node ($\lambda_1 > 0$). The case when $\gamma \neq 0$ is sometimes called an improper node, and is characterized by

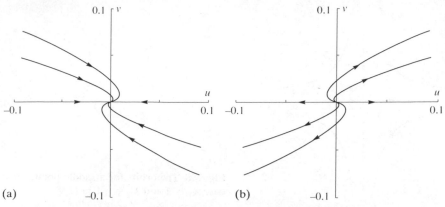

Fig. 5.6 The orbits for an improper node.
(a) The stable case when $\lambda_1 = -1$ and $\gamma = 1$ in (2.11).
(b) The unstable case when $\lambda_1 = 1$ and $\gamma = 1$ in (2.11).

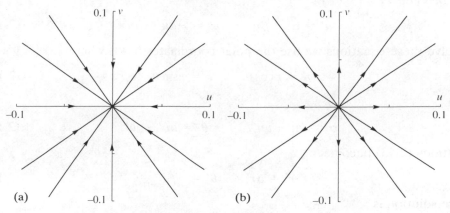

Fig. 5.7 The orbits for a proper node.
(a) The stable case when $\lambda_1 = -1$ and $\gamma = 0$ in (2.11).
(b) The unstable case when $\lambda_1 = 1$ and $\gamma = 0$ in (2.11).

a single dominant orbit, here the u-axis, which gathers all other orbits to it entering, or leaving, the origin, while the case when $\gamma = 0$ is sometimes called a proper node, and has no dominant direction.

(ii) **Saddle points**: Here the eigenvalues are real and have opposite signs, that is, $\lambda_1 \lambda_2 < 0$. The canonical form is (2.6) and equations (2.5) become (2.9) with solution (2.10). The orbits are sketched in figure 5.8 for $\lambda_2 < 0 < \lambda_1$. In this case the origin is called a **saddle point** and is unstable. It is characterized by two dominant orbits. One, here the v-axis, from which all orbits (except the u-axis) emanate as $t \to -\infty$, and the other, here the u-axis, which gathers all orbits (except the v-axis) as $t \to \infty$. We call the u-axis the unstable axis, and the v-axis the stable axis.

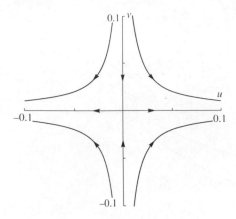

Fig. 5.8 The orbits for a saddle point, for the case $\lambda_1 = 1$ and $\lambda_2 = -1$ in (2.9).

(iii) **Spiral points**: Here the eigenvalues λ_1 and λ_2 are a complex-conjugate pair, $\alpha \pm i\beta$, with $\alpha \neq 0$ and $\beta \neq 0$. The canonical form is (2.8) and equations (2.5) become

$$u' = \alpha u + \beta v, \qquad v' = -\beta u + \alpha v. \tag{2.13}$$

To solve these equations we use the polar coordinates (r, θ), where

$$u = r\cos\theta, \qquad v = r\sin\theta. \tag{2.14}$$

Then

$$rr = uu' + vv', \qquad r^2\theta' = uv' - vu'. \tag{2.15}$$

Equations (2.13) become

$$r' = \alpha r, \qquad \theta' = -\beta, \tag{2.16}$$

whose solution is

$$r = r_0 e^{\alpha t}, \qquad \theta = \theta_0 - \beta t, \tag{2.17}$$

where r_0 and θ_0 are arbitrary constants. The orbits are sketched in figure 5.9 for $\alpha > 0$), which describes a **stable spiral point**, and in figure 5.10 for $\alpha > 0$ (and $\beta > 0$) which describes an **unstable spiral point**. A spiral point is sometimes also called a **focus**.

(iv) **Centre**: Here both eigenvalues are pure imaginary, and form a complex-conjugate pair $\pm i\beta$ with $\beta \neq 0$. The canonical form is again (2.8), but now $\alpha = 0$. The solution is obtained as in case (c) above and, hence, is obtained by putting $\alpha = 0$ in (2.17), so that

$$r = r_0, \qquad \theta = \theta_0 - \beta t. \tag{2.18}$$

The orbits are circles, and are sketched in figure 5.11 (with $\beta > 0$). The origin is a **centre**, and is uniformly, but not asymptotically stable, for the linear equation (2.2). Note, however, that unlike the three previous cases (i), (ii) and (iii),

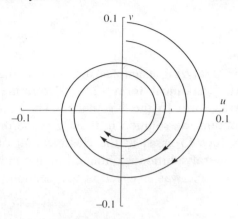

Fig. 5.9 The orbits for a stable spiral point, for the case when $\alpha = -0.15$ and $\beta = 1$ in (2.13).

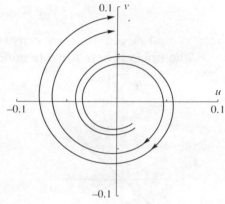

Fig. 5.10 The orbits for an unstable spiral point, for the case when $\alpha = 0.15$ and $\beta = 1$ in (2.13).

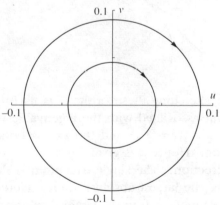

Fig. 5.11 The orbits for a centre, for the case when $\alpha = 0$ and $\beta = 1$ in (2.13).

the principle of linearized stability cannot be used in this case, where we are regarding the linear equation (2.2) as the first variational equation of a nonlinear system.

To illustrate this system of classification, we consider the following examples. First, suppose that

$$x' = -3x + 2y,$$

$$y' = \quad x - 2y.$$

In analyzing examples such as this there are two options. The first is to obtain the canonical form (2.5), determine the orbits in the u-v plane, and then transform to the x-y plane. The second option, and the one used here, is to obtain the solution in the x-y plane directly and use the invariance of the eigenvalues to assist in the classification of the orbits. For the example above, the general solution is obtained by the method described in section 2.5. Indeed, this example was analyzed in that section, and the general solution is

$$x = c_1 e^{-t} - 2c_2 e^{-4t},$$

$$y = c_1 e^{-t} + c_2 e^{-4t},$$

where c_1 and c_2 are arbitrary constants. The eigenvalues are -1 and -4 and, hence, the origin is here a stable node. The orbits are sketched in figure 5.12.

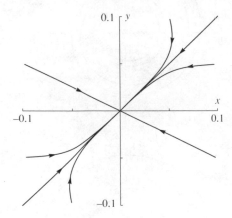

Fig. 5.12 The orbits for $x' = -3x + 2y$, $y' = x - 2y$.

In sketching these orbits it is useful to determine the dominant orbit, which is that associated with the eigenvalue which has the smallest absolute value. Here this is $\lambda_1 = -1$, and the axis is obtained by putting $c_2 = 0$ in the general solution. Hence it is given by $x = y$. As $t \to \infty$ all orbits, except one, approach this direction. The single exception is the axis associated with the eigenvalue which has the largest absolute value. Here this is $\lambda_2 = -4$, and the axis is obtained by putting $c_1 = 0$ in the general solution. Hence it is given by $x + 2y = 0$.

Second, consider

$$x' = \quad 5x - 4y,$$

$$y' = -8x + \quad y.$$

Here the matrix A_0 is

$$\begin{bmatrix} 5 & -4 \\ -8 & 1 \end{bmatrix},$$

which has eigenvalues 9 and -3 and, hence, the origin is a saddle point. The corresponding eigenvectors are

$$\begin{bmatrix} 1 \\ -1 \end{bmatrix}, \quad \begin{bmatrix} 1 \\ 2 \end{bmatrix},$$

and the general solution is

$$x = c_1 e^{9t} + c_2 e^{-3t},$$
$$y = -c_1 e^{9t} + 2c_2 e^{-3t},$$

where c_1 and c_2 are arbitrary constants. The orbits are sketched in figure 5.13.

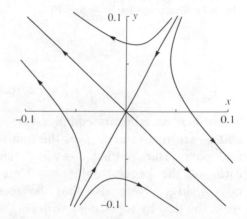

Fig. 5.13 The orbits for $x' = 5x - 4y$, $y' = -8x + y$.

Here the unstable axis is that associated with the positive eigenvalue ($\lambda_1 = 9$), and is obtained by putting $c_2 = 0$ in the general solution, so that it is given by $x + y = 0$. The stable axis is that associated with the negative eigenvalue ($\lambda_2 = -3$), and is obtained by putting $c_1 = 0$ in the general solution, so that it is given by $y = 2x$.

Third, consider

$$x' = 5x + 10y,$$
$$y' = -x - y.$$

This example was analyzed in section 2.5 and the general solution is

$$x = e^{2t}\{c_1(3\cos t - \sin t) + c_2(3\sin t + \cos t)\},$$
$$y = e^{2t}\{-c_1\cos t - c_2\sin t\}.$$

Here the eigenvalues are $2 \pm i$ and, hence, the origin is an unstable focus. The orbits are sketched in figure 5.14. In obtaining these sketches it is useful to observe that the general solution can also be written in the form

$$x = r_0 e^{2t}\{3\cos(t - t_0) - \sin(t - t_0)\},$$
$$y = r_0 e^{2t}\{-\cos(t - t_0)\}.$$

The simplest method of obtaining this is to recall that a complex-valued solution

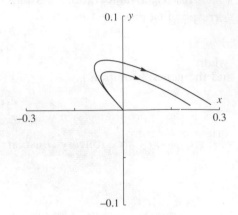

Fig. 5.14 The orbits for $x' = 5x + 10y$, $y' = -x - y$.

is

$$\begin{bmatrix} x \\ y \end{bmatrix} = ce^{2t}(\cos t + \mathrm{i}\sin t)\begin{bmatrix} 3+\mathrm{i} \\ -1 \end{bmatrix},$$

where c is an arbitrary complex-valued constant. Putting $c = c_1 - \mathrm{i}c_2$, where c_1 and c_2 are real-valued, gives the first form of the general solution, on taking the real part. But, putting $c = r_0 e^{-\mathrm{i}t_0}$ and taking the real part, gives the second form of the general solution. Note that this shows that $c_1 = r_0\cos t_0$ and $c_2 = r_0\sin t_0$. Note also that the second form could also be simply obtained from the first by replacing t with $t - t_0$ and putting $c_2 = 0$. Now we find that

$$y = -r_0 e^{2t}\cos(t - t_0), \qquad x + 3y = -4r_0 e^{2t}\sin(t - t_0),$$

so that

$$y^2 + \tfrac{1}{4}(x + 3y)^2 = r_0^2 e^{4t}.$$

Hence, the orbits are elliptical spirals whose 'major and minor axes' are $x + 3y = 0$ and $y = 0$, respectively. These last relations are also useful in determining the sense of time t increasing.

Finally, consider the equation for a simple harmonic oscillator with linear damping,

$$u'' + \nu u' + \omega^2 u = 0, \tag{2.19}$$

where ν is a constant. In applications ν is usually small and positive, but here it is instructive to allow ν to take all real values. To put this equation into the form (2.2), let

$$x = u, \qquad y = \frac{u'}{\omega}.$$

Then (2.19) becomes

$$x' = \omega y, \qquad y' = -\omega x - \nu y.$$

The general solution is

$$x = r_0 e^{-\frac{1}{2}\nu t} \cos \sigma(t - t_0),$$

$$y = r_0 e^{-\frac{1}{2}\nu t} \left(-\frac{\sigma}{\omega} \sin \sigma(t - t_0) - \frac{\nu}{2\omega} \cos \sigma(t - t_0) \right),$$

where

$$\sigma = \sqrt{\omega^2 - \tfrac{1}{4}\nu^2}.$$

This is the appropriate form when $\nu^2 < 4\omega^2$ so that σ is real and positive. The origin is a centre for $\nu = 0$, a stable focus for $\nu > 0$ and an unstable focus for $\nu < 0$. The orbits are sketched for $\nu > 0$ in figure 5.15.

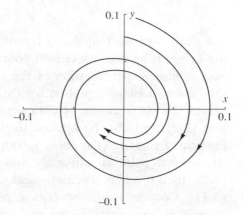

Fig. 5.15 The orbits for the simple harmonic oscillator with linear damping (2.19), for the case when $\nu = 0.2$ and $\sigma = 1$ ($\omega = 1.005$).

The solution is that for a damped oscillation of frequency σ, and damping rate $\frac{1}{2}\nu$. For $\nu^2 > 4\omega^2$ the general solution is

$$x = e^{-\frac{1}{2}\nu t}\{c_1 e^{\rho t} + c_2 e^{-\rho t}\},$$

$$y = e^{-\frac{1}{2}\nu t}\left\{ \left(\frac{\rho}{\omega} - \frac{\nu}{2\omega} \right) c_1 e^{\rho t} - \left(\frac{\rho}{\omega} + \frac{\nu}{2\omega} \right) c_2 e^{-\rho t} \right\},$$

where

$$\rho = \sqrt{\tfrac{1}{4}\nu^2 - \omega^2}.$$

The origin is a stable node for $\nu > 0$ and an unstable node for $\nu < 0$; note that $\rho < \frac{1}{2}|\nu|$. For $\nu > 0$ the solution is sometimes called an overdamped oscillation, since $x(t)$ has at most one zero and decays to zero monotonically as $t \to \infty$.

5.3 Nonlinear perturbations of plane, autonomous systems

We now consider the nonlinear, plane autonomous system (2.1). Our intention is to study the behaviour of the orbits of (2.1) in the vicinity of a critical point. By virtue of the transformation (1.2), there is no loss of generality in choosing the critical point to be at the origin. In this case the equation adopts the form (1.3), which, in our present notation, becomes

$$x' = ax + by + F(x, y), \qquad y' = cx + dy + G(x, y), \tag{3.1}$$

where

$$\frac{|F|}{r}, \frac{|G|}{r} \to 0 \quad (\text{as } r \to 0). \tag{3.1}$$

Here $r = \sqrt{x^2 + y^2}$, and we suppose further that $F(x, y)$ and $G(x, y)$ are continuous, with continuous partial derivatives, for all x and y, in accord with the corresponding hypotheses for $f(x, y)$ and $g(x, y)$. The constants a, b, c and d are given by

$$A_0 = \begin{bmatrix} a & c \\ c & d \end{bmatrix} = \begin{bmatrix} f_x & f_y \\ g_x & g_y \end{bmatrix} \quad (\text{at } x = 0). \tag{3.2}$$

We shall suppose that the determinant of A_0 (i.e. $ad - bc$) is not zero, so that the origin is an isolated critical point for (3.1). We shall regard the system (3.1) as a nonlinear perturbation of the linear system (2.2), which is, of course, just the first variational equation for (3.1). Our interest now is to establish to what extent the classification of critical points, given in section 5.2 for the linear system (2.2), can also be applied to the nonlinear system (3.1). The principle of linearized stability (theorems 4.4 and 4.5 of section 4.3) already ensures that, if the critical point is uniformly and asymptotically stable for the linear system (2.2), then it is also uniformly and asymptotically stable for the nonlinear system (3.1). Conversely, if the critical point is unstable for the linear system (2.2), then it is also unstable for the nonlinear system (3.1). Here we wish to go further and consider how the topology of the orbits for the linear system (2.2) are affected by a nonlinear perturbation of the form (3.1).

In analyzing the topology of the orbits for the nonlinear system (3.1), it is useful to introduce the polar coordinates (r, θ), where

$$x = r \cos \theta, \qquad y = r \sin \theta. \tag{3.3}$$

Then, since

$$rr' = xx' + yy', \qquad r^2 \theta' = xy' - yx', \tag{3.4}$$

the equations (3.1) become

$$r' = r\{a \cos^2 \theta + (b + c) \cos \theta \sin \theta + d \sin^2 \theta\} + M(r, \theta),$$

$$\theta' = \{c \cos^2 \theta + (d - a) \cos \theta \sin \theta - b \sin^2 \theta\} + N(r, \theta), \tag{3.5}$$

where

$$\frac{|M|}{r}, |N| \to 0 \quad (\text{as } r \to 0).$$

Here, the last property follows from the relations

$$M = \cos \theta \, F(r \cos \theta, r \sin \theta) + \sin \theta \, G(r \cos \theta, r \sin \theta),$$

$$rN = \cos \theta \, G(r \cos \theta, r \sin \theta) - \sin \theta \, F(r \cos \theta, r \sin \theta). \tag{3.6}$$

Note that M and N are periodic functions of θ with period 2π. Before proceeding, we must extend the classification of critical points to the nonlinear system (3.1), since so far our classification has only been made for the linear system (2.2). Naturally, the definitions which follow are based on our analysis of the linear system in section 5.2.

Definition: (i) For the nonlinear system (3.1), the origin is said to be a stable (unstable) **node** if there exist constants α_0 and β_0 $(0 \leqslant \alpha_0, \beta_0 < \pi)$ such that, in some sufficiently small neighbourhood of the origin, as $t \to \infty$ $(t \to -\infty)$, $r \to 0$ for all orbits, and there are an infinite number of orbits for which $\theta \to \alpha_0$ and an infinite number of orbits for which $\theta \to \alpha_0 + \pi$, with just two remaining orbits, for one of which $\theta \to \beta_0$ and for the other $\theta \to \beta_0 + \pi$. Usually $\alpha_0 \neq \beta_0$, but if $\alpha_0 = \beta_0$ then the origin is said to be a stable (unstable) improper node. On the other hand, if, in some sufficiently small neighbourhood of the origin, as $t \to \infty$ $(t \to -\infty)$, $r \to 0$ for all orbits and, for every angle γ_0 there is an orbit such that $\theta \to \gamma_0$, then the origin is said to be a stable (unstable) proper node.

(ii) For the nonlinear system (3.1), the origin is said to be a **saddle point** if there exist constants α_0 and β_0 $(0 \leqslant \alpha_0, \beta_0 < \pi; \; \alpha_0 \neq \beta_0)$ such that, in some sufficiently small neighbourhood of the origin, as $t \to \infty$, there is a unique orbit for which $r \to 0$ and $\theta \to \alpha_0$ and a unique orbit for which $r \to 0$ and $\theta \to \alpha_0 + \pi$, while, as $t \to -\infty$, there is a unique orbit for which $r \to 0$ and $\theta \to \beta_0$ and a unique orbit for which $r \to 0$ and $\theta \to \beta_0 + \pi$. The first two orbits together form a curve called the stable manifold, and the last two orbits together form a curve called the unstable manifold. All the other orbits tend away from the origin as $t \to \infty$ or $t \to -\infty$ and, further, any orbit which starts sufficiently close to the stable (unstable) manifold will tend away from this manifold as $t \to \infty$ $(t \to -\infty)$.

(iii) For the nonlinear system (3.1), the origin is said to be a stable (unstable) **spiral point** if, in some sufficiently small neighbourhood of the origin, $r \to 0$ and $|\theta| \to \infty$ as $t \to \infty$ $(t \to -\infty)$ for all orbits.

(iv) For the nonlinear system (3.1), the origin is said to be a **centre** if, within every arbitrarily small neighbourhood of the origin, there exists at least one closed orbit (i.e. a periodic solution) surrounding the origin.

The theorems which follow will establish that the classification system described in section 5.2 for the linear system (2.2) is preserved by the nonlinear perturbation (3.1) for the cases of nodes, saddle points and spiral points. However, a centre for the linear system (2.2) becomes either a spiral point or a centre for the nonlinear system (3.1).

We shall begin with the case (iii) when the origin is a spiral point for the linear system (2.2).

Theorem 5.1: If the origin is a stable (unstable) spiral point for the linear system (2.2), then it is a stable (unstable) spiral point for the nonlinear system (3.1).

Proof: By a linear transformation of the dependent variables x and y, we may suppose that the matrix A_0 (3.2) is in canonical form (see (2.4) and (2.5)). Clearly such a linear transformation retains the nonlinear property of the nonlinear perturbation terms F and G, namely, that $|F|/r$ and $|G|/r \to 0$ as $r \to 0$. Since the origin is a spiral point for the linear system (2.2), the canonical form for A_0 is (2.8) and, hence, we suppose that the nonlinear system (3.1) has the form, for $\alpha \neq 0$ and $\beta \neq 0$,

$$x' = \alpha x + \beta y + F(x, y), \qquad y' = -\beta x + \alpha y + G(x, y). \tag{3.7}$$

In terms of the polar coordinates (r, θ), this system becomes (see (3.3), (3.4) and (3.5)),

$$r' = \alpha r + M(r, \theta), \qquad \theta' = -\beta + N(r, \theta), \tag{3.8}$$

where

$$\frac{|M|}{r}, |N| \to 0, \quad (\text{as } r \to 0).$$

It is sufficient to consider the stable case when $\alpha < 0$, since the transformation $t \to -t$ enables results for the case $\alpha < 0$ as $t \to \infty$ to be transposed into analogous results for the case $\alpha > 0$ as $t \to -\infty$.

Now, by the principle of linearized stability (theorem 4.4 of section 4.3) the zero solution of (3.7) is uniformly and asymptotically stable, since the eigenvalues of the matrix A_0 are $\alpha \pm i\beta$. Hence, in some sufficiently small neighbourhood of the origin, $r \to 0$ as $t \to \infty$ for all orbits. Next, given any $\epsilon > 0$, there exists a $\delta > 0$ such that $0 \leqslant r \leqslant \delta$ implies that $|N| \leqslant \epsilon$ (and also $|M| \leqslant \epsilon r$). But, since $r \to 0$ as $t \to \infty$, there exists a t_0 such that $t \geqslant t_0$ implies that $0 \leqslant r \leqslant \delta$. Indeed, it can be shown that

$$r \leqslant r_0 \exp\{(\alpha + \epsilon)(t - t_0)\},$$

although we shall not need to use this result explicitly. Now, using the equation for θ in (3.8), we see that

$$\theta = \theta_0 - \beta(t - t_0) + \int_{t_0}^{t} N(r(s), \theta(s)) \, ds.$$

Hence, for $t \geqslant t_0$ we can conclude that

$$(-\beta - \epsilon)(t - t_0) \leqslant \theta - \theta_0 \leqslant (-\beta + \epsilon)(t - t_0). \tag{3.9}$$

Now we choose ϵ so that $0 < \epsilon < |\beta|$, and then it follows from (3.9) that $\theta \to -\infty$ (or $+\infty$) as $t \to \infty$ according as $\beta > 0$ (or $\beta < 0$).

Next we consider case (iv) when the origin is a centre for the linear system (2.2).

Theorem 5.2: If the origin is a centre for the linear system (2.2), then it is either a spiral point (stable or unstable), or a centre, for the nonlinear system (3.1).

Proof: The formulation is similar to that for theorem 5.1 and, hence, we can suppose that the nonlinear system (3.1) has the form (3.7), where $\beta \neq 0$ but now $\alpha = 0$. Hence, here equation (3.8) becomes

$$r' = M(r, \theta), \qquad \theta' = -\beta + N(r, \theta). \tag{3.10}$$

Next, given any $\eta > 0$, there exists a $\gamma > 0$ such that $0 \leqslant r \leqslant \gamma$ implies that $|N| \leqslant \eta$ (and also $|M| \leqslant \eta r$). Thus, in the neighbourhood of the origin $0 \leqslant r \leqslant \gamma$, it follows that

$$(-\beta - \eta) < \theta' < (-\beta + \eta). \tag{3.11}$$

We now choose η so that $0 < \eta < |\beta|$, and then (3.11) shows that θ' has the same sign as $-\beta$ and, hence, θ is a monotonic function of t, provided, of course, that the orbit being considered remains within the neighbourhood $0 \leqslant r \leqslant \gamma$. Thus we can replace the pair of equations (3.10) with the single equation

$$\frac{dr}{d\theta} = H(r, \theta) = \frac{M(r, \theta)}{-\beta + N(r, \theta)}, \tag{3.12}$$

which determines the orbit being considered in the form $r = r(\theta)$. Note that, since the denominator in the expression (3.12) for $H(r, \theta)$ is bounded away from zero, it follows that

$$\frac{|H|}{r} \to 0 \quad (\text{as } r \to 0). \tag{3.13}$$

Now consider the unique orbit through the point (r_0, θ_0). From (3.12) it follows that

$$r = r_0 + \int_{\theta_0}^{\theta} H(r, \phi) \, d\phi. \tag{3.14}$$

But, from (3.13), given any $\epsilon > 0$, there exists a $\delta > 0$ such that $0 \leqslant r \leqslant \delta$ implies that $|H| \leqslant \epsilon r$. Hence

$$|r - r_0| \leqslant \epsilon \int_{\theta_0}^{\theta} r(\phi) \, d\phi,$$

or

$$r_0 \exp\{-\epsilon(\theta - \theta_0)\} \leqslant r \leqslant r_0 \exp\{\epsilon(\theta - \theta_0)\}. \tag{3.15}$$

Here the last line follows by imitating the proof of Gronwall's lemma (section 1.3). Now, from (3.11), we see that $\theta' \neq 0$ for any t, and, further, $|\theta'|$ is bounded away from zero. Hence the orbit $r = r(\theta)$ will complete one circuit of the origin in a finite time. From (3.15) we see that, throughout this circuit, $r = r_0 + O(\epsilon)$ and, hence, by choosing r_0 sufficiently small, the orbit remains within the neighbourhood of the origin $0 \leqslant r \leqslant \delta$ for $\theta_0 \leqslant \theta \leqslant \theta_0 + 2\pi$; there is no loss of generality here in supposing that θ is increasing (i.e. $\beta < 0$). Indeed, if we choose ϵ so that $\exp(2\pi\epsilon) < 2$ then it is sufficient to choose $r_0 < \frac{1}{2}\delta$.

Suppose next that the origin is not a centre for the nonlinear system (3.10) (or (3.12)). Then, by decreasing δ if necessary, we can assume that there are no closed orbits in $0 \leqslant r \leqslant \delta$. Let the orbit $r = r(\theta)$ meet the half-line $\theta = \theta_0$ again at $r = r_1$, that is, $r_1 = r(\theta_0 + 2\pi)$. Since there are no closed orbits $r_1 \neq r_0$; with no loss of generality we let $r_0 > r_1$. It now follows that $r(\theta) > r(\theta + 2\pi)$ for all $\theta \geqslant \theta_0$, since any equality would imply the presence of a closed orbit. Now continue the orbit from the point $(r_1, \theta_0 + 2\pi)$ until it meets the half-line $\theta = \theta_0$ again at $r_2 = r(\theta_0 + 4\pi)$. Clearly, $r_1 > r_2$. Continuing this process, we generate the sequence $r_0, r_1, r_2, \ldots, r_n$, where $r_0 > r_1 > r_2 > \ldots > r_n > 0$. Hence, by the monotonic-sequence theorem, there exists $R \geqslant 0$ such that $r_n \to R$ as $n \to \infty$. If $R = 0$, then $r(\theta) \to 0$ as $\theta \to \infty$.

On the other hand, if $R > 0$, then there exists a periodic solution of (3.12), which is the unique orbit through the point (R, θ_0). To show this, form the sequence of orbits $r_n(\theta)$ defined as the unique orbits through the points (r_n, θ_0); clearly, $r_n(\theta) = r(\theta + 2n\pi)$. Thus $r_n(\theta_0) \to R$ and $r_n(\theta_0 + 2\pi) \to R$ as $n \to \infty$. Now, each $r_n(\theta)$ satisfies (3.12) and, hence, analogously to (3.14),

$$r_n(\theta) = r_n + \int_{\theta_0}^{\theta} H(r_n(\phi), \phi)\, d\phi.$$

Further, by analogy with (3.15),

$$r_n \exp\{-\epsilon(\theta - \theta_0)\} \leqslant r_n(\theta) \leqslant r_n \exp\{\epsilon(\theta - \theta_0)\}.$$

Now consider the sequence of functions $r_n(\theta)$ for $\theta_0 \leqslant \theta \leqslant \theta_0 + 2\pi$. Over this interval it follows that

$$r_n \exp(-2\pi\epsilon) \leqslant r_n(\theta) \leqslant r_n \exp(2\pi\epsilon),$$

or

$$R \exp(-2\pi\epsilon) \leqslant r_n(\theta) \leqslant r_0 \exp(2\pi\epsilon).$$

Thus the sequence of functions $r_n(\theta)$ are defined over the annular region $R \exp(-2\pi\epsilon) \leqslant r \leqslant r_0 \exp(2\pi\epsilon)$, $\theta_0 \leqslant \theta \leqslant \theta_0 + 2\pi$ and, in this region, $H(r, \theta)$ will satisfy a Lipschitz condition with respect to r, with constant L say. This is a consequence of the definitions (3.6) for M and N, and the definition (3.12) for H, from which it follows that H has continuous partial derivatives with respect to r and θ; it is crucial here that r be bounded away from zero. Next let $R(\theta)$

be the unique orbit through the point (R, θ_0). Then $r_n(\theta) \to R(\theta)$ uniformly for $\theta_0 \leqslant \theta \leqslant \theta_0 + 2\pi$ as $n \to \infty$. Indeed, since $R(\theta)$ satisfies (3.12),

$$R(\theta) = R + \int_{\theta_0}^{\theta} H(R(\phi), \phi) \, d\phi.$$

Comparing this expression with the corresponding expression for $r_n(\theta)$ and using the Lipschitz condition for $H(r, \theta)$, we find that

$$|r_n(\theta) - R(\theta)| \leqslant |r_n - R| + L \int_{\theta_0}^{\theta} |r_n(\phi) - R(\phi)| \, d\phi,$$

or

$$|r_n(\theta) - R(\theta)| \leqslant |r_n - R| \exp\{(\theta - \theta_0)\}.$$

The last line follows from an application of Gronwall's Lemma (section 1.3). Since $r_n \to R$ as $n \to \infty$, the uniform convergence of $r_n(\theta)$ to $R(\theta)$ is proven. Further, since $r_n(\theta_0)$ and $r_n(\theta_0 + 2\pi) \to R$ as $n \to \infty$, it follows that $R(\theta_0) = R(\theta_0 + 2\pi) = R$ and, hence, $R(\theta)$ is a closed orbit.

But, since our hypothesis is that there are no closed orbits in the neighbourhood of the origin $0 \leqslant r \leqslant \delta$, we must now conclude that $R = 0$. Hence $r(\theta)$, the unique orbit through the point (r_0, θ_0), spirals inward towards the origin as θ increases, and $r \to 0$ as $\theta \to \infty$. Next, let $\hat{r}(\theta)$ be the unique orbit through the point $(\hat{r}_0, \theta_0)$, where $r_1 < \hat{r}_0 < r_0$. Clearly,

$$r_1(\theta) < \hat{r}_0(\theta) < r(\theta) \quad (\text{for } \theta_0 \leqslant \theta \leqslant \theta + 2\pi),$$

where we recall that $r_1(\theta) = r(\theta + 2\pi)$. The result follows since there is a unique orbit through every point. Thus, if $\hat{r}_1 = \hat{r}(\theta_0 + 2\pi)$, we see that $r_2 < \hat{r}_1 < r_1$. Continuing this process, it can be seen that $\hat{r}(\theta)$ also spirals inwards towards the origin as θ increases, and $\hat{r} \to 0$ as $\theta \to \infty$. This is sufficient to establish that, if the origin is not a centre, it is a spiral point.

Next we proceed to show by example that all the options available in theorem 5.2 can be realized. First consider the example

$$x' = y + kxr, \qquad y' = -x + kyr, \tag{3.16}$$

where k is a constant. Here $F = kxr$ and $G = kyr$ and the equation has the required form (3.1). For the corresponding linear system, the origin is a centre, since, in the notation of (3.7), $\alpha = 0$ and $\beta = 1$. In polar coordinates (3.16) becomes

$$r' = kr^2, \qquad \theta' = -1,$$

and so

$$r = (C - kt)^{-1}, \qquad \theta = \theta_0 - t, \tag{3.17}$$

where C is an arbitrary constant. If $r = r_0$ at $t = 0$, then $C = 1/r_0$ and is positive. It follows that the origin is an unstable or stable spiral point, according as $k > 0$ or $k < 0$. If $k = 0$ it is, of course, a centre.

To show that the nonlinear perturbation can preserve a centre, consider the example,

$$x' = y + yr, \qquad y' = -x - xr. \tag{3.18}$$

Again this has the required form (3.1), and is a centre for the corresponding linear equation, with $\alpha = 0$ and $\beta = 1$. In polar coordinates (3.18) becomes

$$r' = 0, \qquad \theta' = -(1+r),$$

and so

$$r = r_0, \qquad \theta = \theta_0 - (1+r_0)t. \tag{3.19}$$

Thus the orbits form a family of circles surrounding the origin, which is thus a centre.

To demonstrate that, when the origin is a centre, the orbits may contain both closed orbits and spirals arbitrarily close to the origin, consider the following example:

$$x' = y + xr^2 \sin \frac{\pi}{r}, \qquad y' = -x + yr^2 \sin \frac{\pi}{r}. \tag{3.20}$$

Again this has the required form (3.1) and is a centre for the corresponding linear equation, with $\alpha = 0$ and $\beta = 1$. In polar coordinates (3.20) becomes

$$r' = r^3 \sin \frac{\pi}{r}, \qquad \theta' = -1. \tag{3.21}$$

Hence the circles $r = 1/k$ $(k = 1, 2, \ldots)$ form a family of closed orbits, which are arbitrarily close to the origin as $k \to \infty$. Also, since r' is of one sign for $1/k < r < 1/(k+1)$, while $\theta = \theta_0 - t$, the orbits between adjacent circles are spirals, which alternate between inward and outward spirals.

Next we turn to case (i) when the origin is a node for the linear system (2.2).

Theorem 5.3: If the origin is a stable (unstable) node for the linear system (2.2), then it is a stable (unstable) node for the nonlinear system (3.1).

Proof: In this case the eigenvalues λ_1 and λ_2 of the matrix A_0 are real-valued, and the canonical form for A_0 is either (2.6), when the eigenvalues are distinct $(\lambda_1 \neq \lambda_2)$, or (2.7), otherwise. For the latter case, when $\lambda_1 = \lambda_2$, the canonical form either has $\gamma \neq 0$, corresponding to the case when the eigenvalue has a geometric multiplicity one, or $\gamma = 0$, when the eigenvalue has a geometric multiplicity two. In the former case, the origin is an improper node for the linear system (see figure 5.6), and remains an improper node for the nonlinear system. In the latter case, the origin is a proper node for the linear system (see figure

5.7), and remains a proper node for the nonlinear system, provided that the conditions on F and G are strengthened to:

$$|F| \text{ and } |G| \text{ are } O(r^{1+\delta}) \quad \text{(as } r \to 0), \tag{3.22}$$

for some constant $\delta > 0$. That the condition (3.22) is necessary in this case is shown by example later in this section. We shall omit the proofs for these exceptional cases when $\lambda_1 = \lambda_2$, and refer the reader to Coddington and Levinson (1955, 377–380, 384–387).

For the case when the eigenvalues λ_1 and λ_2 are distinct ($\lambda_1 \neq \lambda_2$), the canonical form for A_0 is (2.6), and the nonlinear system (3.1) takes the form

$$x' = \lambda_1 x + F(x, y), \qquad y' = \lambda_2 y + G(x, y). \tag{3.23}$$

In terms of the polar coordinates (r, θ), this system becomes (see (3.3), (3.4) and (3.5))

$$r' = r(\lambda_1 \cos^2 \theta + \lambda_2 \sin^2 \theta) + M(r, \theta),$$

$$\theta' = (\lambda_2 - \lambda_1) \cos \theta \sin \theta + N(r, \theta). \tag{3.24}$$

Without loss of generality, we shall suppose that the origin for the linear system is a stable node, with $\lambda_2 < \lambda_1 < 0$ (see figure 5.4). The unstable case can be treated, as in theorem 5.1, through the transformation $t \to -t$.

Now, by the principle of linearized stability (theorem 4.4 of section 4.3) the zero solution of (3.23) is uniformly and asymptotically stable, since the eigenvalues of the matrix A_0 are λ_1 and λ_2. Hence, in some sufficiently small neighbourhood of the origin, $r \to 0$ as $t \to \infty$ for all orbits. Next, given any $\epsilon > 0$, there exists a $\delta > 0$ such that $0 \leqslant r \leqslant \delta$ implies that $|N| \leqslant \epsilon$ (and also $|M| \leqslant \epsilon r$). But, since $r \to 0$ as $t \to \infty$, there exists a t_0 such that $t \geqslant t_0$ implies that $0 \leqslant r \leqslant \delta$. Indeed, it can be shown that

$$r_0 \exp\{(\lambda_2 + \epsilon)(t - t_0)\} \leqslant r \leqslant r_0 \exp\{(\lambda_1 + \epsilon)(t - t_0)\} \tag{3.25}$$

and, further, $r \to 0$ monotonically (since $r' < 0$) and exponentially fast, as $t \to \infty$. Here, of course, we must choose ϵ so that $0 < \epsilon < |\lambda_1| \, (< |\lambda_2|)$.

Next choose γ $(0 < \gamma < \frac{1}{4}\pi)$ and form sectors of angular width 2γ centred on each of the axes (see figure 5.16). For instance, the sector S_1 is located on the positive x-axis and defined by $|\theta| \leqslant \gamma$, while the sector S_2 is located on positive y-axis and defined by $|\theta - \frac{1}{2}\pi| \leqslant \gamma$. Now, for the neighbourhood of the origin $0 \leqslant r \leqslant \delta$, we find from (3.24) that

$$\theta' = (\lambda_2 - \lambda_1) \cos \theta \sin \theta + O(\epsilon). \tag{3.26}$$

Hence, by choosing ϵ sufficiently small, $\theta' < 0$ $(\theta' > 0)$ on $\theta = \gamma$ $(\theta = -\gamma)$ and, hence, any orbit entering the sector S_1 must remain in that sector and approach the origin as $t \to \infty$. Further, by choosing γ sufficiently small, we see that all these orbits approach the origin along the half-line $\theta = 0$. Similarly, any orbit entering the sector S_3 $(|\theta - \pi| \leqslant \gamma)$ will remain in that sector and approach the

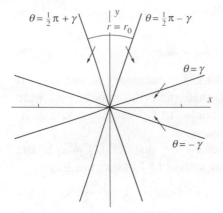

Fig. 5.16 Diagram for theorem 5.3.

origin along the half-line $\theta = \pi$ as $t \to \infty$. By contrast, $\theta' < 0$ ($\theta' > 0$) on $\theta = \frac{1}{2}\pi - \gamma$ ($\theta = \frac{1}{2}\pi + \gamma$) and, hence, no orbit can enter the sector S_2 across these boundaries, at least in the neighbourhood of the origin $0 \leqslant r \leqslant \delta$. Further, for all orbits in the sector $\gamma < \theta < \frac{1}{2}\pi - \gamma$, $\theta' < 0$ and $|\theta'|$ is bounded away from zero. Hence, all these orbits must reach the boundary $\theta = \gamma$ of S_1 in finite time. A similar situation holds for the remaining sectors.

It remains to show that, of all the orbits which commence in the sector S_2 for $0 \leqslant r \leqslant \delta$, there is a unique orbit which stays in the sector and approaches the origin along the half-line $\theta = \frac{1}{2}\pi$ as $t \to \infty$, while all other orbits leave S_2 (and hence eventually enter S_1 or S_3). Consider the circular arc $r = r_0$, for $|\theta - \frac{1}{2}\pi| \leqslant \gamma$, where $0 < r_0 < \delta$ (see figure 5.16). Let P be an arbitrary point on this arc, and consider the unique orbit which commences at P. For P close to the boundary $\theta = \frac{1}{2}\pi - \gamma$ ($\theta = \frac{1}{2}\pi + \gamma$), the orbit will exit S_2 on this boundary. Hence, using the continuity of orbits with respect to their initial conditions and the fact that distinct orbits cannot intersect, it follows that there is at least one value of P such that the corresponding orbit does not exit S_2. By choosing γ sufficiently small, this orbit will approach the origin along the direction $\theta = \frac{1}{2}\pi$ as $t \to \infty$. To show that this orbit is unique, suppose that, on the contrary, there are two such orbits, each of which may be expressed in the form $x = x_i(y)$ ($i = 1, 2$), where $x_i(y)/y \to 0$ as $y \to 0$ through positive values. Let $x(y) = x_1(y) - x_2(y)$, where we suppose that $x(y) \neq 0$ except at $y = 0$. Then it follows from (3.23) that

$$\frac{dx}{dy} = \frac{\lambda_1 x}{\lambda_2 y}\{1 + h(y)\}, \qquad (3.27)$$

where $h(y) \to 0$ as $y \to 0$ through positive values. In obtaining this result we use the properties of F and G defined in (3.1) (and in the sentence immediately following), together with the fact that $x(y)/y \to 0$ as $y \to 0$ through positive values. Then, given any $\sigma > 0$, there exists $\eta > 0$ such that $0 < y < \eta$ implies that $|h(y)| < \sigma$. We can then use (3.27) to show that

$$|x| \leq Cy^{\nu}, \tag{3.28}$$

where $\nu = |\lambda_1|(1+\sigma)/|\lambda_2|$ and C is a positive constant. Further, by choosing σ sufficiently small, we see that $0 < \nu < 1$. Hence $|x|/y \to \infty$ as $y \to 0$ through positive values, which is a contradiction.

Similarly, of all the orbits which commence in the sector S_4 ($|\theta - \frac{3}{2}\pi| < \gamma$), there is a unique orbit which stays in the sector and approaches the origin along the half-line $\theta = \frac{3}{2}\pi$ as $t \to \infty$, while all other orbits leave S_4 (and hence eventually enter S_1 or S_3).

Before proceeding, we give an example to illustrate the necessity to impose the extra conditions (3.22) when considering proper nodes. Thus let,

$$x' = -x - y(\ln r)^{-1}, \qquad y' = -y + x(\ln r)^{-1}. \tag{3.29}$$

Here $F = -y(\ln r)^{-1}$, $G = x(\ln r)^{-1}$ and the equation has the required form (3.1), but F and G do not satisfy the conditions (3.22). For the corresponding linear system (2.2), there is a single eigenvalue, -1, of geometric multiplicity two, so that the origin is a proper (stable) node. In polar coordinates (3.29) becomes

$$r' = -r, \qquad \theta' = (\ln r)^{-1},$$

and so

$$r = r_0 e^{-t}, \qquad \theta = \theta_0 - \ln(t - \ln r_0). \tag{3.30}$$

Hence $\theta \to -\infty$ as $t \to \infty$, and the origin is a (stable) spiral point.

Finally, we turn to case (ii) when the origin is a saddle point for the linear system (2.2).

Theorem 5.4: If the origin is a saddle point for the linear system (2.2), then it is a saddle point for the nonlinear system (2.1).

Proof: The proof is similar to that for theorem 5.3, so we will just sketch the outline. The equations to be considered are again (3.23), or (3.24), but now we suppose that $\lambda_2 < 0 < \lambda_1$, so that the origin is a saddle point for the linear system (see figure 5.8). As in theorem 5.3, the plane is divided into sectors (see figure 5.16). We consider the neighbourhood of the origin $0 \leq r \leq \delta$, such that $|M| \leq \epsilon r$ and $|N| \leq \epsilon$. Then, on the boundary $\theta = \gamma$ ($\theta = -\gamma$) of the sector S_1 we find that $\theta' > 0$ ($\theta' < 0$), and also that $r' > 0$ (since now $\lambda_1 > 0$). Further, $r' > 0$ throughout the sector S_1 and, hence, all orbits entering S_1 will remain in S_1, but cannot reach the origin and, instead, will reach the boundary $r = \delta$.

On the other hand, on the boundary $\theta = \frac{1}{2}\pi - \gamma$ ($\theta = \frac{1}{2}\pi + \gamma$) of the sector S_2, we find that $\theta' < 0$ ($\theta' > 0$), and also that $r' < 0$ (since $\lambda_2 < 0$). Hence, no orbit can enter S_2 across these boundaries. Further, all orbits in the sector

$\gamma < \theta < \frac{1}{2}\pi - \gamma$ will either reach the boundary $r = \delta$ or enter the sector S_1. Finally, just as in the proof of theorem 5.3, it can be shown that of all the orbits which commence in the sector S_2, for $0 \le r \le \delta$, there is a unique orbit which stays in the sector and approaches the origin along the direction $\theta = \frac{1}{2}\pi$ as $t \to \infty$, while all the other orbits leave S_2. Similarly, of all the orbits which commence in the sector S_4, there is a unique orbit which stays in the sector and approaches the origin along the half-line $\theta = \frac{3}{2}\pi$ as $t \to \infty$, while all the other orbits leave S_4.

Thus we have shown that, as $t \to \infty$, there are two orbits which approach the origin along the directions $\theta = \frac{1}{2}\pi$ and $\frac{3}{2}\pi$, while all remaining orbits leave the neighbourhood of the origin. To complete the proof, we make the transformation $t \to -t$ and, by an analogous argument, show that there are two orbits which approach the origin along the directions $\theta = 0$ and π, respectively, as $t \to -\infty$, while all remaining orbits leave the neighbourhood of the origin.

These theorems jointly establish that, in studying the behaviour of the orbits near the critical points of the nonlinear system (2.1), it is sufficient to consider instead the first variational equation for each critical point and, hence, confine attention to linear systems of the form (2.2). The only exception to this procedure is when the linear system predicts the presence of a centre, in which case the nonlinear terms are crucial in deciding whether the origin is in fact a centre, or a spiral point. Indeed, it is precisely because the linear system fails in this case to act as an indicator for the nonlinear system that we shall spend much of the next few chapters developing criteria for the presence of periodic solutions, and describing procedures for determining the outcome of a nonlinear perturbation of a centre for a linear system.

In order to illustrate how the study of critical points is useful in determining the orbits in the phase plane, we consider the following example:

$$x' = x(l - py - kx), \qquad y' = y(-m + qx). \qquad (3.31)$$

Here l, m, p, q and k are positive constants. This is a Lotka–Volterra predator-prey system in which x is the prey and y is the predator. Indeed, we discussed this system in section 5.1 (see (1.4)) when $k = 0$; in that case the system can be solved exactly and the orbits are sketched in figure 5.3. With $k > 0$, exact solutions are not available, but much information about the orbits can be obtained by studying the critical points of (3.31). As we shall see, the term whose coefficient is k can be thought of as a 'dissipative' term, although in the context of population dynamics it is a 'self-limiting' term for the x-population.

In studying the critical points of (3.31), we first observe that (3.31) has the general form (2.1). If (x_0, y_0) is a critical point, let

$$u = x - x_0, \qquad v = y - y_0. \qquad (3.32)$$

Then the first variational equation for (x_0, y_0) is (see (1.2) and (1.3))

$$\begin{bmatrix} u' \\ v' \end{bmatrix} = A_0 \begin{bmatrix} u \\ v \end{bmatrix},$$

where

$$A = \begin{bmatrix} f_x & f_y \\ g_x & g_y \end{bmatrix},$$

and

$$A_0 = A \quad \text{at } x = x_0, y = y_0. \tag{3.33}$$

Here, from (3.31), we see that

$$A = \begin{bmatrix} l - py - 2kx & -px \\ qy & -m + qx \end{bmatrix}. \tag{3.34}$$

The critical points for (3.31) are

$$\text{(i)} \quad x_0 = 0, \qquad y_0 = 0,$$

$$\text{(ii)} \quad x_0 = \frac{l}{k}, \qquad y_0 = 0,$$

$$\text{(iii)} \quad x_0 = \frac{m}{q}, \qquad y_0 = \frac{1}{p}\left(l - \frac{km}{q}\right). \tag{3.35}$$

We shall study each of these critical points in turn.

First consider the critical point (i). For this case the matrix A_0 is given by

$$A_0 = \begin{bmatrix} l & 0 \\ 0 & -m \end{bmatrix}.$$

The eigenvalues are l and $-m$ and, hence, this is a saddle point. The orbits near the critical point are obtained by solving (3.33), and we get

$$u = u_0 e^{lt}, \qquad v = v_0 e^{-mt}.$$

Here u_0 and v_0 are arbitrary constants. Indeed, the linearized system is precisely the same as (1.5) for the case $k = 0$, and the orbits are sketched in figure 5.1 (with a different notation to that being used now).

Next consider the critical point (ii), for which case the matrix A_0 becomes

$$A_0 = \begin{bmatrix} -l & -\dfrac{pl}{k} \\ 0 & -m + \dfrac{ql}{k} \end{bmatrix}.$$

The eigenvalues are $-l$ and $-m + ql/k$. There are two sub-cases to consider. If $ql < km$, both the eigenvalues are negative and this critical point is a stable node. Otherwise, it is a saddle point. The orbits near the critical point are again obtained by solving (3.33), and we get

$$u = u_0 e^{-lt} - \frac{plv_0 e^{\mu t}}{k(l+\mu)}, \qquad v = v_0 e^{\mu t},$$

where

$$\mu = -m + \frac{ql}{k}.$$

Here u_0 and v_0 are arbitrary constants. The orbits are sketched in figure 5.17 for the case $\mu > 0$, when the critical point is a saddle point.

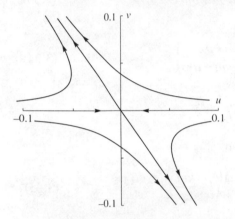

Fig. 5.17 The orbits near the critical point (ii) for (3.31) for the case $l = p = 0.5$, $m = q = 1$ and $k = \frac{1}{4}$.

Finally, we consider the critical point (iii), for which case the matrix A_0 becomes

$$A_0 = \begin{bmatrix} -\dfrac{km}{q} & -\dfrac{pm}{q} \\ \dfrac{1}{p}(ql-km) & 0 \end{bmatrix}.$$

The eigenvalues are λ_1 and λ_2, which are given by the roots of the following quadratic equation for λ:

$$\lambda^2 + \frac{km}{q}\lambda + \frac{m}{q}(ql-km) = 0.$$

Thus, the critical point is a stable spiral point if $ql > km(1+k/4q)$, since then $\lambda_1 = \bar\lambda_2$, with $\mathrm{Re}\,\lambda_1 = -km/2q < 0$. If $km < ql < km(1+k/4q)$, the critical point is a stable node and, if $ql < km$, it is a saddle point. For the first case $(ql > km(1+k/4q))$ the orbits are given by

$$u = r_0 \frac{pm}{q} \exp\left(-\frac{kmt}{2q}\right)\cos\nu(t-t_0),$$

$$v = r_0 \exp\left(-\frac{kmt}{2q}\right)\left(-\frac{km}{2q}\cos\nu(t-t_0) + \nu\sin\nu(t-t_0)\right)$$

where

$$\lambda_1, \lambda_2 = -\frac{km}{2q} \pm i\nu.$$

Here r_0 and t_0 are arbitrary constants, and the orbits are sketched in figure 5.18. The remaining two cases are left as an exercise.

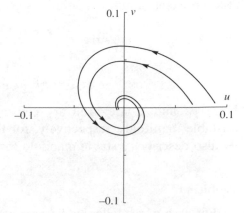

Fig. 5.18 The orbits near the critical point (iii) for (3.31), for the case of figure 5.17.

We are now in a position to consider the orbits in the remainder of the phase plane. Since, in practical applications of (3.31), x and y represent population densities which are intrinsically non-negative, we shall consider only the first quadrant. Further, we shall confine our attention to the case when k is small so that the second critical point (ii) is a saddle point (i.e. $\mu > 0$), and the third critical point (iii) is a stable spiral point, which lies in the first quadrant. With these restrictions, equations (3.31) were integrated numerically, and the results are shown in figure 5.19.

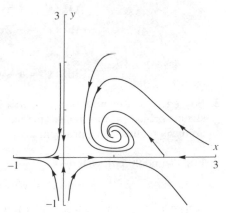

Fig. 5.19 The orbits for equation (3.31) for the case of figure 5.17.

In particular, note the comparison with the local orbits near each critical point, sketched in figures 5.1, 5.17 and 5.18. The results should also be compared with the corresponding orbits for the case $k = 0$, sketched in figure 5.3. It can be seen that the effect of the term whose coefficient is k, is to convert a centre for the case $k = 0$ into a stable spiral point when $k > 0$ (and sufficiently small). It is also interesting to observe here that the value of k ($k = ql/m$ or $\mu = 0$) for which the third critical point (iii) becomes unstable as k increases and moves out of the first quadrant, is precisely the same value for which the second critical point (ii) becomes stable. Finally, we note that, for this example (3.31) there are two exact families of solutions given by

$$x = 0, \qquad y = y_0 e^{-mt} \tag{3.36}$$

and

$$x = l(k + Ce^{-lt})^{-1}, \qquad y = 0. \tag{3.37}$$

Here y_0 and C are arbitrary constants. They correspond to the stable and unstable manifolds, respectively, for the first critical point (i). The second family also describes a stable manifold for the second critical point (ii).

Problems

1 For each of the following linear systems, classify the critical point at the origin, and sketch the orbits in the phase plane.

$$\begin{array}{lll} \text{(a)} \ \ x' = -2x + y, & \text{(b)} \ \ x' = 3x - 2y, & \text{(c)} \ \ x' = \quad x + y, \\ \quad\ \ y' = \quad x - 2y; & \quad\ \ y' = 2x - 2y; & \quad\ \ y' = -5x - 3y; \\ \text{(d)} \ \ x' = \ -x + 5y, & \text{(e)} \ \ x' = \quad 3x + 2y, & \\ \quad\ \ y' = -2x + 5y; & \quad\ \ y' = -2x - y. & \end{array}$$

2 For each of the following nonlinear systems, locate the critical points, classify them, and sketch the orbits near each critical point. Use the information gained to attempt a sketch of the whole phase plane.

$$\begin{array}{ll} \text{(a)} \ \ x' = x - y^2, & y' = 5x - 5y; \\ \text{(b)} \ \ x' = -2x + y^2, & y' = x - 3y + y^2; \\ \text{(c)} \ \ x' = 1 - xy, & y' = x - y^3; \\ \text{(d)} \ \ x' = 1 - y, & y' = x^2 - y^2. \end{array}$$

3 For each of the nonlinear systems in problem 2, use numerical integration to construct orbits in the phase plane, and compare the results with the critical-point analysis requested in problem 2.

4 For the nonlinear system

$$x' = y + kxr^2, \qquad y' = -x + kyr^2,$$

where $r^2 = x^2 + y^2$, show that the origin is the only critical point, and is a stable or unstable spiral point, according as $k < 0$ or $k > 0$.

5 The interaction between two nearly identical species is governed by the equations

$$x' = x(l - kx - py), \qquad y' = y(l - ky - qx),$$

where k, l, p and q are positive constants. For the case when $q > p > k$ (i.e. the 'interaction' terms are stronger than the 'self-limiting' terms and the x-population is stronger than the y-population), find all the critical points, classify them, and sketch the orbits near each critical point. Use the information gained to attempt a sketch of the whole phase plane. Then use numerical integration to construct orbits in the phase plane, and compare the results with the critical-point analysis.

6 The interaction between two competing species is governed by the equations

$$x' = x(l - kx - py), \qquad y' = y(m - ky + qx),$$

where k, l, m, p and q are positive constants. This is a Lotka–Volterra predator-prey system in which x is the prey and y is the predator. This model is an adaptation of example (3.31) of section 5.3, in that the growth rate of the y population (in the absence of the x-population) is here $m - ky$ (instead of $-m$ in (3.31)), indicating that the y-population can exist in the absence of the x-population. Find all the critical points, classify them, and sketch the orbits near each critical point, distinguishing between the cases $kl \gtrless mp$. Use the information gained to attempt a sketch of the whole phase plane. Then use numerical integration to construct orbits in the phase plane, and compare the results with the critical-point analysis.

7 The motion of a simple pendulum with a linear damping is governed by the equation

$$u'' + vu' + \omega^2 \sin u = 0,$$

where $v > 0$. In the x-y phase plane for which $x = u$ and $\omega y = u'$, find all the critical points, classify them, and sketch the orbits near each critical point. Use the information gained to attempt a sketch of the phase plane, and then compare with orbits obtained from a numerical integration.

8 Show that the origin is a centre for the nonlinear system

$$x' = y + 2xy, \qquad y' = -x + xy.$$

[*Hint*: Form an equation for dy/dx and then use separation of variables to obtain an expression for the orbits.]

9 For the nonlinear system

$$x' = -x + y^2, \qquad y' = x - 2y + y^2,$$

locate the critical points, classify them, and sketch the orbits near each critical point. Show that $x = y$ is an orbit and is (for $x > 0$) the unstable manifold through one of the critical points. Also show that $|x - y| \to 0$ as $t \to \infty$ for all the other orbits. [*Hint*: Form an equation for $x - y$.]

CHAPTER SIX
PERIODIC SOLUTIONS OF
PLANE AUTONOMOUS SYSTEMS

6.1 Preliminary results

In this chapter we consider planar systems of the form (2.1) of section 5.2, that is, we let

$$x' = f(x, y), \qquad y' = g(x, y), \tag{1.1}$$

where x, y, f and g are all scalars. Our interest here is in obtaining criteria for the existence, or otherwise, of periodic solutions of (1.1). Here we recall that a periodic solution is such that $x(t+T) = x(t)$ and $y(t+T) = y(t)$ for all t, where the constant T is the period. Clearly, the orbits of periodic solutions are closed curves in the phase plane.

Before proceeding to obtain some general results, let us discuss the following illustrative examples. First, consider

$$x' = y(1+r^2), \qquad y' = -x(1+r^2) \tag{1.2}$$

where $r^2 = x^2 + y^2$. To solve these equations we use the polar co-ordinates (r, θ), where

$$x = r\cos\theta, \qquad y = r\sin\theta. \tag{1.3}$$

Then, recalling that

$$rr' = xx' + yy', \qquad r^2\theta' = xy' - yx', \tag{1.4}$$

we find that

$$r' = 0, \qquad \theta' = -1 - r^2.$$

The solution is

$$r = r_0, \qquad \theta = \theta_0 - (r_0^2 + 1)t.$$

These form a family of periodic solutions, each of which is described by a circle in the phase plane. The orbits are sketched in figure 6.1. On each circle of radius r_0, the period is given by

$$T = \oint \mathrm{d}t = -\int_0^{2\pi} \frac{1}{\theta'}\, \mathrm{d}\theta = \frac{2\pi}{1+r_0^2}.$$

Calling r_0 the amplitude of the solution, note that the period is a function of the

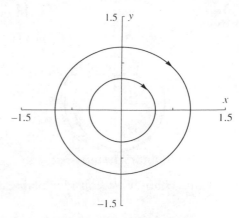

Fig. 6.1 The orbits for equation (1.2).

amplitude. Clearly these solutions are orbitally stable, but not asymptotically orbitally stable.

Next, consider

$$u'' + (u^2 + u'^2 - 1)u' + u = 0, \qquad (1.5)$$

which can be interpreted as an oscillator with a nonlinear damping term, $(u^2 + u'^2 - 1)u'$. Note that the damping is positive (negative) if $u^2 + u'^2 > 1$ (< 1). To put this in the form (1.1), let

$$x = u, \qquad y = u'$$

so that

$$x' = y, \qquad y' = -x - (r^2 - 1)y, \qquad (1.6)$$

where, again, $r^2 = x^2 + y^2$. Transforming to polar co-ordinates (see (1.3) and (1.4)) we get

$$r' = -r(r^2 - 1)\sin^2\theta, \qquad \theta' = -1 - (r^2 - 1)\sin\theta\cos\theta.$$

Here the origin is the only critical point, and is an unstable focus. By inspection we see that $r = 1$, $\theta = \theta_0 - t$ is a periodic solution, with period 2π, and is described by the circle of radius one in the phase plane. Further, it is the only periodic solution, since $r' \leqslant 0$ ($\geqslant 0$) for $r > 1$ (< 1); indeed, $r' < 0$ (> 0) for $r > 1$ (< 1) except on the x-axis ($\theta = 0, \pi$), where $r' = 0$ and $\theta' = -1$, indicating that the orbits cross the x-axis at right-angles. It follows that the periodic solution $r = 1$ is asymptotically orbitally stable. The orbits of (1.6), obtained from a numerical solution, are sketched in figure 6.2.

These two examples indicate that a periodic solution occurs, in general, either as a member of a family of periodic solutions (as in (1.2)), or as an isolated periodic solution (as in (1.6)). The first kind will be considered again briefly in section 6.4, and in more depth in chapter 11. The second kind is the subject of the following definition.

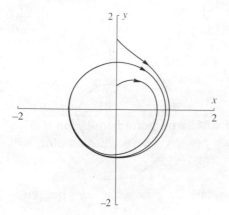

Fig. 6.2 The orbits for equation (1.6).

Definition: A **limit cycle**, C, is the orbit of an isolated periodic solution of (1.1).

Here, by an isolated periodic solution, we mean that there are no other periodic solutions in any arbitrarily small neighbourhood of C (i.e. the set of points x such that $d(x, C) < \epsilon$ for ϵ arbitrarily small, where $d(x, C)$ is the distance of the point x from C). In the remainder of this section, we present some results which may assist in deciding whether or not periodic solutions exist, and in determining their stability.

Theorem 6.1: There are no periodic solutions in any region where $(f_x + g_y)$ is of one sign.

Proof: Suppose that C is the orbit of a periodic solution, enclosing a region S. Then, by Green's theorem in the plane,

$$\iint_S (f_x + g_y) \, dx \, dy = \oint_C f \, dy - g \, dx.$$

But, on the orbit C, (1.1) holds and so

$$\oint_C f \, dy - g \, dx = \oint_C (x' \, dy - y' \, dx) = 0.$$

It follows that $f_x + g_y$ must change sign in S.

As an illustration of this result, consider

$$u'' + \phi(u) u' + \psi(u) = 0,$$

where $\phi(u)$ is always positive for all u (or is always negative for all u). This equation describes a nonlinear oscillator with positive (or negative) damping for all u. Let

$$x = u, \qquad y = u',$$

so that

$$x' = y, \qquad y' = -\psi(x) - \phi(x)\,y.$$

Here

$$f_x + g_y = -\phi(x)$$

and is of one sign everywhere. Hence there can be no periodic solutions. This is to be expected on physical grounds, since positive (negative) damping will clearly cause all non-zero solutions to decay (grow).

Theorem 6.2: The orbit C of a periodic solution must enclose at least one critical point.

Proof: This theorem is a special case of several more general results concerning the index of a critical point, which we develop in the next section. However, it is convenient to prove this result independently here. The proof will be by contradiction. Suppose that the orbit C contains a region S, with no critical points, that is, $f^2 + g^2 \neq 0$ in S. Now let ϕ be the angle between the tangent vector to C and the x-axis (see figure 6.3).

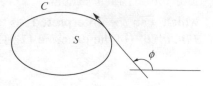

Fig. 6.3 Diagram for theorem 6.2.

Then clearly

$$\oint_C \mathrm{d}\phi = 2\pi.$$

But

$$\tan\phi = \frac{y'}{x'} = \frac{g}{f},$$

so that

$$\mathrm{d}\phi = \frac{f\,\mathrm{d}g - g\,\mathrm{d}f}{f^2 + g^2}.$$

Hence, using Green's theorem in the plane,

$$\oint_C \mathrm{d}\phi = \oint_C \frac{f\,\mathrm{d}g - g\,\mathrm{d}f}{f^2 + g^2} = \iint_S \left\{ \frac{\partial}{\partial f}\left(\frac{f}{f^2 + g^2} \right) + \frac{\partial}{\partial g}\left(\frac{g}{f^2 + g^2} \right) \right\} \mathrm{d}f\,\mathrm{d}g.$$

But the integrand in the integral over S is readily shown to be identically zero, leading to a contradiction.

Note that, in both the examples (1.2) and (1.6), the origin is the only critical point and, indeed, in both cases the periodic solutions enclose the origin. Theorem 6.2 shows that the location of critical points may be an indicator of the presence of periodic solutions, a theme which is developed further in the next section. The next theorem also relates the existence of periodic solutions to the location of critical points.

Theorem 6.3 (Poincaré–Bendixson): Let R be a closed, bounded region containing no critical points of (1.1), and such that there is an orbit p which lies in R for all $t \geq 0$. Then either p is a closed orbit, or it approaches a closed orbit as $t \to \infty$.

Proof: The theorem gives a sufficient condition for the existence of at least one periodic solution of (1.1). While it may seem to be an intuitively obvious result, the proof is beyond the scope of this text (see Coddington and Levinson, 1955, pp. 391–394 or Struble, pp. 179–186).

To illustrate these last two results, consider

$$u'' + F(u, u')u' + u = 0, \tag{1.7}$$

which can be interpreted as an oscillator with a nonlinear damping term, $F(u, u')u'$ (of the example (1.5)). Let

$$x = u, \qquad y = u'$$

so that

$$x' = y, \qquad y' = -x - F(x, y)y.$$

Further, we shall suppose that

$$F(x, y) > 0 \quad (\text{for } r^2 \geq r_2^2),$$

and

$$F(x, y) < 0 \quad (\text{for } r^2 \leq r_1^2),$$

where $r^2 = x^2 + y^2$ and $0 < r_1^2 < r_2^2$. Thus the oscillator has positive damping for r sufficiently large, but negative damping for r sufficiently small. Note that the example (1.5) has $F(x, y) = x^2 + y^2 - 1$ and satisfies these conditions. Here the origin is the only critical point, and is either an unstable focus or an unstable node. Hence, by theorem 6.2, any periodic solution must enclose the origin.

To apply theorem 6.3, we let the region R be the annulus $r_1 \leq r \leq r_2$ (see figure 6.4). Then, using the polar co-ordinates (1.3) and the expressions (1.4), we see that

$$r' = -F(r \cos \theta, r \sin \theta) r \sin^2 \theta, \qquad \theta' = -1 - F(r \cos \theta, r \sin \theta) \sin \theta \cos \theta.$$

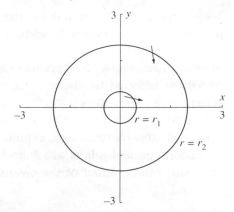

Fig. 6.4 Diagram for the application of theorem 6.3 to (1.7).

Thus, on $r = r_1$, $F < 0$ (> 0), and so $r' \geqslant 0$ ($\leqslant 0$) on $r = r_1$ ($r = r_2$); indeed, $r' < 0$ (> 0) for $r = r_1$ ($r = r_2$) except on the x-axis ($\theta = 0, \pi$), where $r' = 0$ and $\theta' = -1$, indicating that the orbits cross the x-axis at right-angles. Thus, any orbit p which is inside R at $t = 0$, say, must remain inside R for all $t \geqslant 0$. Hence, by theorem 6.3, there exists at least one periodic solution of (1.7) in the region $r_1 < r < r_2$.

Next we turn to consider the stability of the periodic solutions of (1.1). Orbital stability was discussed in section 4.4, and we recall that the first variational equation for the stability of a periodic solution of (1.1) is (4.7) of section 4.4. In our present notation this becomes

$$\begin{bmatrix} \hat{x}' \\ \hat{y}' \end{bmatrix} = A(t) \begin{bmatrix} \hat{x} \\ \hat{y} \end{bmatrix},$$

where

$$A(t) = \begin{bmatrix} f_x & f_y \\ g_x & g_y \end{bmatrix}. \tag{1.8}$$

Here we recall that (1.8) is derived by seeking solutions of (1.1) in the form $(x(t) + \hat{x}, y(t) + \hat{y})$, where $(x(t), y(t))$ is a periodic solution of (1.1). Linearizing (1.1) with respect to $(\hat{x}, \hat{y})$ then gives (1.8), where the matrix $A(t)$ is evaluated for the periodic solution $(x(t), y(t))$ and hence is a periodic function of t. We recall that $(x'(t), y'(t))$ is a solution of (1.8) and, hence, one of the characteristic exponents for (1.8) is zero. However, the principle of linearized stability (here theorem 4.8), then asserts that the remaining characteristic exponents determine the stability. Here there are only two characteristic exponents altogether and, using (1.4) of section 3.1, the second exponent can be determined explicitly. The result is the following theorem.

Theorem 6.4: Let $(x(t), y(t))$ be a periodic solution of (1.1) with period T. Then this solution is asymptotically orbitally stable if $\mu < 0$, where

$$\mu = \frac{1}{T} \int_0^T (f_x + g_y) \, dt. \tag{1.9}$$

Here $(f_x + g_y)$ is evaluated for the periodic solution $(x(t), y(t))$. Conversely, if $\mu > 0$, then this solution is unstable.

Proof: The proof consists in showing that μ is the second characteristic exponent referred to above. We recall from section 3.1 that, if μ is a characteristic exponent, then $\rho = e^{\mu T}$ is a characteristic multiplier (see (1.8) of section 3.1) and ρ is an eigenvalue of the matrix B, defined by (1.3) of section 3.1. Hence, if the characteristic exponents for (1.8) are 0 and μ, the corresponding characteristic multipliers are 1 and ρ. Hence 1 and ρ are the eigenvalues of B and, since the product of the eigenvalues is the determinant of B, we find that

$$\rho = \det B.$$

But, from (1.4) of section 3.1,

$$\det B = \exp\left\{\int_0^T \operatorname{tr} A(t)\, dt\right\},$$

where

$$\operatorname{tr} A(t) = (f_x + g_y), \quad [\text{for } x = (x(t), y(t))].$$

The result (1.9) follows, and then theorem 4.8 of section 4.4 shows that $\mu < 0$ implies asymptotic orbital stability. Conversely, if $\mu > 0$, theorem 4.6 of section 4.3 shows that the periodic solution is unstable.

To illustrate this theorem consider the example (1.6), for which

$$f_x + g_y = -(r^2 - 1) - 2y^2.$$

The periodic solution is

$$x = \cos(t - \theta_0), \qquad y = -\sin(t - \theta_0),$$

the period is 2π and, hence from (1.9),

$$\mu = \frac{1}{2\pi} \int_0^{2\pi} -2\sin^2(t - \theta_0)\, dt,$$

or

$$\mu = -1.$$

Thus, we have confirmed that the solution is asymptotically orbitally stable. By contrast, for the example (1.2), $f_x + g_y = 0$ and theorem 6.4 cannot be used. This is to be expected for this example, as the exact solution shows that the periodic solutions are only orbitally stable, and neither asymptotically orbitally stable nor unstable.

Note that theorem 6.1 together with the result (1.9) shows that the quantity $(f_x + g_y)$ can be used as a measure of the damping associated with the system (1.1). Thus theorem 6.1 can be interpreted as implying that damping which is always positive, or always negative, will prevent the occurrence of periodic solutions. Similarly, theorem 6.4 shows that a periodic solution is asymptotically orbitally stable if the damping is negative when averaged over one period, but is unstable if it is positive when averaged. Note that, by theorem 6.1, $f_x + g_y$ cannot be totally of the same sign in the expression (1.9) for μ, although the example (1.6) just discussed shows that $f_x + g_y$ may take only non-positive (or only non-negative) values on the periodic solution.

6.2 The index of a critical point

A useful indicator of the presence, or otherwise, of a periodic solution is theorem 6.2. As we mentioned in the proof of that theorem, it is a special case of several more general results. Hence we shall now interrupt our discussion of periodic solutions of (1.1) in order to interpolate a brief digression on the subject of the index of a critical point.

Let C be a closed curve in the phase plane, which passes through no critical points of (1.1). Thus, at every point (x, y) on C, the vector $(f(x, y), g(x, y))$ defines a unique direction since $f^2 + g^2 \neq 0$ on C. Indeed, this direction is just that of the unique orbit through (x, y). Let this direction make an angle ϕ with the x-axis (see figure 6.5).

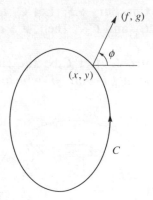

Fig. 6.5 The definition of ϕ for the vector $(f(x,y), g(x,y))$ on the curve C.

Definition: The **index** $I(C)$ of the closed curve C is $\Delta\phi/2\pi$, where $\Delta\phi$ is the total change in the angle ϕ that $(f(x, y), g(x, y))$ makes with the x-axis as the point (x, y) traverses C once in the positive (anti-clockwise) sense.

Clearly $I(C)$ is an integer. To determine the index we first observe that

$$\tan \phi = \frac{g}{f},$$

so that

$$d\phi = \frac{f \, dg - g \, df}{f^2 + g^2}.$$

Then it follows that

$$I(C) = \frac{1}{2\pi} \oint_C d\phi = \frac{1}{2\pi} \oint_C \frac{f\,dg - g\,df}{f^2 + g^2}, \tag{2.1}$$

where we note that

$$df = f_x\,dx + f_y\,dy, \qquad dg = g_x\,dx + g_x\,dy. \tag{2.2}$$

Theorem 6.5: Let C be a closed curve containing no critical points, either inside C or on C. Then the index $I(C) = 0$.

Proof: Let S be the region contained by C. Then $f^2 + g^2 \neq 0$ in S, and Green's theorem in the plane may be applied to (2.1) to show that

$$I(C) = \frac{1}{2\pi} \iint_S \left\{ \frac{\partial}{\partial f}\left(\frac{f}{f^2 + g^2}\right) + \frac{\partial}{\partial g}\left(\frac{g}{f^2 + g^2}\right) \right\} df\,dg.$$

But the integrand is readily shown to be identically zero, and the result follows. Note that, from (2.2),

$$df\,dg = (f_x g_y - f_y g_x)\,dx\,dy.$$

Corollary 6.5: Let C_1 and C_2 be closed curves, and let S be the region between C_1 and C_2. Then, if S contains no critical points, $I(C_1) = I(C_2)$.

Proof: Let C be the closed curve defined by C_1, C_2 and two straight-line segments Γ and $-\Gamma$ which join C_1 to C_2 (see figure 6.6).

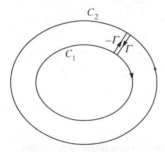

Fig. 6.6 Diagram for corollary 6.5.

Then, since there are no critical points in S, it follows that $I(C) = 0$. But

$$I(C) = I(C_2) + I(-C_1),$$

where we note that the contributions to $I(C)$ of the integrals along Γ and $-\Gamma$ will cancel. But, clearly,

$$I(C_1) = -I(-C_1),$$

and the result follows.

As a consequence of theorem 6.5 and its corollary we can now associate the notion of index with points rather than curves. For instance, if P is any point inside a region S which contains no critical points, then we say that its index is zero, since the index of any closed curve surrounding P, and lying in S, is zero. More pertinently, we define the index of an isolated critical point as follows.

Definition: The **index** $I(P)$ of an isolated critical point P is the index $I(C)$ of any closed curve C surrounding the critical point and containing no other critical point in its interior.

Theorem 6.6: The index of a closed curve C containing k isolated critical points $P_1, \ldots, P_k$ in its interior is given by

$$I(C) = I(P_1) + \ldots + I(P_k). \tag{2.3}$$

Proof: Let C_i $(i = 1, \ldots, k)$ be a closed curve surrounding P_i and containing no other critical point in its interior. Then $I(P_i) = I(C_i)$ $(i = 1, \ldots, k)$. Also, by arguments similar to those used in establishing the corollary to theorem 6.5,

$$I(C) = I(C_1) + \ldots + I(C_2).$$

The result follows.

Theorem 6.7: Let C be the orbit of a periodic solution of (1.1). Then its index $I(C) = 1$.

Proof: Since C is an orbit, the direction field $(f(x, y), g(x, y))$ at any point (x, y) on C coincides with the tanget vector to C (see figure 6.3). Then, clearly,

$$\oint_C d\phi = 2\pi$$

and the result follows immediately from the definition of the index $I(C)$.

It can now be seen that the orbit C of a periodic solution must contain at least one critical point in its interior. Indeed, this is just the content of theorem 6.2, established in section 6.1. Here we see that the result follows from theorems 6.5 and 6.7, since, if C were to enclose no critical points, then $I(C) = 0$ from theorem 6.5, which is contradicted by theorem 6.7. Further, by combining theorems 6.6 and 6.7, the following result holds.

Corollary 6.7: Let C be the orbit of a periodic solution of (1.1), and suppose that the critical points of (1.1) are all isolated. Then the interior of C contains a finite number (≥ 1) of critical points, $P_1, \ldots, P_k$ and

$$I(C) = I(P_1) + \ldots + I(P_k) = 1. \tag{2.4}$$

It remains to relate the index of a critical point to the classification scheme introduced in sections 5.2 and 5.3. In analyzing the index of an isolated critical point, we may, without loss of generality, suppose that the critical point is at the origin. Hence, the system (1.1) adopts the form (3.1) of section 5.3, which we display again here:

$$x' = ax + by + F(x, y), \qquad y' = cx + dy + G(x, y), \qquad (2.5)$$

where

$$\frac{|F|}{r}, \frac{|G|}{r} \to 0 \quad (\text{as } r \to 0).$$

Here $r = \sqrt{x^2 + y^2}$ and we assume that $ad - bc \neq 0$. Then the following theorem holds.

Theorem 6.8: The index of an isolated critical point is -1 if the point is a saddle point and $+1$ otherwise.

Proof: To determine the index we may choose any curve C enclosing the origin and no other critical point. We choose C sufficiently close to the origin so that the nonlinear terms F and G in (2.5) may be neglected. Then, using (2.1) and (2.2), we find that the index is given by

$$\frac{1}{2\pi} \oint_C \frac{ad - bc}{(ax + by)^2 + (cx + dy)^2} (x \, dy - y \, dx).$$

Next we introduce the polar co-ordinates (1.3), and find that this integral becomes

$$\frac{ad - bc}{2\pi} \int_0^{2\pi} \frac{1}{(a \cos \theta + b \sin \theta)^2 + (c \cos \theta + d \sin \theta)^2} \, d\theta.$$

The integral can now be evaluated by standard techniques, and we find that the index is given by

$$\text{sign}(ad - bc).$$

But, if the eigenvalues of the matrix A_0 associated with the linearized part of (2.5) (see (3.2) of section 5.3) are λ_1 and λ_2, then $ad - bc = \lambda_1 \lambda_2$, and the index is $\text{sign}(\lambda_1 \lambda_2)$. Referring now to the classification scheme of section 5.2, we see that the index is $+1$ for nodes (λ_1 and λ_2 real with $\lambda_1 \lambda_2 > 0$), -1 for saddle points (λ_1 and λ_2 real with $\lambda_1 \lambda_2 < 0$), and $+1$ for spiral points and centres ($\lambda_1 = \bar{\lambda}_2$ and so $\lambda_1 \lambda_2 = |\lambda_1|^2 > 0$).

When this theorem is combined with corollary 6.7, we see that the orbit C of a periodic solution must enclose an odd number of critical points (i.e. k in corollary 6.7 is an odd integer), and the number of saddle points enclosed is one less than the number of remaining critical points.

To illustrate how these results may be used to help determine whether or not periodic solutions are present, we reconsider the example (3.31) of section 5.3, which we restate here:

$$x' = x(l-py-kx), \qquad y' = y(-m+qx). \qquad (2.6)$$

It was shown in section 5.3 that there are three critical points (see (3.35) of section 5.3). For $ql < km$ they are a saddle point at the origin, a stable node at $(l/k, 0)$ and a saddle point at $(m/q, (l-km/q)p^{-1})$. Consequently, any periodic solution must enclose the node at $(l/k, 0)$ and exclude the two saddle points. However, since here the x-axis is also an orbit of (2.6) (see (3.36) of section 5.3), clearly no other orbit can cross the x-axis (except at $x = 0$ or $x = l/k$) and, hence, there are no periodic solutions.

For $ql > km$ the critical points are a saddle point at the origin, a saddle point at $(l/k, 0)$ and a stable node, or a stable spiral point, at $(m/q, (l-km/q)p^{-1})$. Consequently, any periodic solution must enclose this last critical point, and exclude the two saddle points. Since the x- and y-axes are also orbits of (2.6) (see (3.36) of section 5.3), any periodic solution must be confined to the first quadrant, where $x, y > 0$. Indeed, all orbits which commence in this quadrant must remain in it. However, in this case it can be shown that there are no periodic solutions. To demonstrate this, we shall use theorem 6.1, but first we use the transformation

$$X = \ln x, \qquad Y = \ln y.$$

Note that this is valid in the first quadrant, where $x, y > 0$. Then (2.6) becomes

$$X' = l - pe^Y - ke^X, \qquad Y' = -m + qe^X.$$

Denoting the right-hand sides by $F(X, Y)$ and $G(X, Y)$, respectively, we see that

$$F_X + G_Y = -ke^X < 0$$

and, hence, by theorem 6.1, there can be no periodic solutions for $k > 0$. Of course, when $k = 0$, we showed in section 5.1 that there is a family of periodic solutions in the first quadrant (see figure 5.3). The orbits for $k > 0$ are sketched in figure 5.19.

6.3 Van der Pol equation

The Van der Pol equation is

$$u'' + \epsilon(u^2 - 1)u' + u = 0, \qquad (3.1)$$

where ϵ is a positive constant. It describes an oscillator with a nonlinear damping term, $\epsilon(u^2 - 1)u'$, whose coefficient is positive for $u^2 > 1$ but is negative for $u^2 < 1$. Interest in this equation arose initially through its application to the modeling of nonlinear electric circuits. In this context it has been extensively studied (see, for instance, Minorsky, 1962, Ch. 4 and Part IV). It is now well-known that it has a single limit cycle which is asymptotically orbitally stable. Our reason for discussing it here is that it serves as a useful example of an equation which has a limit-cycle solution, and it is of interest to examine how the structure of the limit cycle changes as the parameter ϵ is varied.

To put (3.1) into the form (1.1), we use the Lienard transformation

$$x = u, \qquad y = u' - \epsilon(u - \tfrac{1}{3}u^3), \tag{3.2}$$

so that (3.1) becomes

$$x' = y + \epsilon(x - \tfrac{1}{3}x^3), \qquad y' = -x. \tag{3.3}$$

The origin is the only critical point and is an unstable focus for $0 < \epsilon < 2$, and an unstable node for $\epsilon \geq 2$. Hence, by theorem 6.2, any limit cycle must enclose the origin. Further, here $f_x + g_y = \epsilon(1 - x^2)$ and, hence, by theorem 6.1, any limit cycle must include a portion outside the strip $x^2 = 1$, as well as a portion inside this strip. This is self-evident on physical grounds, as the limit cycle must include regions of both positive and negative damping.

This is as far as the general results of section 6.1 can take us, although the Poincaré–Bendixson theorem (6.3) can be used to infer the existence of at least one closed orbit. However, a stronger result is obtained in the following theorem.

Theorem 6.9: There exists a unique limit cycle for (3.3), which is asymptotically orbitally stable.

Proof: First we note that the equations (3.3) are invariant under the transformation $x \to -x$, $y \to -y$. It follows that, if there is a limit cycle, it is symmetric about the origin. Hence, it will be sufficient to consider only the orbits in $x \geq 0$.

Next we observe from (3.3) that

$$x' \gtrless 0, \quad \text{according as } y \gtrless \epsilon(\tfrac{1}{3}x^3 - x),$$

and

$$y' \gtrless 0, \quad \text{according as } x \gtrless 0.$$

Consider an orbit which commences at the point A on the positive y-axis (see figure 6.7). It must move (as t increases) in a direction such that x increases but y decreases until it meets the cubic $y = \epsilon(\tfrac{1}{3}x^3 - x)$, after which it must move in a direction for which x decreases and y decreases until it meets the negative y-axis at the point B. To show the existence of a limit cycle it is sufficient to show that there exists a point A such that $OA = OB$.

Now consider,

$$E = \tfrac{1}{2}(x^2 + y^2), \tag{3.4}$$

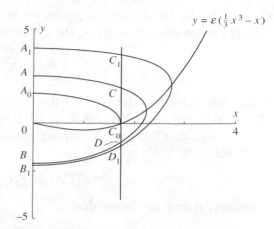

$$y = \varepsilon(\tfrac{1}{3}x^3 - x)$$

Fig. 6.7 Diagram for theorem 6.9, the existence proof for the Van der Pol limit cycle.

which may be considered as a measure of the energy associated with the oscillator. It is also $\tfrac{1}{2}r^2$, where r is the distance from the origin. Then, using (3.3),

$$E' = xx' + yy' = \epsilon x^2(1 - \tfrac{1}{3}x^2). \tag{3.5}$$

We let $E(P)$ denote E evaluated at the point P, with co-ordinates $(x(t), y(t))$, on the orbit from A to B, and define

$$E(PQ) = E(P) - E(Q),$$

or

$$E(PQ) = \int_Q^P dE = \int_Q^P E' \, dt. \tag{3.6}$$

Here E' is evaluated from (3.5). Then, to show the existence of a limit cycle, it is sufficient to show there exists a point A such that $E(BA) = 0$, so that $E(A) = E(B)$, as this is clearly equivalent to $OA = OB$.

Let C_0 be the point $(\sqrt{3}, 0)$, where the cubic $y = \epsilon(\tfrac{1}{3}x^3 - x)$ cuts the x-axis (in $x > 0$), and let the unique orbit through C_0 cut the positive (negative) y-axis at A_0 (B_0). Then, on the orbit from A_0 to B_0 (or any interior orbit) $x^2 < 3$, $E' > 0$ from (3.5) and so $E(B_0A_0) > 0$, or $E(B_0) > E(A_0)$. Hence it is sufficient to consider points A such that A lies above A_0. Next, for the orbit from A to B, let C and D be the points where the orbit intersects $x = \sqrt{3}$ in $y > 0$ and $y < 0$, respectively (see figure 6.7). Now

$$E(CA) = \int_A^C dE = \int_A^C \frac{dE}{dx} \, dx,$$

where

$$\frac{dE}{dx} = \frac{E'}{x'} = \frac{\epsilon x^2(1 - \tfrac{1}{3}x^2)}{y + \epsilon x(1 - \tfrac{1}{3}x^2)}, \tag{3.7}$$

on using (3.3) and (3.5). If we let A_1 be such that $OA_1 > OA$, and we let B_1,

C_1 and D_1 be the analogous points on the orbit through A_1 to the points B, C and D on the orbit through A, then it follows from (3.7) that

$$\left(\frac{\mathrm{d}E}{\mathrm{d}x}\right)_{AC} > \left(\frac{\mathrm{d}E}{\mathrm{d}x}\right)_{A_1C_1} > 0, \tag{3.8}$$

where the subscripts denote that the derivative is evaluated on the orbit AC, or A_1C_1, respectively. The result (3.8) is obtained by observing that, for each fixed x, $\mathrm{d}E/\mathrm{d}x$ (3.7) is a decreasing function of y. Using (3.8), it follows from (3.7) that

$$E(CA) > E(C_1A_1) > 0. \tag{3.9}$$

Similarly, it may be shown that

$$E(BD) > E(B_1D_1) > 0. \tag{3.10}$$

Next, we consider the orbit from C to D, where

$$E(DC) = \int_C^D \mathrm{d}E = \int_C^D \frac{\mathrm{d}E}{\mathrm{d}y}\,\mathrm{d}y,$$

and

$$\frac{\mathrm{d}E}{\mathrm{d}y} = \frac{E'}{y'} = -\epsilon x(1 - \tfrac{1}{3}x^2) > 0, \tag{3.11}$$

on using (3.3) and (3.5). Also

$$E(D_1C_1) = \int_{C_1}^C \frac{\mathrm{d}E}{\mathrm{d}y}\,\mathrm{d}y + \int_C^D \frac{\mathrm{d}E}{\mathrm{d}y}\,\mathrm{d}y + \int_D^{D_1} \frac{\mathrm{d}E}{\mathrm{d}y}\,\mathrm{d}y, \tag{3.12}$$

and the contributions from the first and third portions (i.e. from C_1 to C and from D to D_1) are both negative. On the portion CD, $\mathrm{d}E/\mathrm{d}y$ is a monotonically increasing function of x, and so

$$\left(\frac{\mathrm{d}E}{\mathrm{d}y}\right)_{C_1D_1} > \left(\frac{\mathrm{d}E}{\mathrm{d}y}\right)_{CD} > 0, \tag{3.13}$$

where it is understood here that $\mathrm{d}E/\mathrm{d}y$ (3.11) is being evaluated as a function of x for each fixed y, and the result (3.13) holds over the values of y common to the orbits C_1D_1 and CD. Using (3.13), it follows from (3.11) and (3.12) that

$$E(D_1C_1) < E(DC) < 0. \tag{3.14}$$

Now, since

$$E(BA) = E(BD) + E(DC) + E(CA), \tag{3.15}$$

it follows from (3.9), (3.10) and (3.14) that

$$E(BA) > E(B_1A_1). \tag{3.16}$$

Thus $E(BA)$ is a monotonically decreasing function of A as OA increases, at least for $OA > OA_0$. Also, we have already noted that $E(B_0A_0) > 0$. Next we show that $E(BA) \to -\infty$ as $OA \to \infty$. To demonstrate this, we note from (3.9) and (3.10) that $E(CA)$ and $E(BD)$ are positive and decreasing and, hence, are bounded. On the other hand, from (3.11) $E(DC) \to -\infty$ as $OA \to \infty$, since dE/dy increases without bound as $x \to \infty$. We have now shown that $E(BA)$ is a monotonically decreasing function of OA which is positive at OA_0 and tends to $-\infty$ as $OA \to \infty$. Hence, there exists a *unique* point A such that $E(BA) = 0$, and this is sufficient to establish the existence of a unique limit cycle.

Further, this same argument shows that this limit cycle is asymptotically orbitally stable. Indeed, if A is a point inside the limit cycle, then $E(BA) > 0$ and so $OB > OA$, showing that the orbit is approaching the limit cycle, while, if A is a point outside the limit cycle, then $E(BA) < 0$ and so $OB < OA$ and, again, the orbit is approaching the limit cycle.

To show the structure of the limit cycle, equations (3.3) have been integrated numerically and the results are shown in figure 6.8. Case (a) is for a small value of ϵ ($\epsilon = 0.1$), case (b) for $\epsilon = 1$ and case (c) for a large value ($\epsilon = 5$). The corresponding periodic solutions $x(t)$ are shown in figure 6.9 for the three cases.

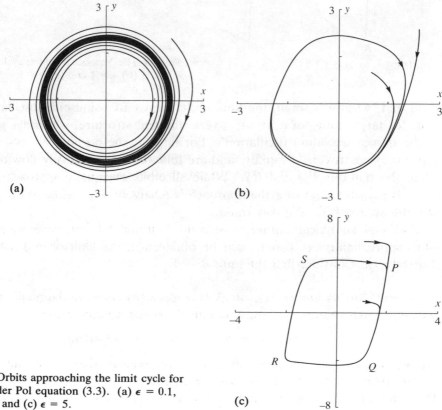

Fig. 6.8 Orbits approaching the limit cycle for the Van der Pol equation (3.3). (a) $\epsilon = 0.1$, (b) $\epsilon = 1$ and (c) $\epsilon = 5$.

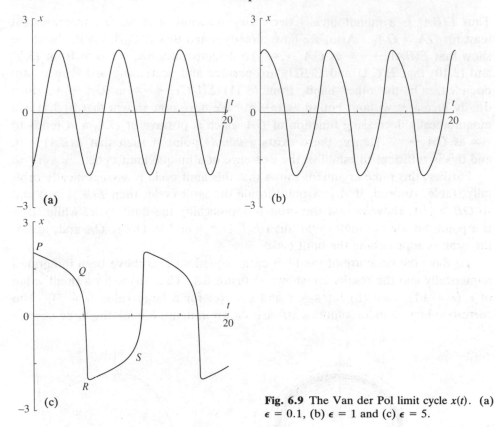

Fig. 6.9 The Van der Pol limit cycle $x(t)$. (a) $\epsilon = 0.1$, (b) $\epsilon = 1$ and (c) $\epsilon = 5$.

Note that, when ϵ is small, the limit cycle is close to a sinusoidal oscillation, but that, for large values of ϵ, it has a very different structure, which has given rise to the name 'relaxation oscillation'. For large ϵ, the portions SP and QR (see figure 6.9) are traversed rapidly, and are followed by a relatively slow relaxation along the portions PQ and RS. While all orbits eventually approach the limit cycle, for small values of ϵ the approach is relatively slow, while for large values of ϵ the approach is relatively rapid.

Although analytical expressions for the Van der Pol limit cycle are not available, approximate expressions can be obtained in the limits $\epsilon \to 0$ and $\epsilon \to \infty$, respectively. Consider first the limit $\epsilon \to 0$.

(i) $\epsilon \to 0$: In this limit (3.1), or (3.3), reduces to a simple harmonic oscillator, whose general solution is readily obtained. Hence we may write

$$x = r_0 \cos t + O(\epsilon), \qquad y = -r_0 \sin t + O(\epsilon), \tag{3.17}$$

where, without loss of generality, we have assumed that $x = r_0$ and $y = 0$ at $t = 0$. Thus, as $\epsilon \to 0$, all solutions are periodic with period 2π, and it follows that the period T of the limit cycle is $T = 2\pi + O(\epsilon)$. Now, recalling the definition of E (3.4) and the result (3.5) for E', we see that

$$\frac{1}{T}\int_0^T E'\,dt = 0,$$

or

$$\frac{1}{T}\int_0^T x^2(1-\tfrac{1}{3}x^2)\,dt = 0. \tag{3.18}$$

This expression is valid for all $\epsilon > 0$ and, in particular, must hold in the limit $\epsilon \to 0$. On substituting (3.17) into (3.18), we find that

$$\frac{1}{2\pi}\int_0^{2\pi} r_0^2 \cos^2 t\,(1-\tfrac{1}{3}r_0^2 \cos^2 t)\,dt = O(\epsilon),$$

or

$$\tfrac{1}{2}r_0^2 - \tfrac{1}{8}r_0^4 = O(\epsilon). \tag{3.19}$$

It follows that, to an error of $O(\epsilon)$, $r_0 = 2$ and, hence, the limit cycle, to this order of approximation, is a circle of radius 2. In effect, the result (3.18) has enabled us to select from all the possible sinusoidal oscillations (3.17) precisely that one which corresponds to the limit cycle. This technique is an illustration of the method of averaging, which is the subject of chapter 9. The same result can be obtained by more straight-forward, but related, perturbation techniques which are described in chapter 7.

Next we recall from (1.9) that the stability of the limit cycle is determined by the sign of the characteristic exponent μ. Here $f_x + g_y = \epsilon(1-x^2)$, and so (1.9) becomes

$$\mu = \frac{1}{T}\int_0^T \epsilon(1-x^2)\,dt. \tag{3.20}$$

Then, using the approximation (1.9), we obtain

$$\mu = -\epsilon + O(\epsilon^2). \tag{3.21}$$

This result confirms that the limit cycle is asymptotically orbitally stable, and also shows that, as $\epsilon \to 0$, the approach to the limit cycle is slow and occurs on a time scale of ϵ^{-1}. Both this result, and that for the limit cycle itself (i.e. (3.19)), are in agreement with the numerical calculations.

(ii) $\epsilon \to \infty$: In this limit it is useful to rescale the variables y and t by putting

$$y = \epsilon\eta, \qquad t = \epsilon\tau. \tag{3.22}$$

The motivation for this is that the time-scale for the limit cycle is ϵ and, since x continues to vary on a scale of unity, it follows from (3.3) that y must also scale with ϵ. In terms of η and τ, (3.3) becomes

$$\frac{dx}{d\tau} = \epsilon^2(\eta + x - \tfrac{1}{3}x^3), \qquad \frac{d\eta}{d\tau} = -x. \tag{3.23}$$

Hence

$$x\frac{dx}{d\eta} = -\epsilon^2(\eta + x - \tfrac{1}{3}x^3). \tag{3.24}$$

Here the small parameter is ϵ^{-2}, but the right-hand side of (3.23), and also the right-hand side of (3.24), is not analytic in ϵ^{-2}. Hence, a regular perturbation expansion in powers of ϵ^{-2} is not possible. Indeed, the limit $\epsilon^2 \to \infty$ is a singular perturbation. Although we shall not be developing any general techniques for singular perturbations in this text, approximate solutions for (3.24) will be obtained here by using heuristic arguments. Thus, as $\epsilon^2 \to \infty$, it is apparent from (3.24) that

$$\text{either} \quad \eta + x - \tfrac{1}{3}x^3 \approx 0, \quad \text{or} \quad \frac{d\eta}{dx} \approx 0. \tag{3.25}$$

Combining these approximations with the information obtained about the limit cycle in the course of proving theorem 6.9, we can deduce that a suitable approximation to the limit cycle as $\epsilon \to \infty$ is obtained by following the cubic $\eta = \tfrac{1}{3}x^3 - x$ from the point P $(x = 2, \eta = \tfrac{2}{3})$ to its turning point Q $(x = 1, \eta = -\tfrac{2}{3})$, and then following the line $\eta = -\tfrac{2}{3}$ until meeting the cubic again at the point R $(x = -2, \eta = -\tfrac{2}{3})$. The circuit is completed by symmetry and, hence, follows the cubic to the next turning point S $(x = -1, \eta = \tfrac{2}{3})$ and then follows the line $\eta = \tfrac{2}{3}$ until meeting the cubic again at the point P. The result is sketched in figure 6.10.

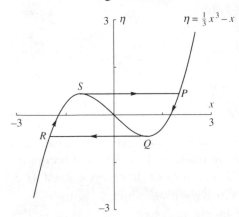

Fig. 6.10 Diagram for the Van der Pol limit cycle in the limit $\epsilon \to \infty$.

To determine an approximation for the period T, we observe that

$$T = \epsilon \oint d\tau = -\epsilon \oint \frac{1}{x}\, d\eta, \tag{3.26}$$

where the integral is around the limit cycle. Clearly, the dominant contributions come from the portions PQ and RS on the cubic and, using the symmetry, it is sufficient to consider just the portion PQ. Here $\eta = \frac{1}{3}x^3 - x$, $d\eta = (x^2 - 1)\,dx$, and so

$$T \approx 2\epsilon \int_1^2 \frac{x^2 - 1}{x}\,dx,$$

or

$$T \approx \epsilon(3 - 2\ln 2). \tag{3.27}$$

On the portions PQ and RS of the limit cycle $x^2 > 1$, and the limit cycle is in a phase of positive damping, where the amplitude relaxes from $|x| \approx 2$ to $|x| \approx 1$. This is followed by a rapid transition on QR and SP, where the limit cycle passes through a region of negative damping ($x^2 < 1$) and overshoots to the points R and P. To analyze this process further, we combine the approximate expressions (3.25) with (3.23). Thus, on the portion PQ, we use the approximation $\eta \approx \frac{1}{3}x^3 - x$ together with the second equation in (3.23), to obtain

$$(x^2 - 1)\frac{dx}{d\tau} \approx -x,$$

or

$$\ln x - \tfrac{1}{2}x^2 \approx \tau + \ln 2 - 2. \tag{3.28}$$

Here we are assuming, without loss of generality, that $\tau = 0$ at the point P. Note that, as $x \to 1$, we find from (3.28) that

$$(x - 1)^2 \approx (\tfrac{3}{2} - \ln 2) - \tau. \tag{3.29}$$

Thus, as $x \to 1$, $\tau \to \frac{3}{2} - \ln 2$, in agreement with our previous approximation for the period (3.27). Next, on the portion QR, $\eta \approx -\frac{2}{3}$ and, combining this approximation with the first equation in (3.23), we obtain

$$\frac{dx}{d\tau} \approx -\tfrac{1}{3}\epsilon^2(x - 1)^2(x + 2),$$

or

$$-\frac{3}{1 - x} + \ln\frac{1 - x}{x + 2} = 3\epsilon^2(\tau - \tau_0). \tag{3.30}$$

Here τ_0 is an arbitrary constant of integration, but, since this approximation must match with (3.28) at the point R ($x \to 1$), clearly $\tau_0 \approx \frac{3}{2} - \ln 2$. As $x \to 2$, we find from (3.30) that

$$1 - x \approx -\frac{1}{\epsilon^2(\tau - \tau_0)}, \tag{3.31}$$

while, as $x \to -2$,

$$x + 2 \approx k\exp\{-3\epsilon^2(\tau - \tau_0)\}, \tag{3.32}$$

where k is a constant. The expressions (3.18) and (3.30) can now be used to construct an approximation to the limit cycle $x(\tau)$, and the result is shown in figure 6.11, where we use the symmetry of the limit cycle to complete the portions RS and SP.

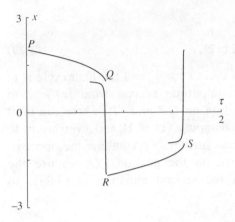

Fig. 6.11 The approximate solution $x(\tau)$, where $t = \epsilon\tau$, for the Van der Pol limit cycle in the limit $\epsilon \to \infty$.

In sketching figure 6.11, we observe from (3.32) that $x \to -2$ exponentially fast as $\epsilon^2(\tau - \tau_0) \to \infty$. Hence the approximation (3.30) joins smoothly to the approximation corresponding to (3.28) for the portion RS. However, from (3.31) we see that the approximation (3.30) is singular as $\tau \to \tau_0$ and, hence, does not join smoothly to the approximation (3.28) as $x \to 1$ (see (3.29) which shows that the approximation (3.28) develops a vertical tangent as $x \to 1$). To correct this, we need to examine the region near the point Q in more detail. Thus, we let

$$1 - x = \delta u, \qquad \eta + \tfrac{2}{3} = \delta^2 v, \qquad \tau - \tau_0 = \delta^2 s, \qquad (3.33)$$

where δ is a small parameter which has yet to be determined. The scaling in (3.33) is determined from the observation that Q is a turning point on the cubic and, hence, $(1-x)^2$ and $\eta + \tfrac{2}{3}$ must scale together, while the second equation of (3.23) implies that $\eta + \tfrac{2}{3}$ and $\tau - \tau_0$ must scale together. We also let $\tau_0 = \tfrac{3}{2} - \ln 2 + \delta^2 s_0$, where $\delta^2 s_0$ is a correction term yet to be determined. Substitution of (3.33) into (3.23) now gives

$$\frac{du}{ds} = \epsilon^2 \delta^3 (-v + u^2 - \tfrac{1}{3}\delta u^3), \qquad \frac{dv}{ds} = -1 + \delta u. \qquad (3.34)$$

Hence we must choose $\delta = \epsilon^{-2/3}$, and so obtain the approximate equations

$$\frac{du}{ds} \approx u^2 - v, \qquad \frac{dv}{ds} \approx -1. \qquad (3.35)$$

It follows from the second of these equations that $v \approx -(s - s_1)$, where s_1 is a constant of integration. Then, to solve the first equation in (3.35), we put

$$u = -\frac{1}{\psi}\frac{d\psi}{ds},$$

and so

$$\frac{d^2\psi}{ds^2} + (s - s_1)\psi \approx 0. \tag{3.36}$$

This is Airy's equation, for which two linearly independent solutions are $\mathrm{Ai}(s_1 - s)$ and $\mathrm{Bi}(s_1 - s)$, which are the Airy functions of the first and second kind, respectively. To find the appropriate solution of (3.36), we first express the matching conditions (3.29) and (3.31) in the present variables (3.33). Thus (3.29) becomes

$$u \approx -\sqrt{-(s + s_0)} \quad (\text{as } s \to -\infty), \tag{3.37}$$

while (3.31) becomes

$$u \approx -\frac{1}{s} \quad (\text{as } s \to 0). \tag{3.38}$$

It follows from (3.37) that $s_1 = -s_0$ and $\psi = \mathrm{Ai}(s_1 - s)$, so that, from (3.36),

$$u = \frac{\mathrm{Ai}'(s_1 - s)}{\mathrm{Ai}(s_1 - s)}. \tag{3.39}$$

In verifying the condition (3.37), we use the result that

$$\mathrm{Ai}(z) \sim \frac{1}{2\sqrt{\pi}z^{1/4}}\exp\{-\tfrac{2}{3}z^{3/2}\} \quad (\text{as } z \to \infty). \tag{3.40}$$

Next we observe from (3.39) that u is singular at the zeros of the Airy function. Since here s is increasing from $-\infty$, we need to consider only the smallest zero, α, such that $\mathrm{Ai}(-\alpha) = 0$, where we note that $\alpha \approx 2.338$. As $s_1 - s \to -\alpha$, (3.39) becomes

$$u \approx \frac{1}{s_1 + \alpha - s}. \tag{3.41}$$

To match with (3.38), we must now choose $s_1 = -\alpha$. This completes our examination of the structure of the limit cycle in the vicinity of the point Q. In particular, we can now improve our estimate of the period (3.27) be observing that the matching from (3.29) to (3.31) occurs over an interval $\delta^2 s_0$ (on the τ-scale), and we have shown that $s_0 = \alpha$. Hence

$$T \approx \epsilon(3 - 2\ln 2 + 2\alpha\epsilon^{-4/3} + \ldots), \tag{3.42}$$

where we can anticipate that the next term is relatively $O(\epsilon^{-2})$.

Next we turn to the approximate calculation of the characteristic exponent μ, here given by (3.20). From (3.22) and (3.23) this is equivalent to

$$\mu T = -\epsilon^2 \oint \frac{1 - x^2}{x}\, d\eta, \tag{3.43}$$

where the integral is around the limit cycle. Then, using the approximations (3.25) and (3.27), we obtain

$$\mu \approx -\epsilon \frac{\frac{3}{2} + 2\ln 2}{3 - 2\ln 2}. \tag{3.44}$$

As anticipated, this confirms that the limit cycle is asymptotically orbitally stable, and also shows that as $\epsilon \to \infty$ the approach to the limit cycle is rapid and occurs on a time scale of ϵ^{-1}. Both this result, and that for the limit cycle itself (i.e. figure 6.10), are in agreement with the numerical calculations (see figure 6.8 (b) and figure 6.9 (b)).

   We have analyzed the Van der Pol limit cycle in considerable detail because it is a typical example of a relaxation oscillation. Indeed, the concepts and methodology developed for the Van der Pol equation can be adapted to study the Lienard equation:

$$u'' + f(u) u' + g(u) = 0. \tag{3.45}$$

To put this into the form (1.1), we let

$$x = u, \qquad y = u' + F(u),$$

where

$$F(u) = \int_0^u f(v) \, dv. \tag{3.46}$$

Then (3.45) becomes

$$x' = y - F(x), \qquad y' = -g(x). \tag{3.47}$$

Suppose further that

  (i) $F(x) = -F(-x)$ for all x;
  (ii) $F(x) = 0$ only for $x = 0, \pm a$, and $F(x) \to \infty$ monotonically as $x \to \infty$ for
        $x > a$;
  (iii) $g(x) = -g(-x)$ for all x;
  (iv) $g(x) > 0$ for all $x > 0$.   (3.48)

Note that the Van der Pol equations (3.3) satisfy these conditions. It may now be shown that there is a unique limit cycle which is asymptotically orbitally stable. The proof is similar to that for theorem 6.9, with

$$E = \tfrac{1}{2} y^2 + \int_0^x g(x') \, dx'$$

replacing (3.4) as a measure of the energy. The details are left to the exercises (see also Jordan and Smith, 1977, section 11.3).

6.4 Conservative systems

Consider the equation

$$u'' + F(u) = 0, \tag{4.1}$$

which can be regarded as describing the motion of a particle moving in a straight line (the u-axis) subject to a force $-F(u)$ per unit mass. To put (4.1) into the form (1.1), we let

$$x = u, \quad y = u', \tag{4.2}$$

so that

$$x' = y, \quad y' = -F(x). \tag{4.3}$$

The equation (4.1) is said to describe a conservative system, since it contains no dissipative terms; note that $(f_x + g_y)$ is identically zero here. In the form (4.3) the equation can be completely integrated. Thus, we let

$$E = \tfrac{1}{2}y^2 + U(x),$$

where

$$U(x) = \int_0^x F(x') \, dx'. \tag{4.4}$$

Here E may be regarded as the energy of the system. In the mechanical analogy, $\tfrac{1}{2}y^2$ is the kinetic energy and $U(x)$ is the potential energy. Now, using (4.3),

$$E' = yy' + F(x)x' = 0. \tag{4.5}$$

Thus E is a constant on each orbit of (4.3). Hence, equation (4.4) determines y explicitly as a function of x for each orbit, and so the entire phase plane can be sketched.

The system (4.3) is a special case of a Hamiltonian system which we introduced briefly in section 4.5 and will discuss fully in chapter 11. Thus, in general, a frictionless autonomous mechanical system can be described by a function $H(p, q)$, called the Hamiltonian, in terms of which the equations of motion are

$$p' = -\frac{\partial H}{\partial q}, \quad q' = \frac{\partial H}{\partial p}, \tag{4.6}$$

where p and q are, in general, m-vectors. We showed in section 4.5 that H is a constant on each orbit of (4.6). Here $m = 1$, we identify x with q, y with p and the Hamiltonian with E, given by (4.4). In this sub-section we shall confine our discussion to the special case (4.3), although it is clear that the results obtained will have their analogues for the more general case (4.6).

As we indicated above, in discussing the solutions of (4.3) it is sufficient to note that the orbits are defined by (4.4), where E is a constant on each orbit.

Whether or not these are closed orbits corresponding to periodic solutions now depends on the structure of the potential-energy function $U(x)$. First we note that critical points for (4.3) are given by $x = x_c$, $y = 0$, where x_c is a stationary point for $U(x)$, that is,

$$U'(x_c) = F(x_c) = 0. \tag{4.7}$$

We shall confine attention to the generic cases when x_c is either a minimum or a maximum for $U(x)$ and $U''(x_c) \neq 0$.

Suppose first that x_c is a minimum. Then, for x near x_c,

$$U(x) \approx U_c + \tfrac{1}{2}\alpha(x - x_c)^2, \tag{4.8}$$

where $U_c = U(x_c)$ and $\alpha = U''(x_c) > 0$. The critical point $(x_c, 0)$ is then a centre, and corresponds to $E = U_c$. For E just greater than U_c, the orbits described by (4.4) are closed curves corresponding to a family of periodic solutions; for E sufficiently close to U_c, the closed curves are approximately ellipses. The situation is sketched in figure 6.12, and is sometimes called motion in a potential well.

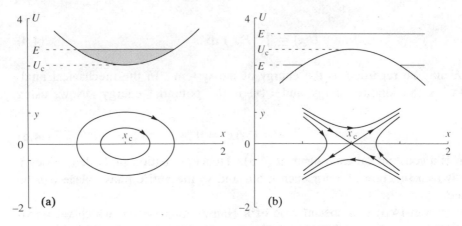

Fig. 6.12 Determination of the orbits for motion near a critical point x_c (here $x_c = 1$) of (4.3). (a) x_c is a minimum of $U(x)$. (b) x_c is a maximum of $U(x)$.

Next let x_c be a maximum. Then, for x near x_c, $U(x)$ is approximately given by (4.7), but now $\alpha < 0$. The critical point $(x_c, 0)$ is now a saddle point. For E close to U_c the orbits are sketched in figure 6.12. Typically the stationary points of $U(x)$ will be a series of alternating maxima and minima. Correspondingly, the critical points of (4.3) will be a series of alternating centres and saddle points, with families of closed orbits surrounding each centre.

As an illustration, consider the following example:

$$u'' + u - \epsilon u^3 = 0, \tag{4.9}$$

which is known as Duffing's equation. Here $F(u) = u - \epsilon u^3$, and so (see (4.4))

$$U(x) = \tfrac{1}{2}x^2 - \tfrac{1}{4}\epsilon x^4. \tag{4.10}$$

There are two cases to consider, depending on whether the parameter $\epsilon > 0$ or < 0. Suppose first that $\epsilon > 0$. Then $U(x)$ has a minimum at $x = 0$ and two symmetrical maximum points at $\pm x_c$, where $x_c = E^{-1/2}$. Correspondingly, there are three critical points, namely, a centre at $(0,0)$ and two saddle points at $(\pm x_c, 0)$. The orbits are sketched in figure 6.13.

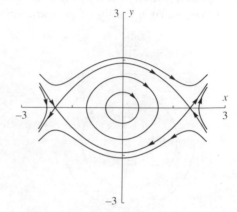

Fig. 6.13 The orbits for Duffing's equation (4.9). Here $\epsilon = \tfrac{1}{4}$ and the orbits shown are for $E = \tfrac{1}{8}, \tfrac{1}{2}, 1$ (the saddle connection), 1.25 and 0.92; the latter two values correspond to orbits outside the saddle connection.

Note that there is a family of periodic orbits surrounding the centre at the origin. This family is bounded above and below by a saddle connection, which are the orbits connecting the saddle points, either $(-x_c, 0)$ to $(x_c, 0)$, or vice-versa. These orbits correspond to $E = E_c = 1/4\epsilon$, which is the value of E for the saddle points. Note that $E = 0$ characterizes the centre at the origin and, hence, the periodic orbits have $0 < E < E_c$. Here, for each saddle connection,

$$y = \pm\sqrt{\tfrac{1}{2}\epsilon}(x_c^2 - x^2),$$

and so

$$x = \pm x_c \tanh(t/\sqrt{2}), \tag{4.11}$$

where, without loss of generality we have chosen $t = 0$ to correspond to $x = 0$. Here the expression for y is obtained from (4.4), and the result for x is then found by integrating the first equation of (4.3). For the periodic orbits, we define the amplitude of the orbit to be A, where $x = \pm A$ when $y = 0$. Then, for each periodic orbit, we find from (4.4) that

$$y^2 = (A^2 - x^2)\{1 - \tfrac{1}{2}\epsilon(x^2 + A^2)\}. \tag{4.12}$$

Substitution of this expression into the first equation of (4.3) then leads to the determination of $x(t)$ in terms of an elliptic function. We shall not give more details but we note that the period T of each orbit is given by

$$T = 4\int_0^A \frac{1}{y}\,dx, \tag{4.13}$$

where we have again used the first equation of (4.3) and also the symmetry of the orbit. On using the above expression for y, this integral can be evaluated in terms of elliptic integrals. In particular, the period is a function of the amplitude (i.e. $T = T(A)$), and it can be shown that T is a monotonically increasing function of A, which tends to infinity as $A \to x_c$.

Next we consider (4.10) when $\epsilon < 0$. In this case $U(x)$ has only one stationary point, which is a minimum at $x = 0$. The orbits are sketched in figure 6.14, and consist of a family of periodic orbits surrounding the centre at the origin.

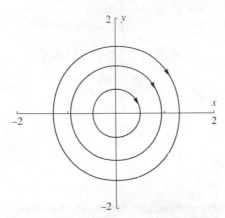

Fig. 6.14 The orbits for Duffing's equation (4.9). Here $\epsilon = -\frac{1}{4}$ and the orbits shown are for $E = \frac{1}{8}$, $\frac{1}{2}$ and 1.

The expressions (4.12) for the orbit, and (4.13) for the period, continue to hold, but now the period T is a monotonically decreasing function of the amplitude A. Note that, in both these cases, the period T tends to 2π and is independent of A as $A \to 0$. Indeed, in this limit (4.9) reduces to the equation for a simple harmonic oscillator.

Next, we consider the example

$$u'' + \sin u = 0, \tag{4.14}$$

which is the equation governing the motion of a simple pendulum. Here $F(u) = \sin u$, and so (see (4.4))

$$U(x) = 1 - \cos x. \tag{4.15}$$

For small amplitudes, $\sin u$ may be expanded as a power series in u and, truncating at the term of $O(u^3)$, we obtain

$$\sin u \approx u - \tfrac{1}{6}u^3 \quad (\text{as } |u| \to 0).$$

With this approximation (4.14) reduces to Duffing's equation (4.9). In general, $U(x)$ is a periodic function of x with period 2π and has minima at $x = 0, \pm 2\pi, \dots$, and maxima at $x = \pm\pi, \pm 3\pi, \dots$. Hence the critical points are periodically placed along the x-axis, with centres at $(0,0)$, $(\pm 2\pi, 0)$, $\dots$, and saddle points at $(\pm\pi, 0)$, $(\pm 3\pi, 0)$, $\dots$. The orbits are sketched in figure 6.15. There is a family of periodic orbits surrounding each centre, bounded above and

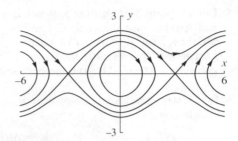

Fig. 6.15 The orbits for the simple pendulum equation (4.14). The orbits shown are for $E = 0.7, 1.4, 2$ (the saddle connection) and 2.5.

below by saddle connections connecting the neighbouring saddle points. The pattern is periodic with period 2π in the x-direction. Note the qualitative resemblance between the orbits for (4.14) and those for Duffing's equation (4.9) when $\epsilon > 0$ (see figure 6.13), provided, of course, that $|u|$ is sufficiently small.

Problems

1 Show that the system of equations

$$x' = y - xF(r), \qquad y' = -x - yF(r),$$

where $r^2 = x^2 + y^2$, has a limit cycle $r = r_0$ for each value of r_0 such that $F(r_0) = 0$ and $F'(r_0) \neq 0$. Show also that the limit cycle is asymptotically orbitally stable if $F'(r_0) > 0$, and unstable if $F'(r_0) < 0$. For the case when

$$F(r) = (r-2)^2(r^2 - 4r + 3),$$

find all limit cycles, determine the stability, and sketch the orbits in the phase plane. [*Hint*: Use the polar co-ordinates (1.3).]

2 Use theorem 6.1 to show that, for each of the following systems, there are no periodic solutions.

    (a) $x' = x + y + x^3 - y^3$, $\qquad y' = -x + 2y - x^2y + y^3$;

    (b) $x' = -xe^y + y^2$, $\qquad y' = x - x^2y$.

3 For the system

$$x' = y - (r^2 - 1)x^3, \qquad y' = -x - (r^2 - 1)x^2y,$$

where $r^2 = x^2 + y^2$, show that there is a unique limit cycle, $r = 1$, which is asymptotically orbitally stable. [*Hint*: Use the polar co-ordinates (1.3).]

4 For each of the following nonlinear systems, locate the critical points, classify them and determine the index. Then use theorems 6.7 and 6.8 to deduce that there are no periodic solutions

    (a) $x' = x(1 - 2y - x^2)$, $\qquad y' = y(-3 + x)$;

    (b) $x' = x(4 - y - x^2)$, $\qquad y' = y(-1 + x)$.

[*Hint*: Use the method described in section 6.2 for the system (2.6), and note also that, for both problems (a) and (b), the x- and y-axes are orbits.]

5 For problem 5 of chapter 5, which describes the interaction between two nearly identical species, use theorems 6.7 and 6.8 to deduce that there are no periodic solutions. Similarly, for problem 6 of chapter 5, which describes the interaction between two competing species, use theorems 6.7 and 6.8 to deduce that there are no periodic solutions. [*Hint*: Use the method indicated for problem 4.]

6 (a) Using the polar co-ordinates (r, θ) (see (1.3)), show that the system (1.1) becomes

$$r' = M(r, \theta), \qquad r\theta' = N(r, \theta),$$

where

$$M(r, \theta) = \cos\theta f(r\cos\theta, r\sin\theta) + \sin\theta g(r\cos\theta, r\sin\theta),$$
$$N(r, \theta) = -\sin\theta f(r\cos\theta, r\sin\theta) + \cos\theta g(r\cos\theta, r\sin\theta).$$

If C is a closed curve which passes through no critical points of (1.1), show that the index $I(C)$ (see (2.1)) is given by

$$I(C) = \frac{1}{2\pi} \oint_C \frac{M\,dN - N\,dM}{M^2 + N^2} + S(C),$$

where $S(C) = 1$, if C encloses the origin, and $S(C) = 0$, otherwise.

(b) If $R = r^{-1}$, show that

$$R' = \hat{M}(R, \theta), \qquad R\theta' = \hat{N}(R, \theta),$$

where

$$\hat{M}(R, \theta) = -R^2 M(R^{-1}, \theta), \qquad \hat{N}(R, \theta) = R^2 N(R^{-1}, \theta).$$

The index of the point at infinity is now defined to be the index of the origin in the R-θ plane. Thus, let $\hat{C}$ be a closed curve surrounding the point $R = 0$ and containing no critical points except possibly $R = 0$. Then, using the result of part (a), the index of the point at infinity is $I(\hat{C})$, where

$$I(\hat{C}) = 1 + \frac{1}{2\pi} \oint_{\hat{C}} \frac{\hat{M}\,d\hat{N} - \hat{N}\,d\hat{M}}{\hat{M}^2 + \hat{N}^2}.$$

Next let C be a closed curve which contains all critical points of (1.1) and the point $r = 0$. Show that

$$I(C) + I(\hat{C}) = 2.$$

This expression provides an alternative formula for $I(\hat{C})$, the index of the point at infinity; note that theorem 6.6 can be used to calculate $I(C)$ when C encloses a finite number of isolated critical points.

(c) Next let $X = R\cos\theta$ and $Y = R\sin\theta$ so that $X = xr^{-2}$ and $Y = yr^{-2}$. Show that

$$X' = \hat{f}(X, Y), \qquad Y' = \hat{g}(X, Y),$$

where

$$\hat{f}(X, Y) = (Y^2 - X^2)f(x, y) - 2XYg(x, y),$$
$$\hat{g}(X, Y) = -2XYf(x, y) + (X^2 - Y^2)g(x, y).$$

Hence, or otherwise, show that

$$I(\hat{C}) = \frac{1}{2\pi} \oint_{\hat{C}} \frac{\hat{f}\, d\hat{g} - \hat{g}\, d\hat{f}}{\hat{f}^2 + \hat{g}^2}.$$

(d) Use the result of part (b) to determine the index of the point at infinity for the nonlinear systems of problem 4.

7 Show that the Raleigh equation

$$v'' + \epsilon(\tfrac{1}{3}v'^3 - v') + v = 0$$

can be transformed into the Van der Pol equation (3.1) by the substitution $u = v'$. Hence, using theorem 6.9, deduce that there is a unique limit cycle which is asymptotically orbitally stable. Use the results for the Van der Pol limit cycle obtained in section 6.3 to describe the Raleigh limit cycle. [*Hint*: Note that the phase plane $x = v'$, $y = -v$ is identical to the Van der Pol phase plane (see (3.3)).]

8 Convert the Lienard equation (see (3.45) and the subsequent discussion)

$$u'' + f(u)u' + g(u) = 0$$

into the planar system

$$x' = y - F(x), \qquad y' = -g(x),$$

where x, y are defined by (3.46). Suppose, further, that the conditions (3.48) hold. Show that there is a unique limit cycle which is asymptotically orbitally stable. [*Hint*: Adapt the proof of theorem 6.9.]

9 (a) For the Lienard equation of problem 8, let

$$E = \tfrac{1}{2}y^2 + G(x),$$

where

$$G(x) = \int_0^x g(x')\, dx'.$$

(b) Show that

$$E' = -g(x)F(x)$$

and, hence, deduce that

$$\frac{1}{T} \int_0^T g(x)F(x)\, dt = 0,$$

where the integrand is evaluated on the limit cycle and T is the period.

(c) If $f(u) = \epsilon(5u^4 - a^4)$ and $g(u) = u$, verify that the conditions (3.48) are satisfied for $\epsilon > 0$. Then adapt the method described in section 6.3 for the Van der Pol equation to show that, as $\epsilon \to 0$, the limit cycle solution is given by

$$u = r_0 \cos t + O(\epsilon),$$

where $r_0^4 = \tfrac{8}{5}a^4$ and $T = 2\pi + O(\epsilon)$.

10 (a) For the conservative system (4.1), verify that the orbits in the phase plane are described by

$$\tfrac{1}{2}y^2 + U(x) = E,$$

where E is a constant, $U(x)$ is defined in (4.4) and x, y are defined in (4.2).

(b) For each of the following systems sketch the orbits in the phase plane:

(i) $U(x) = \cosh x - 1$;   (ii) $U(x) = \tanh^2 x$,

(iii) $U(x) = \tfrac{1}{4}x^4 - \tfrac{1}{2}x^2$;   (iv) $U(x) = x^{-4} - 2x^{-2}$;

(v) $U(x) = x^2 \exp(-x^2)$;   (vi) $U(x) = \tfrac{1}{2}x^2 - \tfrac{1}{4}x^4 + \tfrac{1}{6}x^6$.

11 For the conservative system

$$u'' - u + 2u^3 = 0,$$

show that the orbits in the phase plane $x = u, y = u'$ are given by

$$y^2 = x^2 - x^4 + 2E,$$

where E is a constant. Sketch the orbits, and show that there is a saddle point at $x = 0, y = 0$ ($E = 0$) and centres at $x = \pm 2^{-1/2}, y = 0$ ($E = -\tfrac{1}{8}$). Also show that

$$y^2 = x^2 - x^4, \qquad x = \pm \operatorname{sech} t,$$

are saddle connections connecting the saddle point to itself.

12 For Duffing's equation (4.9),

$$u'' + u - \epsilon u^3 = 0,$$

show that the orbits in the phase plane $x = u, y = u'$ are given by

$$\tfrac{1}{2}y^2 + \tfrac{1}{2}x^2 - \tfrac{1}{4}\epsilon x^4 = E,$$

where E is a constant. Sketch the orbits, and show that for $\epsilon > 0$ there are saddle points at $x = \pm x_c$, where $x_c = \epsilon^{-1/2}$, $y = 0$ ($E = 1/4\epsilon$). Show also that the stable and unstable manifolds through the saddle points are given by

$$x = \pm x_c \tanh(t/\sqrt{2}) \quad \text{or} \quad x = \pm x_c \coth(t/\sqrt{2}).$$

13 For the pendulum equation (4.14)

$$u'' + \sin u = 0,$$

show that the orbits in the phase plane $x = y, y = u'$ are given by

$$\tfrac{1}{2}y^2 + 1 - \cos x = E,$$

where E is a constant. Sketch the orbits, and show that there are centres at $(0,0)$, $(\pm 2\pi, 0)$, ..., and saddle points at $(\pm \pi, 0)$, $(\pm 3\pi, 0)$, ... Show that the saddle connections connecting the saddle points $(\pm \pi, 0)$ are given by

$$\tfrac{1}{2}y^2 = 1 + \cos x \quad \text{and} \quad x = \pi - 4\cot^{-1} e^{\pm t}.$$

CHAPTER SEVEN
PERTURBATION METHODS
FOR PERIODIC SOLUTIONS

7.1 Poincaré–Lindstedt method

In this chapter we will describe procedures for obtaining periodic solutions to equations of the form

$$u'' + u = \epsilon f(u, u'; \epsilon), \tag{1.1}$$

where $f(u, u'; \epsilon)$ is a prescribed function of u, u' and the parameter ϵ, and is defined for all values of these variables. Our interest here is when ϵ is a small parameter, $|\epsilon| \ll 1$. Note that both the Van der Pol equation, (3.1) of section 6.3, and Duffing's equation, (4.1) of section 6.4, have the form of (1.1).

When $\epsilon = 0$, equation (1.1) reduces to

$$u'' + u = 0, \tag{1.2}$$

which has the general solution,

$$u = A_0 \cos(t - t_0), \tag{1.3}$$

where A_0 and t_0 are arbitrary constants. Of course, equation (1.2) is that for a simple harmonic oscillator, and the general solution (1.3) is periodic with period 2π. In the phase plane for which $x = u$ and $y = u'$, the orbits are circles and the origin (the only critical point) is a centre. From this point of view, equation (1.1), for $0 < |\epsilon| \ll 1$, can be regarded as a perturbation of a centre. Our aim is to determine which, if any, of the periodic solutions (1.3) survive this perturbation. In this section we shall describe a perturbation-expansion procedure, the Poincaré–Lindstedt method, for obtaining periodic solutions of (1.1). In the next section we shall give the justification for the perturbation expansion, and in section 7.3 we shall discuss the stability of the periodic solutions obtained.

For simplicity we shall suppose that $f(u, u'; \epsilon)$ is an analytic function of its arguments u, u' and ϵ. This condition is usually met in applications. It now follows from theorem 1.6 of chapter 1 that the solutions of (1.1) are analytic functions of t, ϵ and the initial conditions. In particular, we can expect the solutions to have a convergent power-series expansion in ϵ, and hence we write

$$u = u_0(t) + \epsilon u_1(t) + \epsilon^2 u_2(t) + \dots. \tag{1.4}$$

This series is substituted into (1.1), and the result is a sequence of equations for $u_0, u_1, u_2, \dots$. We shall illustrate the procedure for Duffing's equation ((4.9) of section 6.4), which here we put into the form (1.1), so that

$$u'' + u = \epsilon u^3. \tag{1.5}$$

Substituting (1.4) into (1.5) and equating to zero each coefficient of ϵ, we find that

$$u_0'' + u_0 = 0, \qquad u_1'' + u_1 = u_0^3, \qquad u_2'' + u_2 = 3u_0^2 u_1, \qquad \dots. \tag{1.6}$$

The general solution for u_0 is given by (1.3), and is periodic with period 2π. Since equation (1.1) (or here (1.5)) is autonomous, there is no loss of generality in putting $t_0 = 0$. Then the equation for u_1 becomes

$$u_1'' + u_1 = A_0^3 \cos^3 t. \tag{1.7}$$

To solve (1.7), we use the trigonometric identity

$$\cos^3 t = \tfrac{1}{4} \cos 3t + \tfrac{3}{4} \cos t, \tag{1.8}$$

and then the general solution of (1.7) is readily found to be

$$u_1 = A_1 \cos(t - t_1) - \tfrac{1}{32} A_0^3 \cos 3t + \tfrac{3}{8} A_0^3 t \sin t. \tag{1.9}$$

Again, there is no loss of generality in putting $t_1 = 0$. However, there is, clearly, no non-trivial choice of the constants A_0 and A_1 for which u_1 can be periodic. The difficulty lies with the last term of (1.9) which is said to be **secular**, since it grows without bound as $t \to \infty$. This is contrary to our analysis of Duffing's equation in section 6.4, which implies that, for sufficiently small ϵ (and bounded values of A_0), there is a family of periodic solutions. The source of the difficulty can be located by observing that, from (1.3) and (1.9), we may write (with $t_0 = t_1 = 0$)

$$u_0 + \epsilon u_1 = (A_0 + \epsilon A_1) \cos\{t(1 - \tfrac{3}{8}\epsilon A_0^2)\}$$

$$- \tfrac{1}{32}\epsilon A_0^3 \cos\{3t(1 - \tfrac{3}{8}\epsilon A_0^2)\} + O(\epsilon^2). \tag{1.10}$$

Expanding the right-hand side of (1.10) in powers of ϵ, we see that it agrees with the sum of (1.3) and (1.9) to terms of $O(\epsilon^2)$. But now, to $O(\epsilon)$, there is no secular term. Instead, to this order, the solution is periodic with period $2\pi(1 - \tfrac{3}{8}\epsilon A_0^2)^{-1}$. Indeed, our analysis of Duffing's equation in section 6.4 has already shown that the period of each periodic solution will depend on ϵ (and A_0).

Hence, in seeking to find periodic solutions of (1.1), we must anticipate that the period $T(\epsilon)$ will be a function of ϵ. We let the corresponding frequency be

$$\omega(\epsilon) = \frac{2\pi}{T(\epsilon)}, \tag{1.11}$$

and replace the independent variable t with τ, where

$$\tau = \omega t. \tag{1.12}$$

If now $u = u(\tau; \epsilon)$ then periodic solutions have a fixed period of 2π with respect to τ. Equation (1.1) becomes

$$\omega^2 \frac{d^2u}{d\tau^2} + u = \epsilon f\left(u, \omega \frac{du}{d\tau}; \epsilon\right). \tag{1.13}$$

Since, clearly, $T = 2\pi + O(\epsilon)$, we may put

$$\omega = 1 + \epsilon\kappa, \tag{1.14}$$

and then (1.13) can be put into the form

$$\frac{d^2u}{d\tau^2} + u = \epsilon g\left(u, \frac{du}{d\tau}; \epsilon, \kappa\right),$$

where

$$g\left(u, \frac{du}{d\tau}; \epsilon, \kappa\right) = f\left(u, \omega \frac{du}{d\tau}; \epsilon\right) + \frac{1 - \omega^2}{\epsilon} \frac{d^2u}{d\tau^2}. \tag{1.15}$$

Note that here, on the right-hand side of the expression for g, we may regard $d^2u/d\tau^2$ as a function of u, $du/d\tau$, ϵ and κ from (1.13). Using this convention and recalling (1.14), we see that

$$g\left(u, \frac{du}{d\tau}; \epsilon, \kappa\right) = f\left(u, \frac{du}{d\tau}; 0\right) + 2\kappa u + O(\epsilon). \tag{1.16}$$

We now replace the expansion (1.4) with

$$u = u_0(\tau) + \epsilon u_1(\tau) + \epsilon^2 u_2(\tau) + \dots, \qquad \kappa = \kappa_0 + \epsilon\kappa_1 + \dots, \tag{1.17}$$

and substitute these series into (1.15). We again illustrate the procedure for Duffing's equation (1.5), for which (1.15) becomes

$$\frac{d^2u}{d\tau^2} + u = \epsilon u^3 - \epsilon(2\kappa + \epsilon\kappa^2) \frac{d^2u}{d\tau^2}. \tag{1.18}$$

Substituting (1.17) into (1.18) and equating to zero each coefficient of ϵ, we find that

$$\frac{d^2u_0}{d\tau^2} + u_0 = 0,$$

$$\frac{d^2u_1}{d\tau^2} + u_1 = u_0^3 - 2\kappa_0 \frac{d^2u_0}{d\tau^2},$$

$$\frac{d^2u_2}{d\tau^2} + u_2 = 3u_0^2 u_1 - (2\kappa_1 + \kappa_0^2) \frac{d^2u_0}{d\tau^2} - \kappa_0 \frac{d^2u_1}{d\tau^2},$$

$$\vdots \tag{1.19}$$

The general solution for u_0 is

$$u_0 = A_0 \cos(\tau - \tau_0), \tag{1.20}$$

where A_0 and τ_0 are arbitrary constants, and is periodic with period 2π. Again, there is no loss of generality in putting $\tau_0 = 0$. Using the trigonometric identity (1.8), the equation for u_1 is

$$\frac{d^2u_1}{d\tau^2} + u_1 = \tfrac{1}{4}A_0^3 \cos 3\tau + 2A_0(\kappa_0 + \tfrac{3}{8}A_0^2)\cos \tau. \tag{1.21}$$

The general solution of (1.21) is readily found to be

$$u_1 = A_1 \cos(\tau - \tau_1) - \tfrac{1}{32}A_0^3 \cos 3\tau + A_0(\kappa_0 + \tfrac{3}{8}A_0^2)\tau \sin \tau. \tag{1.22}$$

Again, there is no loss of generality in putting $\tau_1 = 0$. However, the secular term in (1.22) can now be removed by choosing κ_0 appropriately, namely,

$$\kappa_0 = -\tfrac{3}{8}A_0^2. \tag{1.23}$$

The consequent solution for u_1 is periodic in τ, with period 2π. Of course, this was to be expected, as it was the basis of the transformation (1.12). Note that the expression for $u_0 + \epsilon u_1$ obtained from (1.20) and (1.22) is identical, to $O(\epsilon^2)$, with (1.10). Also, it is useful to observe that the secular term in (1.22) arises from the forcing term proportional to $\cos \tau$ on the right-hand side. Indeed, it is clear that, if the right-hand side of the equation for u_1 (or that for $u_2, \ldots$) is expressed in terms of the Fourier variables $\{\cos n\tau, \sin n\tau; \ n = 0, 1, 2, 3, \ldots\}$, then the secular-producing terms are $\cos \tau$ and $\sin \tau$. Setting the coefficients of these 'resonant' terms to zero in equations such as (1.21) is clearly more efficient than first finding the general solution, and then identifying the secular terms.

At this stage the constant A_1 has not been determined. It may be chosen to satisfy some auxiliary condition. Here, for instance, we could impose the condition that

$$u = A, \qquad \frac{du}{d\tau} = 0 \qquad (\text{at } \tau = 0). \tag{1.24}$$

This defines the constant A as the **amplitude** of the periodic solution. In the phase plane for which $x = u$ and $y = u'$, A is the distance from the origin of the point where the orbit cuts the x-axis. Here we can put $A = A_0$ and then $A_1 = \tfrac{1}{32}A_0^3$. With u_1 now determined, we can proceed to consider the equation for u_2. Removal of the coefficients of the resonant, or secular-producing, terms on the right-hand side will then determine κ_2. We shall leave this calculation for the exercises. Clearly, at least in principle, this procedure can be continued indefinitely. We emphasise, however, that often there are diminishing returns the further the expansion is pursued. In many cases, the most significant information is obtained at the lowest orders. For instance, here κ_0 (1.23) is found at the $O(\epsilon)$ stage of the expansion, and shows that the frequency ω (1.11) is decreased (increased) accordingly as ϵ is increased (decreased) from zero.

For another illustrative example, we consider the Van der Pol equation ((3.1) of section 6.3), or

$$u'' + u = \epsilon(1 - u^2)u'. \tag{1.25}$$

Thus, (1.15) becomes

$$\frac{d^2 u}{d\tau^2} + u = \epsilon(1-u^2)\omega\frac{du}{d\tau} - \epsilon(2\kappa+\epsilon\kappa^2)\frac{d^2 u}{d\tau^2}. \tag{1.26}$$

Substituting the expansions (1.17) into (1.26) and equating to zero each coefficient of ϵ, we find that

$$\frac{d^2 u_0}{d\tau^2} + u_0 = 0, \qquad \frac{d^2 u_1}{d\tau^2} + u_1 = \frac{d}{d\tau}(u_0 - \tfrac{1}{3}u_0^3) - 2\kappa_0\frac{d^2 u_0}{d\tau^2}, \qquad \dots. \tag{1.27}$$

The general solution for u_0 is again (1.20) (with $\tau_0 = 0$) and, again using the trigonometric identity (1.8), the equation for u_1 is

$$\frac{d^2 u_1}{d\tau^2} + u_1 = A_0(\tfrac{1}{4}A_0^2 - 1)\sin\tau + \tfrac{1}{4}A_0^3\sin 3\tau + 2\kappa_0 A_0 \cos\tau. \tag{1.28}$$

Removing the coefficients of the resonant, or secular-producing, terms (i.e. $\cos\tau$ and $\sin\tau$), we find that

$$A_0^2 = 4 \quad \text{and} \quad \kappa_0 = 0, \tag{1.29}$$

where we have rejected the trivial solution $A_0 = 0$. Thus here there is a unique value of A_0 (viz. $A_0 = 2$, since $A_0 = -2$ gives the same orbit in the phase plane) and, hence, there is a unique limit cycle. The approximate solution obtained here agrees, of course, with the analysis of section 6.3 (see (3.19) of section 6.3). The general solution of u_1 is now (with $A_0 = 2$)

$$u_1 = A_1\cos\tau + B_1\sin\tau - \tfrac{1}{4}\sin 3\tau \tag{1.30}$$

and, to satisfy the initial condition (1.24), we must put $A = 2 + \epsilon A_1$ and $B_1 = \tfrac{3}{4}$. Note that, at this stage, the period is $2\pi + O(\epsilon^2)$. To determine κ_1 and A_1, it is necessary to consider the equation for u_2. We shall leave this calculation for the exercises, but note that the result is $A_1 = 0$ and $\kappa_1 = -\tfrac{1}{16}$.

7.2 Poincaré–Lindstedt method, continued

In the previous section, we showed that, to find periodic solutions of (1.1), we must transform this equation, using (1.12), into the form (1.15), which for convenience we restate here:

$$\frac{d^2 u}{d\tau^2} + u = \epsilon g\left(u, \frac{du}{d\tau}; \epsilon, \kappa\right). \tag{2.1}$$

Here we observe that the right-hand side is given by (1.15), but using (1.13), it can be put into the form

$$g\left(u, \frac{du}{d\tau}; \epsilon, \kappa\right) = \frac{1}{\omega^2}f\left(u, \omega\frac{du}{d\tau}; \epsilon\right) + \frac{\omega^2-1}{\epsilon\omega^2}u,$$

where

$$\omega = 1 + \epsilon\kappa. \tag{2.2}$$

In particular, we recall that (see (1.16)),

$$g\left(u, \frac{du}{d\tau}; \epsilon, \kappa\right) = f\left(u, \frac{du}{d\tau}; 0\right) + 2\kappa u + O(\epsilon). \tag{2.3}$$

Associated with (2.1), we impose the initial condition (1.24), or

$$u = A, \qquad \frac{du}{d\tau} = 0 \qquad (\text{at } \tau = 0). \tag{2.4}$$

Because (2.1) is autonomous, there is no loss of generality in choosing this initial condition. In effect we have chosen the origin of the time variable τ to be at the point where the orbit in the phase plane, for which $x = u$ and $y = u'$, cuts the x-axis. Note that, from (1.12), $u' = \omega\,du/d\tau$. Our aim now is to establish that, for sufficiently small values of $|\epsilon|$, there exist values of κ and A such that (2.1) has a periodic solution which has a period 2π with respect to τ. The frequency of this solution is then ω (2.2), its period is $T = 2\pi/\omega$ with respect to t, and its amplitude is A.

We shall suppose that $f(u, u'; \epsilon)$ is an analytic function of its arguments u, u' and ϵ, and is defined for all values of these variables. Then, from (2.2), $g(u, du/d\tau; \epsilon, \kappa)$ is an analytic function of its arguments u, $du/d\tau$, ϵ and κ. It then follows from theorem 1.6 of chapter 1 that the initial value problem (2.4) for (2.1) has a unique solution:

$$u = u(\tau; \kappa, A, \epsilon), \tag{2.5}$$

which is an analytic function of τ and the parameters κ, A and ϵ. By theorem 1.7 of chapter 1, the solution exists for all $\tau \geqslant 0$, provided $|u|$ and $|du/d\tau|$ are bounded. Next, since this function (2.5) satisfies (2.1) and (2.4), it follows that (see (4.14) of section 2.4)

$$u = A\cos\tau + \epsilon \int_0^\tau g\left(u, \frac{du}{ds}; \epsilon, \kappa\right)\sin(\tau - s)\,ds,$$

and

$$\frac{du}{d\tau} = -A\sin\tau + \epsilon \int_0^\tau g\left(u, \frac{du}{ds}; \epsilon, \kappa\right)\cos(\tau - s)\,ds. \tag{2.6}$$

Note that, in the analysis of section 2.4 leading to the result (4.14), it was assumed that the right-hand side of (2.1) is known explicitly. However, the result clearly remains true, when, as here, the right-hand side is only known implicitly, being a function of u and $du/d\tau$. Equations (2.6) can be regarded as a pair of integro-differential equations equivalent to (2.1) and (2.4).

Next we define the functions $P(\kappa, A; \epsilon)$ and $Q(\kappa, A; \epsilon)$ by

$$\epsilon P(\kappa, A; \epsilon) = u(2\pi; \kappa, A, \epsilon) - u(0; \kappa, A, \epsilon),$$

$$\epsilon Q(\kappa, A; \epsilon) = \frac{du}{d\tau}(2\pi; \kappa, A, \epsilon) - \frac{du}{d\tau}(0; \kappa, A, \epsilon). \tag{2.7}$$

Using the result (2.6), it follows that

$$P(\kappa, A; \epsilon) = -\int_0^{2\pi} g\left(u, \frac{du}{ds}; \epsilon, \kappa\right) \sin s \ ds,$$

$$Q(\kappa, A; \epsilon) = \int_0^{2\pi} g\left(u, \frac{du}{ds}; \epsilon, \kappa\right) \cos s \ ds. \tag{2.8}$$

Here, in the integrand, $u = u(s; \kappa, A, \epsilon)$ (see (2.5)). Then a necessary and sufficient condition for $u(\tau; \kappa, A, \epsilon)$ to be a periodic function with period 2π with respect to τ, is

$$P(\kappa, A; \epsilon) = 0, \qquad Q(\kappa, A; \epsilon) = 0. \tag{2.9}$$

This result was established in theorem 2.4 of section 2.4, or, more intuitively, it is easily seen to be a consequence of (2.7). Note that, in the phase plane, the orbit is at the point $(A, 0)$ at the time $\tau = 0$, and after a time $\tau = 2\pi$ is, from (2.7), located at the point $(A + \epsilon P, \epsilon \omega Q)$ (see figure 7.1).

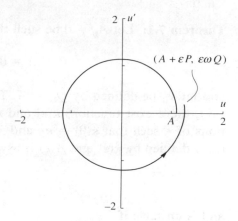

Fig. 7.1 The definition of P and Q (2.9).

Hence the conditions (2.9) are precisely the conditions that the orbit be closed. The equations (2.9) are now regarded as two equations for κ and A as functions of ϵ. To determine when there may be a solution, we consider the limit $\epsilon \to 0$, where we note that P and Q are analytic functions of κ, A and ϵ. Thus, as $\epsilon \to 0$, we let $\kappa \to \kappa_0$, $A \to A_0$ and $u \to u_0$, where

$$u_0 = A_0 \cos \tau. \tag{2.10}$$

This last result is, of course, an immediate consequence of (2.1) as $\epsilon \to 0$. Further, as $\epsilon \to 0$, $P \to P_0$ and $Q \to Q_0$, where

$$P_0(\kappa_0, A_0) = P(\kappa_0, A_0; 0) = -\int_0^{2\pi} g\left(u_0, \frac{du_0}{ds}; 0, \kappa_0\right)\sin s \ ds,$$

$$Q_0(\kappa_0, A_0) = Q(\kappa_0, A_0; 0) = \int_0^{2\pi} g\left(u_0, \frac{du_0}{ds}; 0, \kappa_0\right)\cos s \ ds. \qquad (2.11)$$

Note that, since u_0 is known explicitly from (2.10), the functions P_0, and Q_0 are known explicitly in terms of κ_0 and A_0. Further, using (2.3), we see that

$$P_0 = P_0(A_0) = -\int_0^{2\pi} f\left(u_0, \frac{du_0}{ds}; 0\right)\sin s \ ds$$

and

$$Q_0 = \hat{Q}_0(A_0) + 2\pi\kappa_0 A_0,$$

where

$$\hat{Q}_0(A_0) = \int_0^{2\pi} f\left(u_0, \frac{du_0}{ds}; 0\right)\cos s \ ds. \qquad (2.12)$$

Note that, as the terminology implies, P_0 and Q_0 are functions of A_0 alone. We are now in a position to establish the following theorem.

Theorem 7.1: Let $A_0 \neq 0$ be such that

$$P_0(A_0) = 0, \qquad \frac{dP_0}{dA_0}(A_0) \neq 0, \qquad (2.13)$$

and let κ_0 be defined by $Q_0 = 0$. Then, for sufficiently small $|\epsilon|$ ($0 \leq |\epsilon| < \epsilon_0$, say), there exist functions $\kappa(\epsilon)$ and $A(\epsilon)$ satisfying (2.9), which are analytic functions of ϵ such that $\kappa(0) = \kappa_0$ and $A(0) = A_0$. Further, the periodic solution of (2.1) defined by $\kappa(\epsilon)$ and $A(\epsilon)$ is asymptotically orbitally stable if

$$\epsilon \frac{dP_0}{dA_0} < 0,$$

and is unstable if

$$\epsilon \frac{dP_0}{dA_0} > 0.$$

Proof: First note that, if A_0 is determined from (2.13), then κ_0 is determined as a consequence of $Q_0 = 0$ (see (2.12)). Thus we are assured of the existence of a solution to (2.9) when $\epsilon = 0$; that is, $P_0(\kappa_0, A_0) = Q_0(\kappa_0, A_0) = 0$. Since P and Q are analytic functions of ϵ, κ and A, the implicit function theorem for functions of two variables now guarantees a solution of (2.9) for $\kappa(\epsilon)$ and $A(\epsilon)$ when $0 \leq |\epsilon| < \epsilon_0$, say, which are analytic functions of ϵ and such that $\kappa(0) = \kappa_0$ and $A(0) = A_0$, provided that the Jacobian of the functions P and Q is not zero when $\epsilon = 0$, that is,

$$\frac{\partial(P,Q)}{\partial(\kappa,A)} = \det\begin{bmatrix} \dfrac{\partial P}{\partial \kappa} & \dfrac{\partial P}{\partial A} \\[2mm] \dfrac{\partial Q}{\partial \kappa} & \dfrac{\partial Q}{\partial A} \end{bmatrix} \neq 0 \quad (\text{as } \epsilon \to 0). \tag{2.14}$$

From (2.12) we see that this condition reduces to

$$\det\begin{bmatrix} 0 & \dfrac{\mathrm{d}P_0}{\mathrm{d}A_0} \\[3mm] 2\pi A_0 & 2\pi\kappa_0 + \dfrac{\mathrm{d}\hat{Q}_0}{\mathrm{d}A_0} \end{bmatrix} \neq 0,$$

or

$$2\pi A_0 \frac{\mathrm{d}P_0}{\mathrm{d}A_0} \neq 0. \tag{2.15}$$

But this is just the required condition (2.13) and, hence, the existence is proven. Note that the theorem establishes the existence of a periodic solution for each solution A_0 of $P_0(A_0) = 0$.

The result concerning the stability will be discussed in the next section. Here, however, we shall present a heuristic argument which gives a geometrical meaning to the stability criteria. The periodic orbit commences at the point $(A, 0)$ in the phase plane and, after one period, returns to the same point. Consider a nearby orbit which commences at the point $(A^*, 0)$ close to A. After one period this will have passed around the origin and reached the point $(A^* + \epsilon P^*, \epsilon\omega Q^*)$, where $P^* = P(\kappa, A^*; \epsilon)$ and $Q^* = Q(\kappa, A^*; \epsilon)$, as defined in (2.8) (see figure 7.2). Clearly, the periodic orbit is asymptotically orbitally stable if $A^* \gtrless A$ implies that $\epsilon P^* \lessgtr 0$. In the limit $\epsilon \to 0$, this is equivalent to $\epsilon\,\mathrm{d}P_0/\mathrm{d}A_0 < 0$. Conversely, the periodic orbit is unstable if $A^* \gtrless A$ implies that $\epsilon P^* \gtrless 0$, which in the limit $\epsilon \to 0$ is equivalent to $\epsilon\,\mathrm{d}P_0/\mathrm{d}A_0 > 0$.

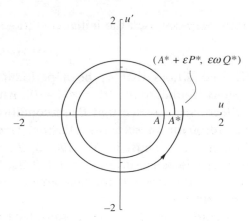

Fig. 7.2 The definition of P^* and Q^* in theorem 7.1.

As an alternative to this heuristic argument (and to the more general discussion of the next section), we can use theorem 6.4 of section 6.1 to determine the stability. Thus we recall that the stability is decided by the sign of the characteristic exponent μ, defined by (1.9) of section 6.1. Here this becomes

$$\mu = \frac{1}{T} \int_0^T \epsilon f_y(x,y;\epsilon) \, dt, \tag{2.16}$$

where we recall that T is the period, $x = u$, $y = u'$ and the integrand is evaluated for the periodic solution. In the limit $\epsilon \to 0$ let $\mu \to \mu_0$, and then

$$\mu_0 = \frac{\epsilon}{2\pi} \int_0^{2\pi} f_y(u_0, u_0'; 0) \, dt, \tag{2.17}$$

where $u_0 = A_0 \cos t$ (see (2.10)). It can now be shown that

$$2\pi\mu_0 = \epsilon \frac{dP_0}{dA_0}, \tag{2.18}$$

where we recall that $P_0(A_0)$ is defined by (2.12). The details are left to the exercises. The stability results in theorem 7.1 now follows from (2.18) and theorem 6.4 of section 6.1.

Next, to illustrate the relationship between theorem 7.1 and the Poincaré–Lindstedt method of section 7.1, we apply the perturbation procedure (1.17) to (2.1). We find that

$$\frac{d^2 u_0}{d\tau^2} + u_0 = 0, \qquad \frac{d^2 u_1}{d\tau^2} + u_1 = g\left(u_0, \frac{du_0}{d\tau}; 0, \kappa_0\right), \qquad \ldots. \tag{2.19}$$

Thus u_0 is given by (2.10), and the general solution for u_1 is (compare (2.6))

$$u_1 = A_1 \cos\tau + \int_0^\tau g\left(u_0, \frac{du_0}{ds}; 0, \kappa_0\right) \sin(\tau - s) \, ds. \tag{2.20}$$

Here we have used the initial conditions (2.4), and expanded

$$A = A_0 + \epsilon A_1 + \ldots.$$

The conditions that u_1 be a periodic function of τ with period 2π are now readily found to be $P_0 = Q_0 = 0$, where P_0 and Q_0 are defined by (2.11). Indeed, we now see that these conditions are precisely those needed to remove the resonant, or secular-producing, terms from the right-hand side of the equation for u_1. Thus, if $g(u_0, du_0/d\tau; 0, \kappa_0)$ is expressed in terms of the Fourier variables $\{\cos n\tau, \sin n\tau; n = 0, 1, 2, 3, \ldots\}$, then the conditions $P_0 = Q_0 = 0$ are just the requirement for the removal of the coefficients of the resonant terms, $\cos\tau$ and $\sin\tau$.

At this point let us again consider the Van der Pol equation,

$$u'' + u = \epsilon(1 - u^2) u'. \tag{2.21}$$

Here, from (2.12),

$$P_0(A_0) = -\int_0^{2\pi} (1 - u_0^2)\frac{du_0}{ds}\sin s \; ds,$$

or

$$P_0(A_0) = \pi A_0(1 - \tfrac{1}{4}A_0^2), \tag{2.22}$$

where we recall that u_0 is given by (2.10). Hence the limit cycle is given by $A_0^2 = 4$ and, further,

$$\frac{dP_0}{dA_0} = -2\pi \quad \text{(for } A_0^2 = 4\text{)}, \tag{2.23}$$

so that the limit cycle is asymptotically orbitally stable for $\epsilon > 0$. Further, it is readily shown from (2.12) that

$$\hat{Q}_0(A_0) = \int_0^{2\pi} (1 - u_0^2)\frac{du_0}{ds}\cos s \; ds = 0 \tag{2.24}$$

and, hence, here $\kappa_0 = 0$. These results of course agree with the results obtained for the Van der Pol equation as $\epsilon \to 0$ in section 6.3 (and also in section 7.1), and provide an independent proof of the existence of the Van der Pol limit cycle, at least for sufficiently small values of ϵ.

Next we consider an example for which the theorem establishes the existence of two limit cycles. Thus, let

$$u'' + u = -\epsilon(\tfrac{1}{2} - \tfrac{5}{2}u^2 + u^4)u'. \tag{2.25}$$

Then, from (2.12),

$$P_0(A_0) = \int_0^{2\pi} (\tfrac{1}{2} - \tfrac{5}{2}u_0^2 + u_0^4)\frac{du_0}{ds}\sin s \; ds,$$

or

$$P_0(A_0) = -\tfrac{1}{8}\pi A_0(4 - 5A_0^2 + A_0^4). \tag{2.26}$$

Hence the limit cycles are given by $A_0^2 = 1$ and $A_0^2 = 4$, or $A_0 = 1$ and $A_0 = 2$ (since the solutions $A_0 = -1$ and $A_0 = -2$ correspond, respectively, to the same orbits in the phase plane). Here

$$\frac{dP_0}{dA_0} = -\tfrac{1}{8}\pi(4 - 15A_0^2 + 5A_0^4),$$

or

$$\frac{dP_0}{dA_0} = \begin{cases} \tfrac{3}{4}\pi & (A_0^2 = 1), \\ -3\pi & (A_0^2 = 4). \end{cases} \tag{2.27}$$

Thus the limit cycle defined by $A_0 = 1$ is unstable, whereas the limit cycle defined by $A_0 = 2$ is asymptotically orbitally stable. Here the origin in the phase plane (where $x = u$ and $y = u'$) is a stable focus (for $|\epsilon| < 4$). The orbits are sketched in figure 7.3, which are obtained from a numerical integration of (2.25).

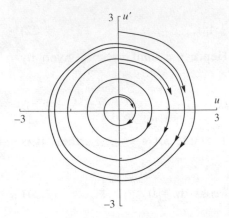

Fig. 7.3 The orbits for equation (2.25) with $\epsilon = 0.1$.

On the other hand, for Duffing's equation

$$u'' + u = \epsilon u^3, \tag{2.28}$$

we find from (2.12) that

$$P_0(A_0) = -\int_0^{2\pi} u_0^3 \sin s \; ds = 0 \quad \text{(for all } A_0),$$

$$\hat{Q}_0(A_0) = \int_0^{2\pi} u_0^3 \cos s \; ds = \tfrac{3}{4}\pi A_0^3,$$

so that

$$\kappa_0 = -\tfrac{3}{8}A_0^2. \tag{2.29}$$

Note that this result for κ_0 agrees with that obtained in the previous section (see (1.23) of section 7.1). However, theorem 7.1 cannot be used here since $P_0(A_0)$ is identically zero for all A_0 and, hence, dP_0/dA_0 is also identically zero. Of course, this is consistent with our analysis of Duffing's equation in section 6.4, where it was shown that there is a family of periodic solutions characterized by a suitable amplitude parameter. Indeed, in terms of the amplitude A, periodic solutions exist for $\epsilon A^2 < 1$. Further, these periodic solutions are even functions of τ (i.e. $u(\tau) = u(-\tau)$ for all τ) and, hence, here (see (2.8))

$$P(\kappa, A; \epsilon) = -\int_0^{2\pi} u^3 \sin s \; ds = 0 \quad \text{(for all } A). \tag{2.30}$$

Thus the role of the perturbation expansion in this case is only to determine κ and, hence, the period, as a function of the amplitude A, and the parameter ϵ, from the equation $Q(\kappa, A; \epsilon) = 0$. Similar comments apply for any equation of the form

$$u'' + u = \epsilon f(u; \epsilon), \tag{2.31}$$

where the right-hand side is independent of u'. This equation corresponds to a conservative system, since, in the terminology of section 6.4, the orbits are given by

$$\tfrac{1}{2}y^2 + U(x) = E,$$

where

$$U(x) = \tfrac{1}{2}x^2 - \epsilon \int_0^x f(x'; \epsilon) \, dx'. \tag{2.32}$$

Here $x = u$, $y = u'$ and the energy E is a constant. Clearly, for sufficiently small ϵ, equation (2.29) defines a set of closed orbits, corresponding to a family of periodic solutions. These are readily seen to be even functions of t, or τ, and hence, again, $P(\kappa, A; \epsilon)$ (2.8) is identically zero.

7.3 Stability

To study the stability of the periodic solutions of (1.1), determined by the perturbation methods of the previous two sections, we shall use the principle of linearized stability discussed in section 4.3. Thus, if $u(t)$ is a periodic solution of (1.1), the first step is to obtain the first variational equation (see (3.4) of section 4.3) corresponding to this solution. This is achieved by replacing u in (1.1) with $u(t) + v$ and linearizing with respect to v. The result is

$$v'' + v = \epsilon \left(\frac{\partial f}{\partial u}(u, u'; \epsilon) v + \frac{\partial f}{\partial u'}(u, u'; \epsilon) v' \right). \tag{3.1}$$

Here the arguments of $\partial f / \partial u$ and $\partial f / \partial u'$ are $u(t)$ and $u'(t)$ and, hence, equation (3.1) is a linear equation for v with periodic coefficients of period T. The characteristic exponents of this equation now determine the stability of $u(t)$. To find these characteristic exponents, we recall the general procedure for linear equations with periodic coefficients (see section 3.1 and section 3.2). Thus, let v^1 and v^2 be the two linearly independent solutions of (3.1) which satisfy the initial conditions:

$$v^1(0) = 1, \quad v^2(0) = 0, \quad v^{1\prime}(0) = 0, \quad v^{2\prime}(0) = 1. \tag{3.2}$$

Then form the fundamental matrix

$$X = \begin{bmatrix} v^1 & v^2 \\ v^{1\prime} & v^{2\prime} \end{bmatrix},$$

where we note that $X(0) = E$. Now the characteristic multipliers ρ are the eigenvalues of the matrix B, where

$$B = X(T). \tag{3.4}$$

The characteristic exponents μ are then given by the relation $\rho = e^{\mu T}$.

Now we recall from section 4.4 that, since (1.1) is an autonomous equation, $v = u'(t)$ is a solution of (3.1). Indeed, this is readily verified by differentiation of (1.1) with respect to t. If we recall that $u(t)$ satisfies the initial conditions

$$u = A, \qquad u' = 0 \qquad (\text{at } t = 0), \tag{3.5}$$

then we see that v^2 is proportional to u', so that

$$v^2 = Ku', \tag{3.6}$$

where K is a constant whose value will not be needed in the sequel, although it may be noted that $K = -A^{-1} + O(\epsilon)$. Next we observe that the periodic solution $u(t)$ of (1.1) also depends on the parameters ϵ and A, so that we may write $u = u(t; \epsilon, A)$. In order for this solution to be periodic of period $T = 2\pi/\omega$, we then showed that the parameters ω and A must depend on ϵ, so that $\omega = \omega(\epsilon)$ and $A = A(\epsilon)$. Here we have chosen to work in terms of the variable t, rather than $\tau = \omega t$ (1.12), so that the parameter ω does not appear initially in the construction of the solution $u(t; A, \epsilon)$, as it is not contained in either the equation (1.1) or the initial condition (3.5). Indeed, in the course of proving theorem 7.1, we showed that $u(t; A, \epsilon)$ was first constructed as the unique solution of (1.1) subject to the initial condition (3.5); insisting that the solution be periodic of period T then determined $\omega = \omega(\epsilon)$ and $A = A(\epsilon)$ *a posteriori*. Thus $u(t; A, \epsilon)$ satisfies (1.1), and (3.5), at a stage when A is still independent of ϵ. Then, since equation (1.1) does not contain any explicit dependence on the parameter A, differentiation of (1.1) with respect to A shows that $\partial u/\partial A$ is a solution of (3.1). This result is, of course, just a special case of the result (3.5) of section 4.3. Further, differentiation of (3.5) with respect to A shows that

$$v^1 = \frac{\partial u}{\partial A}. \tag{3.7}$$

Here it is understood that after differentiation with respect to A, we let $\omega = \omega(\epsilon)$ and $A = A(\epsilon)$.

Thus, using (3.6) and (3.7), we see that the matrix B, defined by (3.3) and (3.4), has been completely determined in terms of the periodic solution $u(t)$. Also we note that v^2 (see (3.6)) is a periodic function of t with period T, and from (3.2) it then follows that $v^2(T) = 0$ and $v^{2\prime}(T) = 1$. Thus B is given by

$$B = \begin{bmatrix} \dfrac{\partial u}{\partial A}(T) & 0 \\[2ex] \dfrac{\partial u'}{\partial A}(T) & 1 \end{bmatrix}. \tag{3.8}$$

Hence the characteristic multipliers ρ, being the eigenvalues of B, are given by

$$\rho = 1, \qquad \frac{\partial u}{\partial A}(T). \tag{3.9}$$

Here we note that the first characteristic multiplier is unity, and the corresponding characteristic exponent is zero. Of course, this result is true for the periodic solutions of any autonomous equation and was established in section 4.4, where it was shown to be consequence of the fact that u' is a solution of the first variational equation (3.1). Theorem 4.8 of section 4.4 then shows that the remaining characteristic multiplier, $\partial u/\partial A(T)$, determines the stability. Next, on differentiating (2.7) with respect to A, it follows that

$$\frac{\partial u}{\partial A}(T) = 1+\epsilon\frac{\partial P}{\partial A}, \qquad \frac{\partial u'}{\partial A}(T) = \epsilon\frac{\partial Q}{\partial A}, \tag{3.10}$$

where we have also used the initial conditions (3.10). Hence the characteristic multiplier determining stability is given by

$$\rho = 1+\epsilon\frac{\partial P}{\partial A}, \tag{3.11}$$

and we deduce that the periodic solution is asymptotically orbitally stable if

$$\epsilon\frac{\partial P}{\partial A} < 0$$

(i.e. $\rho < 1$ for sufficiently small $|\epsilon|$), but is unstable if

$$\epsilon\frac{\partial P}{\partial A} > 0$$

(i.e. $\rho > 1$ for sufficiently small $|\epsilon|$). This result is in fact stronger than that stated in theorem 7.1, which it implies on taking the limit $\epsilon \to 0$.

Problems

1 For the Van der Pol equation

$$u''+\epsilon(u^2-1)u'+u = 0,$$

use the Poincaré–Lindstedt method to construct the limit cycle and its period as $\epsilon \to 0$. Show that the amplitude $A = 2+\alpha_1\epsilon+O(\epsilon^2)$ and the period is $T = 2\pi\{1+T_2\epsilon^2+O(\epsilon^4)\}$, and find the coefficients α_1 and T_2. Here the amplitude is defined so that $u(0) = A$ and $u'(0) = 0$.

2 For each of the following equations, use the Poincaré–Lindstedt method to find all the limit cycles as $\epsilon \to 0$, and determine the stability. In each case show that the origin is a focus, and sketch the orbits in the x-y phase plane, where $x = u$ and $y = u'$.

    (a) $u''+\epsilon(5u^4-1)u'+u = 0$;   (b) $u''+\epsilon(u^2-1)u'+u-\epsilon u^3 = 0$;

    (c) $u''+\epsilon(u^4-10u^2+8)u'+u = 0$;   (d) $u''+\epsilon\{\cos(u^2+u'^2)\}u'+u = 0$,

For equation (a), compare the result with the analysis of problem 9 (b) of chapter 6. For equation (d), note that the limit cycle solutions are given exactly by $u = A\cos t$, where $\cos A^2 = 0$.

3 For each of the equations in problem 2, use numerical integration to construct orbits in the *x-y* phase plane, where $x = u$ and $y = u'$. Compare the results with the perturbation analysis in problem 2.

4 (a) For Duffing's equation

$$u'' + u - \epsilon u^3 = 0,$$

use the Poincaré–Lindstedt method to find the periodic solutions as $\epsilon \to 0$. Show that the period is

$$T = 2\pi\{1 + T_1 \epsilon + T_2 \epsilon^2 + O(\epsilon^3)\},$$

and determine the coefficients T_1 and T_2.

  (b) Show that in the *x-y* phase plane, where $x = u$ and $y = u'$, the periodic solutions of Duffing's equation are given by

$$y^2 = (A^2 - x^2)\{1 - \tfrac{1}{2}\epsilon(x^2 + A^2)\},$$

provided that $\epsilon A^2 < 1$, where A is the amplitude (see the discussion in section 6.4). Also show that the period is given by

$$T = 4 \int_0^A \frac{1}{y} \, dx.$$

Use these expressions to confirm the analysis of part (a).

5 For each of the following equations, use the Poincaré–Lindstedt method to find the periodic solutions as $\epsilon \to 0$. In each case also sketch the orbits in the *x-y* phase plane, where $x = u$ and $y = u'$, without restriction on ϵ.

(a) $u'' + u - \epsilon u^2 = 0,$ (b) $u'' + u - \epsilon(u^3 - u^5) = 0.$

For equation (b), compare the result with the analysis of problem 10 (b) (vi) of chapter 6.

6 Consider the system

$$x' = y + \epsilon F(x, y; \epsilon), \qquad y' = -x + \epsilon G(x, y; \epsilon),$$

which is an alternative formulation to (1.1) of a perturbation of a centre. Use the polar co-ordinates (r, θ), where $x = r\cos\theta$ and $y = r\sin\theta$, to show that this system is equivalent to

$$r' = \epsilon M(r, \theta; \epsilon), \qquad r\theta' = -r + \epsilon N(r, \theta; \epsilon),$$

where

$$M(r, \theta; \epsilon) = \cos\theta \, F(r\cos\theta, r\sin\theta; \epsilon) + \sin\theta \, G(r\cos\theta, r\sin\theta; \epsilon),$$

$$N(r, \theta; \epsilon) = -\sin\theta \, F(r\cos\theta, r\sin\theta; \epsilon) + \cos\theta \, G(r\cos\theta, r\sin\theta; \epsilon).$$

To find periodic solutions by the Poincaré–Lindstedt method, adapt the discussion of section 7.2. Thus, if the period is $T = 2\pi/\omega$, let $\omega = 1 + \epsilon\kappa$ and $\tau = \omega t$, so that

$$\frac{dr}{d\tau} = \frac{\epsilon}{\omega} M(r, \theta; \epsilon), \qquad r\frac{d\theta}{d\tau} = -\frac{r}{\omega} + \frac{\epsilon}{\omega} N(r, \theta; \epsilon).$$

Imposing the initial conditions $r(0) = A$ and $\theta(0) = 0$, these equations define $r = r(\tau; \epsilon, \kappa, A)$ and $\theta = \theta(\tau; \epsilon, \kappa, A)$. Show that a necessary and sufficient condition for the existence of a periodic solution is that

$$P(\kappa, A; \epsilon) = 0, \qquad Q(\kappa, A; \epsilon) = 0,$$

where

$$P(\kappa, A; \epsilon) = \int_0^{2\pi} M(r, \theta; \epsilon) \, d\tau, \qquad Q(\kappa, A; \epsilon) = 2\pi\kappa + \int_0^{2\pi} \frac{N(r, \theta; \epsilon)}{r} \, d\tau.$$

Further, show that as $\epsilon \to 0$, $'A \to A_0$, $\kappa \to \kappa_0$ and $r \to A_0$, $\theta \to -\tau$. Hence deduce that periodic solutions are given by

$$P_0(A_0) = \int_0^{2\pi} M(A_0, -\tau; 0) \, d\tau = 0$$

and

$$Q_0(\kappa_0, A_0) = 2\pi\kappa_0 + \frac{1}{A_0} \int_0^{2\pi} N(A_0, -\tau; \epsilon) \, d\tau = 0.$$

Finally, show that the periodic solutions are asymptotically orbitally stable if

$$\epsilon \frac{dP_0}{dA_0} < 0,$$

and unstable if

$$\epsilon \frac{dP_0}{dA_0} > 0.$$

7 For each of the following systems, use the method described in problem 6 to find all the limit cycles as $\epsilon \to 0$, and determine the stability. In each case show that the origin is a focus, and sketch the orbits in the x-y phase plane.

$$\text{(a)} \quad x' = y + \epsilon(x - \tfrac{1}{3}x^3), \qquad y' = -x;$$

$$\text{(b)} \quad x' = y + \epsilon(x - \tfrac{1}{3}x^3), \qquad y' = -x - \epsilon y^3;$$

$$\text{(c)} \quad x' = y + \epsilon(x - \tfrac{1}{3}x^3), \qquad y' = -x + \epsilon y.$$

Note that the system (a) is the Van der Pol equation (see (3.3) of section 6.3), and, in each case, it is possible to eliminate y and obtain a single second order equation for x, to which the Poincaré–Lindstedt method of section 7.2 may be applied.

8 Use the expression (2.12) for $P_0(A_0)$ to show that μ_0, as given by (2.18), agrees with the expression (2.17) for μ_0.

CHAPTER EIGHT
PERTURBATION METHODS
FOR FORCED OSCILLATIONS

8.1 Non-resonant case

In this chapter we will describe perturbation techniques for obtaining periodic solutions to equations of the form

$$u'' + \sigma^2 u = F_0 \cos t + \epsilon f(u, u', t; \epsilon), \tag{1.1}$$

where $f(u, u', t; \epsilon)$ is a prescribed function of u, u', t and the parameter ϵ, is defined for all values of these variables, and is a periodic function of the variable t with period 2π. When $F_0 = 0$ and $f(u, u', t; \epsilon)$ contains no explicit dependence on t, this equation is identical to that considered in the previous chapter (see (1.1) of section 7.1, and observe that the constant σ in (1.1) can then be removed by replacing t with σt). Thus, here our aim is to determine the effect of a periodic forcing term, $F_0 \cos t$, on a system which, in the absence of any forcing, describes a nonlinear oscillator. Other periodic forcing functions could also be considered by the same techniques to be employed here, but we shall not give any details.

When $\epsilon = 0$ and $F_0 = 0$, the equation (1.1) reduces to that for a simple harmonic oscillator of frequency σ and period $2\pi/\sigma$, while the frequency of the forcing term, $F_0 \cos t$, is unity and the period is 2π. One of the main features of the analysis to be presented is the interplay between these two frequencies, the natural frequency σ and the forcing frequency, 1. Indeed, when $\epsilon = 0$, the general solution of (1.1) is readily obtained and, denoting u by u_0 when $\epsilon = 0$, we find that

$$u_0 = C_0 \cos \sigma(t - t_0) + \frac{F_0}{\sigma^2 - 1} \cos t \quad \text{(for } \sigma^2 \neq 1), \tag{1.2}$$

where C_0 and t_0 are arbitrary constants. This describes a natural oscillation of amplitude C_0, and a forced oscillation of amplitude $R_0 = F_0/|\sigma^2 - 1|$. If σ is a rational number, say $\sigma = p/q$, where p and q are relatively prime integers, then the solution (1.2) is a periodic function of period $2\pi q$ for any choice of C_0 and t_0. Further, for the solution to have the same period as the forcing term, namely 2π, σ must be an integer. On the other hand, if σ is an irrational number, then the solution (1.2) can be a periodic function only if $C_0 = 0$, in which case the period is automatically 2π, the period of the forcing function. Next,

190

$$u_0 = C_0 \cos(t - t_0) + \tfrac{1}{2} F_0 t \sin t \quad \text{(for } \sigma^2 = 1\text{)}. \tag{1.3}$$

Now, the forced response (i.e. the term proportional to F_0) is secular, and grows without bound as t increases. This is called the 'resonant' case, although we shall see later that the effect of the nonlinear, perturbing term $\epsilon f(u, u', t; \epsilon)$ in (1.1) is to cause the case $\sigma = n$ for any integer $n = 0, 1, 2, \ldots$, to be also classified as a resonant case.

In this, and the next section, we shall discuss the non-resonant case. First we shall describe the perturbation-expansion procedure, and then, in the next section, we will give the justification. To illustrate the method we shall use the following equation:

$$u'' + \sigma^2 u = F_0 \cos t + \epsilon u^3, \tag{1.4}$$

which is Duffing's equation subjected to a periodic forcing. Before proceeding, we note that this equation arises in the description of the forced motion of a pendulum. Thus, if θ is the angle between the pendulum and the vertical, where the pendulum is constrained to move in a vertical plane, then the equation of motion is

$$\theta'' + \sigma^2 \sin \theta = \theta_0 \cos t, \tag{1.5}$$

where $\sigma^2 = g/l$ and l is the length of the pendulum. Now put $\theta = \delta u$ and $\theta_0 = \delta F_0$, so that (1.5) becomes

$$u'' + \sigma^2 u = F_0 \cos t + \epsilon f(u),$$

where

$$\epsilon f(u) = \sigma^2 \left(u - \frac{\sin \delta u}{\delta} \right). \tag{1.6}$$

Choosing $\epsilon = \tfrac{1}{6} \sigma^2 \delta^2$, we see that $f(u) = u^3 + O(\epsilon)$ and, hence, (1.6) reduces to (1.4) as $\epsilon \to 0$.

Assuming that $\sigma \neq n$ for any integer $n = 0, 1, 2, \ldots$, we now seek to solve (1.4) by a power-series expansion in powers of ϵ. Thus we let

$$u = u_0 + \epsilon u_1 + \epsilon^2 u_2 + \ldots . \tag{1.7}$$

The series is substituted into (1.4), and we equate to zero each coefficient of ϵ. Thus we obtain

$$u_0'' + \sigma^2 u_0 = F_0 \cos t, \quad u_1'' + \sigma^2 u_1 = u_0^3, \quad u_2'' + \sigma^2 u_2 = 3u_0^2 u_1, \quad \ldots . \tag{1.8}$$

The solution for u_0 is given by (1.2). For simplicity we concentrate our attention on the forced oscillation, and put $C_0 = 0$. Thus u_0 is periodic with period 2π. The equation for u_1 is now, using the trigonometric identity (1.8) of section 7.1,

$$u_1'' + \sigma^2 u_1 = \frac{F_0^3}{(\sigma^2-1)^3}(\tfrac{1}{4}\cos 3t + \tfrac{3}{4}\cos t). \tag{1.9}$$

The general solution is

$$u_1 = C_1 \cos\sigma(t-t_1) + \frac{F_0^3}{(\sigma^2-1)^3}\left(\frac{\cos 3t}{4(\sigma^2-9)} + \frac{3\cos t}{4(\sigma^2-1)}\right). \tag{1.10}$$

Again, to ensure that u_1 is periodic with period 2π, we must put $C_1 = 0$. The procedure can clearly be continued indefinitely. There are several points to notice. First, the reason for the restriction $\sigma \neq n$, for any integer n, is now apparent. At the lowest order, $\sigma \neq 1$ in order to avoid a resonance between the forcing term with frequency 1 and the natural oscillation of frequency σ. At the next order, $\sigma \neq 3$ as well, in order now to avoid a resonance between a forcing term with frequency 3 (the third harmonic of the forcing frequency) and the natural oscillation. Clearly, the restriction $\sigma \neq n$ is to avoid a resonance between the natural oscillation and any harmonic of the forcing term. Here only the odd harmonics appear ($n = 1, 3, 5, \dots$), but in the general case (1.1) we would expect all values of n to occur. Next, the response grows as σ approaches any integer. Indeed, since we expect the expansion to be valid for $|\epsilon|$ sufficiently small, we see from (1.10) that, if σ is close to 1, then $|\epsilon| \ll (\sigma^2-1)^3/F_0^2$ in order that $|\epsilon u_1| \ll |u_0|$. An analogous condition must hold if σ is close to 3, etc.. Clearly, although the expansion converges for $\sigma \neq n$ and $|\epsilon|$ sufficiently small (see theorem 8.1 below), its utility is greatly reduced when σ is close to any resonant value (i.e. the integers $n = 0, 1, 2, \dots$).

8.2 Non-resonant case, continued

We now return to consider the general case (1.1), and, for simplicity, we shall suppose that $f(u, u', t; \epsilon)$ is an analytic function of its arguments u, u', t and ϵ. Associated with (1.1) we impose the initial conditions

$$u = A + \frac{F_0}{\sigma^2-1}, \qquad u' = B, \tag{2.1}$$

where A and B are constants yet to be determined. This particular choice of initial condition is motivated by the form of the general solution (1.2) when $\epsilon = 0$. It now follows from theorem 1.6 of chapter 1 that the initial-value problem (2.1) for (1.1) has a unique solution

$$u = u(t; A, B, \epsilon), \tag{2.2}$$

which is an analytic function of t and the parameters ϵ, A and B. By theorem 1.7 of chapter 1, the solution exists for all $t \geqslant 0$, provided that $|u|$ and $|u'|$ are bounded. Our aim now is to establish that, for sufficiently small values of $|\epsilon|$, there exist values of A and B such that (1.1) has a periodic solution of period 2π. The method of proof is analogous to that of section 7.2 for the existence of periodic solutions in the absence of any forcing.

Since the function (2.2) satisfies (1.1) and (2.1), it follows that (see (4.14) of section 2.4)

$$u = A\cos\sigma t + \frac{B}{\sigma}\sin\sigma t + \frac{F_0\cos t}{\sigma^2-1} + \frac{\epsilon}{\sigma}\int_0^t f(u,u',s;\epsilon)\sin\sigma(t-s)\,ds,$$

$$u' = -\sigma A\sin\sigma t + B\cos\sigma t - \frac{F_0\sin t}{\sigma^2-1} + \epsilon\int_0^t f(u,u',s;\epsilon)\cos\sigma(t-s)\,ds. \quad (2.3)$$

Here, in the integral terms on the right-hand side, u is a function of s (and also of the parameters ϵ, A and B). Equations (2.3) can be regarded as a pair of integro-differential equations equivalent to (1.1) and (2.1).

Next we define the functions $P(A,B;\epsilon)$ and $Q(A,B;\epsilon)$ by

$$P(A,B;\epsilon) = u(2\pi;A,B;\epsilon) - u(0;A,B,\epsilon),$$

$$Q(A,B;\epsilon) = u'(2\pi;A,B;\epsilon) - u'(0;A,B,\epsilon). \quad (2.4)$$

Using the result (2.3), it follows that

$$P(A,B;\epsilon) = A(\cos 2\pi\sigma - 1) + \frac{B}{\sigma}\sin 2\pi\sigma + \frac{\epsilon}{\sigma}\int_0^{2\pi} f(u,u',s;\epsilon)\sin\sigma(2\pi-s)\,ds,$$

$$Q(A,B;\epsilon) = -\sigma A\sin 2\pi\sigma + B(\cos 2\pi\sigma - 1) + \epsilon\int_0^{2\pi} f(u,u',s;\epsilon)\cos\sigma(2\pi-s)\,ds.$$

$$(2.5)$$

Then a necessary and sufficient condition for $u(t;\epsilon,A,B)$ to be a periodic function of period 2π with respect to t is (see theorem 2.4 of section 2.4)

$$P(A,B;\epsilon) = 0, \qquad Q(A,B;\epsilon) = 0. \quad (2.6)$$

These equations are regarded as a pair of equations for A and B as functions of ϵ. We are now in a position to establish the following theorem.

Theorem 8.1: For sufficiently small $|\epsilon|$ $(0 \leqslant |\epsilon| < \epsilon_0$, say), there exist functions $A(\epsilon)$ and $B(\epsilon)$, satisfying (2.6), which are analytic functions of ϵ such that $A(0) = 0$ and $B(0) = 0$, provided that $\sigma \neq n$ for $n = 0, 1, 2, \ldots$. Further, the periodic solution of (2.1) defined by $A(\epsilon)$ and $B(\epsilon)$ is uniformly and asymptotically stable if

$$\epsilon\int_0^{2\pi} \frac{\partial f}{\partial u'}(u_0,u_0',s;\epsilon)\,ds < 0, \quad (2.7)$$

and is unstable if this inequality is reversed, where u_0 is given by (1.2) with $C_0 = 0$, that is,

$$u_0 = \frac{F_0}{\sigma^2-1}\cos t. \quad (2.8)$$

Proof: To determine when (2.6) have a solution, we consider the limit $\epsilon \to 0$, and then observe that a solution is obtained by letting $A, B \to 0$ simultaneously. Since P and Q are analytic functions of A and B, the implicit function theorem for functions of two variables now guarantees a solution of (2.6) for $A(\epsilon)$ and $B(\epsilon)$ when $0 \leqslant |\epsilon| < \epsilon_0$, say, which are analytic functions of ϵ and are such that $A(0) = 0$ and $B(0) = 0$, provided that the Jacobian of the functions A and B is not zero when $\epsilon = 0$, that is,

$$\frac{\partial(P, Q)}{\partial(A, B)} = \det \begin{bmatrix} \dfrac{\partial P}{\partial A} & \dfrac{\partial P}{\partial B} \\[2mm] \dfrac{\partial Q}{\partial A} & \dfrac{\partial Q}{\partial B} \end{bmatrix} \neq 0 \quad (\text{as } \epsilon \to 0). \tag{2.9}$$

From (2.5) this condition reduces to

$$\det \begin{bmatrix} \cos 2\pi\sigma - 1 & \dfrac{1}{\sigma} \sin 2\pi\sigma \\[2mm] -\sigma \sin 2\pi\sigma & \cos 2\pi\sigma - 1 \end{bmatrix} \neq 0,$$

or

$$2(1 - \cos 2\pi\sigma) \neq 0,$$

or

$$4 \sin^2 \pi\sigma \neq 0. \tag{2.10}$$

Since $\sigma \neq n$ for any integer $n = 0, 1, 2, \ldots$, this condition is satisfied and the existence is proven.

The relationship of this theorem to the expansion procedure described in section 8.1 is readily obtained by observing that the theorem establishes the existence of a periodic solution $u(t; \epsilon)$ of (1.1) which is an analytic function of ϵ and which reduces to (1.2) (with $C_0 = 0$) when $\epsilon \to 0$.

To study the stability of this solution we shall use the principle of linearized stability (see section 4.3). Thus, if $u(t)$ is the periodic solution of (1.1), the first step is to obtain the first variational equation (see (3.4) of section 4.3) corresponding to this solution. This is achieved by replacing u in (1.1) with $u(t) + v$ and linearizing with respect to v. The result is

$$v'' + \sigma^2 v = \epsilon \left(\frac{\partial f}{\partial u}(u, u', t; \epsilon) v + \frac{\partial f}{\partial u'}(u, u', t; \epsilon) v' \right). \tag{2.11}$$

Here the arguments of $\partial f/\partial u$ and $\partial f/\partial u'$ are $u(t)$ and $u'(t)$ and, hence, equation (2.11) is a linear equation for v with periodic coefficients of period 2π. The characteristic exponents of this equation now determine the stability of $u(t)$. To find these exponents, we let v^1 and v^2 be two linearly independent solutions of (3.1), and form the fundamental matrix

$$X = \begin{bmatrix} v^1 & v^2 \\ v^{1\prime} & v^{2\prime} \end{bmatrix}. \tag{2.12}$$

The initial conditions are chosen so that $X(0) = E$, and then the characteristic multipliers ρ are the eigenvalues of $X(2\pi)$, while the characteristic exponents μ are given by $\rho = e^{2\pi\mu}$.

Next we note that $u(t; A, B, \epsilon)$ is first constructed as the unique solution of (1.1) subject to the initial condition (2.1); insisting that the solution be periodic of period 2π then determines $A(\epsilon)$ and $B(\epsilon)$ *a posteriori*. Thus $u(t; A, B, \epsilon)$ satisfies (1.1) and (2.1) at a stage when A and B are still independent of ϵ. Then, since equation (1.1) does not contain any explicit dependence on the parameters A and B, differentiation of (1.1) with respect to A and B shows that $\partial u/\partial A$ and $\partial u/\partial B$ are solutions of (2.11). This result is, of course, just a special case of the result (3.5) of section 4.3. Further, differentiation of (2.1) with respect to A and B shows that

$$v^1 = \frac{\partial u}{\partial A}, \qquad v^2 = \frac{\partial u}{\partial B}. \tag{2.13}$$

Here it is understood that, after differentiation with respect to A and B, we let $A = A(\epsilon)$ and $B = B(\epsilon)$. Thus the matrix $X(2\pi)$ is given by

$$X(2\pi) = \begin{bmatrix} \dfrac{\partial u}{\partial A}(2\pi) & \dfrac{\partial u}{\partial B}(2\pi) \\ \dfrac{\partial u'}{\partial A}(2\pi) & \dfrac{\partial u'}{\partial B}(2\pi) \end{bmatrix}. \tag{2.14}$$

But, on differentiating (2.4) with respect to A and B, we see that

$$\frac{\partial u}{\partial A}(2\pi) = 1 + \frac{\partial P}{\partial A}, \qquad \frac{\partial u}{\partial B}(2\pi) = \frac{\partial P}{\partial B},$$

$$\frac{\partial u'}{\partial A}(2\pi) = \frac{\partial Q}{\partial A}, \qquad \frac{\partial u'}{\partial B}(2\pi) = 1 + \frac{\partial Q}{\partial B}, \tag{2.15}$$

where we have also used the initial condition (2.1). Hence the characteristic multipliers ρ, being the eigenvalues of $X(2\pi)$, are given by

$$(\rho - 1)^2 - (\rho - 1)\left(\frac{\partial P}{\partial A} + \frac{\partial Q}{\partial B}\right) + \frac{\partial(P, Q)}{\partial(A, B)} = 0. \tag{2.16}$$

Then, by theorem 4.6 of section 4.3, there is uniform and asymptotic stability if both solutions of the quadratic equation (2.16) are such that $|\rho| < 1$, so that $\operatorname{Re}\mu < 0$. Otherwise, there is instability if either of the solutions is such that $|\rho| > 1$.

To analyze the solutions of (2.16), we let

$$2\alpha = \frac{\partial P}{\partial A} + \frac{\partial Q}{\partial B}, \qquad \beta = \frac{\partial(P, Q)}{\partial(A, B)}. \tag{2.17}$$

The solutions of (2.16) are then given by

$$\rho = 1 + \alpha \pm \sqrt{\alpha^2 - \beta}. \tag{2.18}$$

It can now be shown that $|\rho| < 1$ if

$$\beta > 0 \quad \text{and} \quad 2\alpha + \beta < 0 < 4 + 4\alpha + \beta, \tag{2.19}$$

and $|\rho| > 1$ if any of these inequalities are reversed. Hence (2.19) provides the conditions determining the stability. The situation is sketched in figure 8.1.

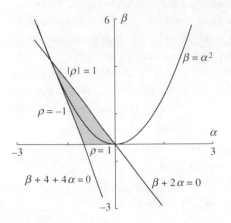

Fig. 8.1 The stability diagram corresponding to (2.19) for non-resonant forced oscillations. The shaded area is the stable zone.

Note that the stability boundaries are $2\alpha + \beta = 0$ for $-2 < \alpha < 0$, along which $|\rho| = 1$ and the two values for ρ are complex conjugates, $\beta = 0$ for $-1 < \alpha < 0$, along which $\rho = 1$ and $1 + 2\alpha$, and $4 + 4\alpha + \beta = 0$ for $-1 < \alpha < -2$, along which $\rho = -1$ and $3 + 2\alpha$.

To proceed further, we next consider the limit $\epsilon \to 0$. From (2.5) we see that

$$\frac{\partial P}{\partial A} = \cos 2\pi\sigma - 1 + \frac{\epsilon}{\sigma} \int_0^{2\pi} \left(\frac{\partial f}{\partial u} \frac{\partial u}{\partial A} + \frac{\partial f}{\partial u'} \frac{\partial u'}{\partial A} \right) \sin \sigma(2\pi - s) \, ds,$$

$$\frac{\partial P}{\partial B} = \frac{\sin 2\pi\sigma}{\sigma} + \frac{\epsilon}{\sigma} \int_0^{2\pi} \left(\frac{\partial f}{\partial u} \frac{\partial u}{\partial B} + \frac{\partial f}{\partial u'} \frac{\partial u'}{\partial B} \right) \sin \sigma(2\pi - s) \, ds,$$

$$\frac{\partial Q}{\partial A} = -\sigma \sin 2\pi\sigma + \epsilon \int_0^{2\pi} \left(\frac{\partial f}{\partial u} \frac{\partial u}{\partial A} + \frac{\partial f}{\partial u'} \frac{\partial u'}{\partial A} \right) \cos \sigma(2\pi - s) \, ds,$$

$$\frac{\partial Q}{\partial B} = \cos 2\pi\sigma - 1 + \epsilon \int_0^{2\pi} \left(\frac{\partial f}{\partial u} \frac{\partial u}{\partial B} + \frac{\partial f}{\partial u'} \frac{\partial u'}{\partial B} \right) \cos \sigma(2\pi - s) \, ds. \tag{2.20}$$

Here, from (2.3),

$$\frac{\partial u}{\partial A} = \cos \sigma s + O(\epsilon), \qquad \frac{\partial u}{\partial B} = \frac{\sin \sigma s}{\sigma} + O(\epsilon),$$

$$\frac{\partial u'}{\partial A} = -\sigma \sin \sigma s + O(\epsilon), \qquad \frac{\partial u'}{\partial B} = \cos \sigma s + O(\epsilon). \tag{2.21}$$

Substituting the relations (2.21) into (2.20), and recalling the definitions (2.17), we now find that

$$2\alpha = 2(\cos 2\pi\sigma - 1) + \epsilon \int_0^{2\pi} \left(\frac{\partial f}{\partial u} \frac{\sin 2\pi\sigma}{\sigma} + \frac{\partial f}{\partial u'} \cos 2\pi\sigma \right) ds + O(\epsilon^2),$$

$$\beta = 2(1 - \cos 2\pi\sigma) + \epsilon \int_0^{2\pi} \left(-\frac{\partial f}{\partial u} \frac{\sin 2\pi\sigma}{\sigma} + \frac{\partial f}{\partial u'}(1 - \cos 2\pi\sigma) \right) ds + O(\epsilon^2),$$

$$(2.22)$$

where here the arguments of $\partial f/\partial u$ and $\partial f/\partial u'$ are u_0 and u_0', and we recall that u_0 is defined by (2.8). It follows that

$$2\alpha + \beta = \epsilon \int_0^{2\pi} \frac{\partial f}{\partial u'}(u_0, u_0', s; \epsilon)\, ds + O(\epsilon^2). \tag{2.23}$$

Thus, in the limit $\epsilon \to 0$, $2\alpha \to -4\sin^2 \pi\sigma$ and $\beta \to 4\sin^2 \pi\sigma$, and, since $\sigma \neq n$ for any integer n, we see that the characteristic multipliers are located on the stability boundary $2\alpha + \beta = 0$ for $-2 \leqslant \alpha < 0$. In this limit the characteristic exponents are $\mu = \pm i\sigma$. To resolve the stability, we must consider the $O(\epsilon)$ terms in (2.22) or, more directly, in (2.23). Initially, suppose that $\alpha \neq -2$. Then, from (2.19), uniform and asymptotic stability holds if $2\alpha + \beta < 0$ (since $\beta > 0$ and $4 + 4\alpha + \beta > 0$), and so (2.23) gives the stability criterion (2.7). It can be shown that, if the characteristic exponents are μ, then

$$\mathrm{Re}\,\mu = \frac{\epsilon}{4\pi} \int_0^{2\pi} \frac{\partial f}{\partial u'}(u_0, u_0', s; \epsilon)\, ds + O(\epsilon^2),$$

$$\mathrm{Im}\,\mu = \pm i\sigma + O(\epsilon). \tag{2.24}$$

Note that special care is needed when $\sigma = n + \frac{1}{2}$ for any integer n ($n = 0, 1, 2, \ldots$), since then $\alpha = -2$ and, in the limit $\epsilon \to 0$, the characteristic multipliers are located on the intersection of the stability boundaries $2\alpha + \beta = 0$ and $4 + 4\alpha + \beta = 0$. In this special case, $\beta = \alpha^2 + O(\epsilon^2)$ and $4 + 4\alpha + \beta = O(\epsilon^2)$, and so the criterion $2\alpha + \beta < 0$ cannot be immediately applied. Instead, (2.18) can be used directly to obtain (2.24), and the stability criterion (2.7) again follows.

In applying the stability criterion (2.7) to the forced Duffing's equation (1.4), we see that $\partial f/\partial u'$ is identically zero and, hence, to $O(\epsilon)$, the stability cannot be determined. In this case it is necessary to continue the calculation of α and β up to at least the terms of $O(\epsilon^2)$, and then use the stability criterion (2.19). Alternatively, we can add a small linear damping term, $-\nu u'$, to the right-hand side of (1.4), where ν (> 0) is $O(\epsilon)$. Thus (1.4) becomes

$$u'' + \sigma^2 u = F_0 \cos t + \epsilon u^3 - \nu u', \tag{2.25}$$

and, using the expansion (1.7), we find that u_0 is given by (2.8), and u_1 is now given by (in place of (1.10))

$$u_1 = \frac{F_0^3}{(\sigma^2 - 1)^3}\left(\frac{\cos 3t}{4(\sigma^2 - 9)} + \frac{3\cos t}{4(\sigma^2 - 1)} \right) + \frac{\hat{\nu} F_0}{(\sigma^2 - 1)^2} \sin t, \tag{2.26}$$

where $\nu = \epsilon \hat{\nu}$. The stability criterion (2.7) is now simply $2\pi\nu > 0$ and, hence, as expected, the solution is uniformly and asymptotically stable if $\nu > 0$.

In general, the addition of a suitable damping term, such as $-\nu u'$, is a useful device when (2.7) initially fails to give any useful information about the stability. Indeed, it is clear that the stability criterion (2.7) is, in essence, a measure of the amount of dissipation associated with the equation (1.1). To illustrate this, consider the forced Van der Pol equation

$$u'' + \sigma^2 u = F_0 \cos t + \epsilon(1 - u^2)u'. \tag{2.27}$$

This has the required form for the application of theorem 8.1 and, hence, for $\sigma \neq n$ for $n = 0, 1, 2, \ldots$, there is a periodic solution of period 2π which is given by the expansion (1.7), where the first term u_0 is just (2.8). The stability criterion (2.7) becomes

$$\epsilon \int_0^{2\pi} \left(1 - \frac{F_0^2}{(\sigma^2 - 1)^2} \cos^2 s\right) ds < 0,$$

or

$$\epsilon \pi \left(2 - \frac{F_0^2}{(\sigma^2 - 1)^2}\right) < 0. \tag{2.28}$$

Hence, for $\epsilon > 0$, this solution is uniformly and asymptotically stable for $2(\sigma^2 - 1)^2 < F_0^2$, and is unstable if this inequality is reversed. Thus, here the forced oscillation is stable when either the forcing frequency is sufficiently close to the natural frequency σ, or when the forcing amplitude $|F_0|$ is sufficiently large. A possible physical interpretation of this result is that, because the limit-cycle solution of the unforced Van der Pol equation is stable, the forced oscillations are corresponding stable if the forcing frequency is sufficiently close to the frequency of the limit cycle, or when the forcing amplitude is sufficiently large to overcome any discrepancy between the forcing frequency and the natural frequency. Alternatively, recalling that the response amplitude $R_0 = F_0/|\sigma^2 - 1|$, we note that the stability criterion (2.28) is equivalent to $R_0^2 > 2$.

8.3 Resonant case

We now consider equation (1.1) when the natural frequency σ is close to one of the resonant values, $\sigma = n$ for $n = 0, 1, 2, \ldots$. We shall consider only the case $n = 1$. The theory for the other cases is similar, and some aspects of the higher-order resonances will be taken up in the exercises. Thus, here we suppose that

$$\sigma^2 = 1 + \epsilon\kappa, \tag{3.1}$$

where κ is a new parameter which measures how closely the resonance is tuned. Next we recall the solution (1.2) (or (2.8)) for the forced oscillation when $\epsilon = 0$. Then the amplitude of the forced oscillation is $F_0(\sigma^2 - 1)^{-1}$ and, if we wish to keep the amplitude of the response bounded with respect to ϵ when (3.1) holds, then, clearly, we must ensure that F_0 is $O(\epsilon)$. Hence, we put

$$F_0 = \epsilon G_0. \tag{3.2}$$

With the conditions (3.1) and (3.2), equation (1.1) becomes

$$u'' + u = \epsilon g(u, u', t; \epsilon),$$

where

$$g(u, u', t; \epsilon) = G_0 \cos t - \kappa u + f(u, u', t; \epsilon). \tag{3.3}$$

Now, when $\epsilon \to 0$, equation (3.3) reduces to the equation for a simple harmonic oscillator, whose general solution is

$$u_0 = A_0 \cos t + B_0 \sin t, \tag{3.4}$$

where we again denote u by u_0 when $\epsilon = 0$. Here A_0 and B_0 are arbitrary constants and, for any value of these constants, the solution (3.4) for u_0 is a periodic function of t with period 2π. The main aim of the subsequent analysis is now to find out how the perturbation $\epsilon g(u, u', t; \epsilon)$ on the right-hand side of (3.3) determines A_0 and B_0.

As in the previous two sections, we shall suppose that $f(u, u', t; \epsilon)$ is an analytic function of its arguments u, u', t and ϵ. However, in contrast to (2.1), we now impose the initial conditions

$$u = A, \qquad u' = B \qquad \text{(at } t = 0), \tag{3.5}$$

where A and B are constants yet to be determined. Here we anticipate that, as $\epsilon \to 0$, $A, B \to A_0, B_0$, respectively. Next, it follows from theorem 1.6 of chapter 1 that the initial-value problem (3.5) for (3.3) has a unique solution

$$u = u(t; A, B, \epsilon), \tag{3.6}$$

which is an analytic function of t and the parameters ϵ, A and B. Also, by theorem 1.7 of chapter 1, the solution exists for all $t \geq 0$ such that $|u|$ and $|u'|$ are bounded. Our aim now is to establish that, for sufficiently small values of $|\epsilon|$, there exist values of A and B such that (1.1) has a periodic solution of period 2π. The method of proof is analogous to that of theorems 7.1 and 8.1. Thus, since (3.6) satisfies (3.3) and (3.5), it follows that

$$u = A \cos t + B \sin t + \epsilon \int_0^t g(u, u', s; \epsilon) \sin(t - s) \, \mathrm{d}s,$$

$$u' = -A \sin t + B \cos t + \epsilon \int_0^t g(u, u', s; \epsilon) \cos(t - s) \, \mathrm{d}s. \tag{3.7}$$

Here, in the integral terms on the right-hand side, u is a function of s (and also of the parameters A, B and ϵ).

Next we define the functions $P(A,B;\epsilon)$ and $Q(A,B;\epsilon)$ by

$$\epsilon P(A,B;\epsilon) = u(2\pi;A,B,\epsilon) - u(0;A,B,\epsilon),$$

$$\epsilon Q(A,B;\epsilon) = u'(2\pi;A,B,\epsilon) - u'(0;A,B,\epsilon). \tag{3.8}$$

Using the result (3.7), it follows that

$$P(A,B;\epsilon) = -\int_0^{2\pi} g(u,u',s;\epsilon)\sin s \, ds,$$

$$Q(A,B;\epsilon) = \int_0^{2\pi} g(u,u',s;\epsilon)\cos s \, ds. \tag{3.9}$$

Then a necessary and sufficient condition for $u(t;\epsilon,A,B)$ to be a periodic function of period 2π with respect to t is

$$P(A,B;\epsilon) = 0, \qquad Q(A,B;\epsilon) = 0. \tag{3.10}$$

These equations are regarded as a pair of equations for A and B as functions of ϵ. To determine when there may be a solution, we consider the limit $\epsilon \to 0$, where we note that P and Q are analytic functions of A, B and ϵ. Thus, as $\epsilon \to 0$, we let $A \to A_0$, $B \to B_0$, $u \to u_0$ (given by (3.4)) and then $P \to P_0$ and $Q \to Q_0$, where

$$P_0(A_0,B_0) = P(A_0,B_0;0) = -\int_0^{2\pi} g(u_0,u_0',s;0)\sin s \, ds,$$

$$Q_0(A_0,B_0) = Q(A_0,B_0;0) = \int_0^{2\pi} g(u_0,u_0',s;0)\cos s \, ds. \tag{3.11}$$

Note that, since u_0 is known explicitly from (3.4), the functions P_0 and Q_0 are known explicitly in terms of A_0 and B_0. In this same limit (3.10) reduces to

$$P_0(A_0,B_0) = 0, \qquad Q_0(A_0,B_0) = 0. \tag{3.12}$$

We are now in a position to establish the following theorem.

Theorem 8.2: Let A_0 and B_0 satisfy (3.12), where

$$\frac{\partial(P_0,Q_0)}{\partial(A_0,B_0)} = \det \begin{bmatrix} \dfrac{\partial P_0}{\partial A_0} & \dfrac{\partial P_0}{\partial B_0} \\[2ex] \dfrac{\partial Q_0}{\partial A_0} & \dfrac{\partial Q_0}{\partial B_0} \end{bmatrix} \neq 0. \tag{3.13}$$

Then, for sufficiently small ϵ ($0 \leq |\epsilon| < \epsilon_0$, say), there exist functions $A(\epsilon)$ and $B(\epsilon)$ satisfying (3.10), which are analytic functions of ϵ such that $A(0) = A_0$ and $B(0) = B_0$. Further, the periodic solution defined by $A(\epsilon)$ and $B(\epsilon)$ is uniformly and asymptotically stable if

$$\frac{\partial(P_0, Q_0)}{\partial(A_0, B_0)} > 0 \quad \text{and} \quad \epsilon\left(\frac{\partial P_0}{\partial A_0} + \frac{\partial Q_0}{\partial B_0}\right) < 0, \tag{3.14}$$

and is unstable if either of these inequalities is reversed.

Proof: We are assuming that a solution to (3.10) exists when $\epsilon = 0$ (i.e. A_0 and B_0 satisfy (3.12)). Since P and Q are analytic functions of ϵ, κ and A, the implicit-function theorem for functions of two variables now guarantees a solution of (3.10) for $A(\epsilon)$ and $B(\epsilon)$ when $0 \leqslant |\epsilon| < \epsilon_0$, say, which are analytic functions of ϵ and are such that $A(0) = A_0$ and $B(0) = B_0$, provided that the Jacobian of the functions P and Q is not zero when $\epsilon = 0$. But this is precisely the condition (3.13), and existence is proven.

The stability analysis is similar to that given in the proof of theorem 8.1 in section 8.2. Thus here, if $u(t)$ is the periodic solution of (3.3), we set $u = u(t) + v$ in (3.3), and then the first variational equation for v is

$$v'' + v = \epsilon\left(\frac{\partial g}{\partial u}(u, u', t; \epsilon)v + \frac{\partial g}{\partial u'}(u, u', t; \epsilon)v'\right). \tag{3.15}$$

Here the arguments of $\partial g/\partial u$ and $\partial g/\partial u'$ are $u(t)$ and $u'(t)$ and, hence, (3.15) is a linear equation for v with periodic coefficients of period 2π. The characteristic multipliers ρ, which determine the stability, are the eigenvalues of the matrix $X(2\pi)$, where X is the fundamental matrix for (3.15) and, as in the proof of theorem 8.1, is given by (2.12) with $X(0) = E$. Further, just as in the proof of theorem 8.1, the solutions v^1 and v^2 of (3.15) are given by (2.13), that is,

$$v^1 = \frac{\partial u}{\partial A}, \qquad v^2 = \frac{\partial u}{\partial B}.$$

Hence, once again, $X(2\pi)$ is given by (2.14). Now, however, in place of (2.15), we find from (3.8) that

$$\frac{\partial u}{\partial A}(2\pi) = 1 + \epsilon\frac{\partial P}{\partial A}, \qquad \frac{\partial u}{\partial B}(2\pi) = \epsilon\frac{\partial P}{\partial B},$$

$$\frac{\partial u'}{\partial A}(2\pi) = \epsilon\frac{\partial Q}{\partial A}, \qquad \frac{\partial u'}{\partial B}(2\pi) = 1 + \epsilon\frac{\partial Q}{\partial B}. \tag{3.16}$$

In effect, we have replaced P and Q in (2.15) by ϵP and ϵQ, in agreement with the difference between (2.4) and (3.8). Hence now, in place of (2.16), the characteristic multipliers ρ are given by

$$(\rho - 1)^2 - \epsilon(\rho - 1)\left(\frac{\partial P}{\partial A} + \frac{\partial Q}{\partial B}\right) + \epsilon^2\frac{\partial(P, Q)}{\partial(A, B)} = 0. \tag{3.17}$$

As in the proof of theorem 8.1 (see (2.19)), it can be shown that $|\rho| < 1$ if

$$\epsilon^2\frac{\partial(P, Q)}{\partial(A, B)} > 0,$$

$$\epsilon\left(\frac{\partial P}{\partial A} + \frac{\partial Q}{\partial B}\right) + \epsilon^2\frac{\partial(P, Q)}{\partial(A, B)} < 0$$

and

$$4+2\epsilon\left(\frac{\partial P}{\partial A}+\frac{\partial Q}{\partial B}\right)+\epsilon^2\frac{\partial(P,Q)}{\partial(A,B)} > 0. \tag{3.18}$$

Conversely, if any of these inequalities are reversed, then $|\rho| > 1$. If we now take the limit $\epsilon \to 0$, we obtain the stability criterion (3.14). Indeed, it follows from (3.17) that

$$\rho = 1+\epsilon\rho_1+O(\epsilon^2),$$

where

$$\rho_1 = \alpha_0 \pm \sqrt{\alpha_0^2-\beta_0}$$

and

$$2\alpha_0 = \left(\frac{\partial P_0}{\partial A_0}+\frac{\partial Q_0}{\partial B_0}\right), \qquad \beta_0 = \frac{\partial(P_0,Q_0)}{\partial(A_0,B_0)}. \tag{3.19}$$

For uniform and asymptotic stability we require that $\mathrm{Re}(\epsilon\rho_1) < 0$, and for instability $\mathrm{Re}(\epsilon\rho_1) > 0$. The stability criterion (3.14) then follows from (3.19). Note that the characteristic exponents μ are given by $\rho = e^{2\pi\mu}$ and, hence,

$$2\pi\mu = \epsilon\rho_1+O(\epsilon^2).$$

The stability boundaries are $\alpha_0 = 0$ for $\beta_0 > 0$, along which $\mathrm{Re}\,\rho_1 = 0$, and $\beta_0 = 0$ for $\epsilon\alpha_0 < 0$, along which $\rho_1 = 0$ and $2\alpha_0$. The situation is sketched in figure 8.2 (for $\epsilon > 0$). Note that the stability boundary $\beta_0 = 0$ also corresponds to the failure of the criterion (3.13), and we shall show below that it corresponds to the existence of multiple solutions of (3.12). Also, by an analysis similar to that of the previous section (see, e.g. (2.20) and the following discussion), it may be shown that

$$2\alpha_0 = \int_0^{2\pi} \frac{\partial f}{\partial u'}(u_0, u_0', s; \epsilon)\,\mathrm{d}s. \tag{3.20}$$

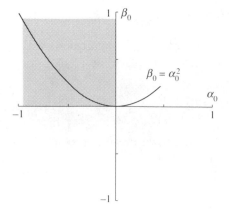

Fig. 8.2 The stability diagram corresponding to (3.19) for resonant forced oscillations. The shaded area is the stable zone.

Hence the second stability criterion in (3.14), $2\epsilon\alpha_0 < 0$, is analogous to the stability criterion (2.7) obtained for the non-resonant case. Hence it can be regarded as a measure of the amount of dissipation associated with the system.

Next, we substitute (3.3) into (3.11) to obtain

$$P_0(A_0, B_0) = \pi\kappa B_0 + \hat{P}_0(A_0, B_0),$$

$$Q_0(A_0, B_0) = \pi G_0 - \pi\kappa A_0 + \hat{Q}_0(A_0, B_0),$$

where

$$\hat{P}_0(A_0, B_0) = -\int_0^{2\pi} f(u_0, u_0', s; 0)\sin s \; ds,$$

and

$$\hat{Q}_0(A_0, B_0) = \int_0^{2\pi} f(u_0, u_0', s; 0)\cos s \; ds. \tag{3.21}$$

Here the dependence of P_0 and Q_0 on the external parameters κ and G_0 has been explicitly displayed, since here $\hat{P}_0$ and $\hat{Q}_0$ depend solely on A_0 and B_0. We now see that (3.12) can be regarded as a pair of equations for A_0 and B_0 in terms of κ and G_0. To analyze these equations it is customary to put

$$A_0 = R_0\cos\phi_0, \qquad B_0 = R_0\sin\phi_0, \tag{3.22}$$

and then to use (3.12) to determine R_0 and ϕ_0 in terms of κ and G_0. Note that, using (3.22), (3.4) becomes

$$u_0 = R_0\cos(t - \phi_0) \tag{3.23}$$

and, hence, R_0 is called the **response amplitude**, and ϕ_0 is the phase of the response relative to the forcing. Mostly, interest centres on R_0, and it is useful to sketch R_0 as a function of κ, with G_0 an external parameter defining a family of curves. Such plots are called **response diagrams** as they display the amplitude R_0 of the response as a function $R_0(\kappa)$ of the tuning parameter κ. We now show that turning points on the response curves, where $d\kappa/dR_0 = 0$, correspond to $\beta_0 = 0$ and, hence, to a stability boundary (provided, of course, that $\epsilon\alpha_0 < 0$). Some typical response diagrams, namely, those for the forced Duffing's equation, are shown in figure 8.3 (for the case without dissipation) and figure 8.4 (for the case with dissipation). It is intuitively clear that turning points correspond to the coalescence of distinct branches of the response curves; at such points the implicit-function theorem fails, since the Jacobian in (3.13) vanishes, that is, β_0 (3.19) is zero. To demonstrate that $\beta_0 = 0$ is equivalent to $d\kappa/dR_0 = 0$, we differentiate (3.12) with respect to R_0, using (3.21) to isolate the dependence of P_0 and Q_0 on κ. Hence, we find that

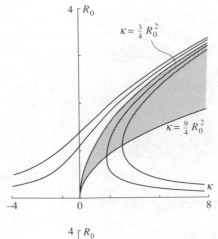

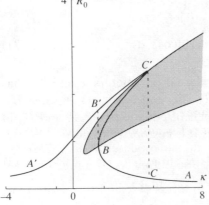

Fig. 8.3 The response diagram (4.4) for forced resonant oscillations of Duffing's equation (4.1). The curves shown are for $G_0 = 1, 2$. The shaded area is the unstable zone.

Fig. 8.4 The response diagram (4.9) for forced resonant oscillations of the damped Duffing's equation (4.6). The curve shown is for $\hat{\nu} = 0.4$ and $G_0 = 1$. The shaded area is the unstable zone.

$$
\begin{bmatrix}
\dfrac{\partial P_0}{\partial A_0} & \dfrac{\partial P_0}{\partial B_0} \\[2ex]
\dfrac{\partial Q_0}{\partial A_0} & \dfrac{\partial Q_0}{\partial B_0}
\end{bmatrix}
\begin{bmatrix}
\cos \phi_0 \\[1ex]
\sin \phi_0
\end{bmatrix}
= \frac{\mathrm{d}\kappa}{\mathrm{d}R_0}
\begin{bmatrix}
-\sin \phi_0 \\[1ex]
\cos \phi_0
\end{bmatrix}. \tag{3.24}
$$

Since $R_0 \neq 0$, it follows immediately that $\mathrm{d}\kappa/\mathrm{d}R_0 = 0$ is equivalent to the vanishing of the determinant of the 2×2 matrix on the left-hand side of (3.24), which, of course, is just β_0.

Next we substitute the transformations (3.22) and (3.23) into (3.21), and form the expressions

$$
M_0(R_0, \phi_0) = P_0 \cos \phi_0 + Q_0 \sin \phi_0,
$$

$$
N_0(R_0, \phi_0) = -P_0 \sin \phi_0 + Q_0 \cos Q_0,
$$

where

$$
M_0 = \pi G_0 \sin \phi_0 - \int_0^{2\pi} f(u_0, u_0', s; 0) \sin(s - \phi_0) \, \mathrm{d}s
$$

and

$$N_0 = \pi G_0 \cos\phi_0 - \pi\kappa R_0 + \int_0^{2\pi} f(u_0, u_0', s; 0)\cos(s - \phi_0)\, ds. \qquad (3.25)$$

Now change the variable in the integral terms in (3.25) from s to $s + \phi_0$. Since the integrands are periodic functions of period 2π in s, it follows that (see the lemma in section 3.1)

$$M_0 = \pi G_0 \sin\phi_0 - \int_0^{2\pi} f(u_0, u_0', s + \phi_0; 0)\sin s\, ds,$$

$$N_0 = \pi G_0 \cos\phi_0 - \pi\kappa R_0 + \int_0^{2\pi} f(u_0, u_0', s + \phi_0; 0)\cos s\, ds,$$

where here

$$u_0 = R_0 \cos s. \qquad (3.26)$$

Clearly, the equations (3.12) are equivalent to

$$M_0 = 0, \qquad N_0 = 0. \qquad (3.27)$$

The equations (3.26) and (3.27) are usually easier to use than the original version, namely (3.12). In particular if $f = f(u, u'; \epsilon)$ is autonomous (i.e. does not contain any explicit dependence on t), then the integral terms in (3.26) are independent of ϕ_0; in this case elimination of ϕ_0 to obtain R_0 as a function of κ is easily achieved. Next it may be shown from (3.25) that

$$2\alpha_0 = \frac{1}{R_0}\frac{\partial}{\partial R_0}(R_0 M_0) + \frac{1}{R_0}\frac{\partial N_0}{\partial \phi_0},$$

and

$$\beta_0 = \det\begin{bmatrix} \dfrac{\partial M_0}{\partial R_0} & \dfrac{\partial N_0}{\partial R_0} \\[2ex] \dfrac{1}{R_0}\dfrac{\partial M_0}{\partial \phi_0} - \dfrac{N_0}{R_0} & \dfrac{1}{R_0}\dfrac{\partial N_0}{\partial \phi_0} + \dfrac{M_0}{R_0} \end{bmatrix}. \qquad (3.28)$$

The simplest method of obtaining these results is to observe that the transformation from A_0, B_0 to R_0, ϕ_0, and of P_0, Q_0 to M_0, N_0, is, in effect, a transformation from Cartesian to polar co-ordinates. Standard vector formulae then lead to the required result.

8.4 Resonant oscillations for Duffing's equation

We now reconsider the forced Duffing's equation (1.4) for the case when the natural frequency σ is close to the resonant value, 1. Using the transformations

(3.1) and (3.2), we see that (1.4) becomes

$$u'' + u = \epsilon(G_0 \cos t - \kappa u + u^3). \qquad (4.1)$$

This has the required form (3.3) and, hence, we may apply theorem 8.2 and the accompanying discussion. Note that here $f(u, u', t; \epsilon)$ in (1.1) is just u^3. Thus, with u_0 given by (3.4), we find that $P_0(A_0, B_0)$ and $Q_0(A_0, B_0)$ (3.11) are here given by

$$P_0(A_0, B_0) = \pi \kappa B_0 - \tfrac{3}{4} \pi B_0 (A_0^2 + B_0^2),$$

$$Q_0(A_0, B_0) = \pi G_0 - \pi \kappa A_0 + \tfrac{3}{4} \pi A_0 (A_0^2 + B_0^2). \qquad (4.2)$$

In obtaining these expressions it is useful to recall the alternative form (3.21) for P_0 and Q_0. Before seeking solutions of (3.12) (i.e. $P_0 = 0$ and $Q_0 = 0$), we first employ the transformation (3.22) from A_0, B_0 to R_0, ϕ_0. Then, using (3.25) and (3.26), we find that

$$M_0(R_0, \phi_0) = \pi G_0 \sin \phi_0,$$

$$N_0(R_0, \phi_0) = \pi G_0 \cos \phi_0 - \pi \kappa R_0 + \tfrac{3}{4} \pi R_0^3. \qquad (4.3)$$

We now determine R_0 and ϕ_0 from (3.27) (i.e. $M_0 = 0$ and $N_0 = 0$). It follows that $\phi_0 = 0, \pi$ and

$$R_0(\kappa - \tfrac{3}{4} R_0^2) = \pm G_0, \qquad (4.4)$$

where the alternate signs refer to $\phi_0 = 0$ and π. The response diagram is plotted in figure 8.3, where, without loss of generality, we may suppose that $G_0 > 0$.

In discussing the response curves, we first note that, when $G_0 = 0$, the solutions of (4.4) are either $R_0 = 0$ or $\kappa = \tfrac{3}{4} R_0^2$. The latter solution can be recognised as the nonlinear frequency correction for the periodic solution of the unforced Duffing's equation (see section 7.1). Then, as $G_0 = 0$ increases, the response curves emerge from these curves for $G_0 = 0$. Those curves to the right (left) of the parabola $\kappa = \tfrac{3}{4} R_0^2$ correspond to $\phi_0 = 0$ (π) and, hence, are in (out of) phase with the forcing. The turning points on these curves (with G_0 fixed) are obtained from (4.4) by calculating $d\kappa/dR_0$. We find that

$$R_0 \frac{d\kappa}{dR_0} = -\kappa + \tfrac{9}{4} R_0^2$$

and, hence, the turning points where $d\kappa/dR_0 = 0$ are given by the parabola $\kappa = \tfrac{9}{4} R_0^2$. Note that, to the right of each turning point, there are three solutions for R_0 for each value of κ (with G_0 fixed), but only one solution to the left of the turning point. The response diagram can be usefully compared to that for the corresponding linear equation, that is, when the term ϵu^3 in (4.1) is omitted. For the linear equation the response curves are $R_0 = G_0/|\kappa|$, and there is only one value of R_0 for each value of κ, but $R_0 \to \infty$ as $\kappa \to 0$. The nonlinear term

ϵu^3 is needed to resolve this singularity, and, indeed, from figure 8.3, we see that R_0 is finite for all (finite) values of κ. Indeed, the effect of the nonlinear term can be interpreted as bending the linear response curves to the right, with the consequence that multiple solutions are introduced.

Next we turn to the question of stability, and calculate α_0 and β_0 either from (3.19) using (4.2), or from (3.28) using (4.3). We find that

$$\alpha_0 = 0, \qquad \beta_0 = \pi^2(\kappa - \tfrac{3}{4}R_0^2)(\kappa - \tfrac{9}{4}R_0^2). \tag{4.5}$$

First we observe that $\beta_0 \neq 0$ except when $\kappa = \tfrac{3}{4}R_0^2$, which corresponds to $G_0 = 0$, or when $\kappa = \tfrac{9}{4}R_0^2$, which corresponds to the turning points, where $d\kappa/dR_0 = 0$. Uniform and asymptotic stability requires $\epsilon\alpha_0 < 0$ and $\beta_0 > 0$, and, otherwise, there is instability (see theorem 8.2). Hence we can infer that the response is unstable for $\tfrac{3}{4}R_0^2 < \kappa < \tfrac{9}{4}R_0^2$, which is the zone between the two parabolas. Outside this zone $\beta_0 > 0$ but $\alpha_0 = 0$, and so we cannot use the stability criterion (3.14). We shall show below that the introduction of a small amount of dissipation will allow us to infer the stability of the response curves in the zone outside the parabolas (i.e. $\kappa < \tfrac{3}{4}R_0^2$ or $\kappa > \tfrac{9}{4}R_0^2$). It is interesting to note that

$$\frac{dR_0}{dG_0}(\kappa - \tfrac{9}{4}R_0^2) = \pm 1,$$

where the derivative is taken with κ fixed, and so dR_0/dG_0 is positive in the stable zone, but negative in the unstable zone. This is in agreement with the intuitive notion that, when an increase (decrease) in R_0 requires an increase (decrease) in G_0, then stability will ensue, since a small perturbation in R_0 could not be sustained. On the other hand, when an increase (decrease) in R_0 requires a decrease (increase) in G_0, then we expect instability.

To resolve the ambiguity concerning α_0, we follow the same device used in section 8.2, and include a small damping term in (4.1). We add a term $-\nu u'$ to the right-hand side of (4.1), where ν (> 0) is $O(\epsilon)$. Putting $\nu = \epsilon\hat{\nu}$, (4.1) becomes

$$u'' + u = \epsilon(G_0\cos t - \kappa u - \hat{\nu}u' + u^3). \tag{4.6}$$

In place of (4.2) we now obtain

$$P_0(A_0, B_0) = \pi\kappa B_0 - \pi\hat{\nu}A_0 - \tfrac{3}{4}\pi B_0(A_0^2 + B_0^2),$$

$$Q_0(A_0, B_0) = \pi G_0 - \pi\kappa A_0 - \pi\hat{\nu}B_0 + \tfrac{3}{4}\pi A_0(A_0^2 + B_0^2), \tag{4.7}$$

and, in place of (4.3), we get

$$M_0(R_0, \phi_0) = \pi G_0\sin\phi_0 - \pi\hat{\nu}R_0,$$

$$N_0(R_0, \phi_0) = \pi G_0\cos\phi_0 - \pi\kappa R_0 + \tfrac{3}{4}\pi R_0^3. \tag{4.8}$$

We can determine R_0 and ϕ_0 from (3.27) (i.e. $M_0 = 0$ and $N_0 = 0$). Eliminating

ϕ_0, we find that

$$R_0^2\{\hat{\nu}^2 + (\kappa - \tfrac{3}{4}R_0^2)^2\} = G_0^2. \tag{4.9}$$

The response diagram is plotted in figure 8.4 for a case when the friction coefficient is not too large $(\hat{\nu}^2 < \tfrac{3}{4}(\tfrac{3}{4}G_0^2)^{2/3})$. Comparing this with the case $\hat{\nu} = 0$ in figure 8.3, we can see that the main qualitative effect of dissipation is to prevent the response curves from reaching to infinity and, instead, for a given G_0, to provide a continuous curve. To obtain the turning points, we calculate $\mathrm{d}\kappa/\mathrm{d}R_0$ and get

$$R_0(\kappa - \tfrac{3}{4}R_0^2)\frac{\mathrm{d}\kappa}{\mathrm{d}R_0} = -\{\hat{\nu}^2 + (\kappa - \tfrac{3}{4}R_0^2)(\kappa - \tfrac{9}{4}R_0^2)\}. \tag{4.10}$$

The turning points correspond to $\mathrm{d}\kappa/\mathrm{d}R_0 = 0$ and, hence, are given by the zeros of the right-hand side of (4.10). There are two turning points, and, referring to figure 8.4, we see that there are three solutions for R_0 for each value of κ between the turning points, but just one solution elsewhere.

To determine the stability, we re-calculate α_0 and β_0 and, in place of (4.50), we now get

$$\alpha_0 = -\pi\hat{\nu}, \qquad \beta_0 = \pi^2\{\hat{\nu}^2 + (\kappa - \tfrac{3}{4}R_0^2)(\kappa - \tfrac{9}{4}R_0^2)\}. \tag{4.11}$$

Since $\epsilon\alpha_0 = -\pi\nu < 0$ for $\nu > 0$, stability is now determined by the sign of β_0, where we note that β_0 has the opposite sign to the right-hand side of (4.10), and, as required, β_0 is zero at the turning points where $\mathrm{d}\kappa/\mathrm{d}R_0 = 0$. Hence, there is instability for the section of the response curve between the turning points, and stability elsewhere.

We are now in a position to discuss the **jump phenomenon**. Suppose that the forcing amplitude G_0 is held constant, but the forcing frequency is varied slowly. By a change of time variable this is equivalent to a slow variation of the tuning parameter κ. Assume that κ is varied so slowly that the response of the oscillator can be described by (3.23) with a corresponding slow variation of the response amplitude R_0, provided that R_0 is on a stable branch of the response curve. Now, for a fixed G_0, vary κ so that the response moves from the point A on the response curve in the direction of κ decreasing. Since this branch is stable, the response is $R_0(\kappa)$ until the turning point B is reached (see figure 8.4). But then, recalling that the branch BC' is unstable, a further decrease in κ will cause the response to jump to the point B', which lies on a stable branch of the response curve, and then move to the point A'. Next reverse the process by slowly increasing κ from the point A'. The response $R_0(\kappa)$ will remain on this branch until the turning point C' is reached, when there is a jump to the point C. A further increase in κ will cause the response to return to the point A. The portion $B'C'CB$ of the response curve forms a hysteresis loop, where the segments BB' and $C'C$ represent a rapid change in response amplitude as κ is slowly varied, while on the other segments the response amplitude varies slowly.

8.5 Resonant oscillations for Van der Pol's equation

Let us now consider the forced Van der Pol equation (2.27) for the case when the natural frequency σ is close to the resonant value, 1. Using the transformations (3.1) and (3.2), we see that (2.27) becomes

$$u'' + u = \epsilon\{G_0 \cos t - \kappa u + (1 - u^2)u'\}. \tag{5.1}$$

This has the required form (3.3) and, hence, we may apply theorem 8.2 and the accompanying discussion. Note that here $f(u, u', t; \epsilon)$ in (1.1) is $(1 - u^2)u'$. The development parallels that for the forced Duffing's equation in section 8.5. Thus, with u_0 given by (3.4), we find that $P_0(A_0, B_0)$ and $Q_0(A_0, B_0)$ (3.11) are here given by

$$P_0(A_0, B_0) = \pi\kappa B_0 + \pi A_0 - \tfrac{1}{4}\pi A_0(A_0^2 + B_0^2),$$

$$Q_0(A_0, B_0) = \pi G_0 - \pi\kappa A_0 + \pi B_0 - \tfrac{1}{4}\pi B_0(A_0^2 + B_0^2). \tag{5.2}$$

Next we employ the transformation (3.22) from A_0, B_0 to R_0, ϕ_0 and find that $M_0(R_0, \phi_0)$ and $N_0(R_0, \phi_0)$ ((3.25) and (3.26)) are here given by

$$M_0 = \pi G_0 \sin \phi_0 + \pi R_0 - \tfrac{1}{4}\pi R_0^3, \qquad N_0 = \pi G_0 \cos \phi_0 - \pi\kappa R_0. \tag{5.3}$$

We can determine R_0 and ϕ_0 from (3.27) (i.e. $M_0 = 0$ and $N_0 = 0$). Eliminating ϕ_0, we see that

$$R_0^2\{\kappa^2 + (1 - \tfrac{1}{4}R_0^2)\} = G_0^2. \tag{5.4}$$

The response diagram is plotted in figure 8.5.

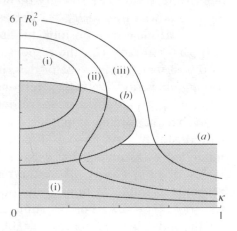

Fig. 8.5 The response diagram (5.4) for forced resonant oscillations of the Van der Pol equation (5.1). The curves shown are for $G_0 = 0.6$ (i), 0.85 (ii) and 1.2 (iii), corresponding, respectively, to the cases (i) $0 < G_0^2 < \tfrac{16}{27}$, (ii) $\tfrac{16}{27} < G_0^2 < \tfrac{32}{27}$ and (iii) $G_0^2 > \tfrac{32}{27}$. The shaded area is the unstable zone, whose boundary is either (a) ($\alpha_0 = 0$) or (b) ($\beta_0 = 0$) (see (5.6)). The curves are symmetric in κ and, hence, only $\kappa \geqslant 0$ is shown.

In discussing the response curves, first note that they are symmetric with respect to κ. Next, when $G_0 = 0$, the solutions of (5.4) are $R_0 = 0$ for all κ, or $\kappa = 0$ and $R_0 = 2$. This latter solution corresponds to the limit-cycle solution of the unforced Van der Pol equation (see section 6.3) in the limit $\epsilon \to 0$. Then, as $|G_0|$ increases, the response curves emerge from those for $G_0 = 0$. The turning points are found from the relation

$$R_0 \kappa \frac{\mathrm{d}\kappa}{\mathrm{d}R_0} = -\{\kappa^2 + (1 - \tfrac{1}{4}R_0^2)(1 - \tfrac{3}{4}R_0^2)\}. \tag{5.5}$$

For $0 < G_0^2 < \tfrac{16}{27}$ there are two disjoint branches to the response curves. One branch forms a closed loop about the point $\kappa = 0$, $R_0 = 2$ and lies in the zone $\tfrac{4}{3} < R_0^2 < \tfrac{16}{3}$: it has a single turning point (where $\mathrm{d}\kappa/\mathrm{d}R_0 = 0$) in $\kappa > 0$ (and one in $\kappa < 0$) which is such that $\tfrac{4}{3} < R_0^2 < 4$. The other branch lies in the zone $0 < R_0^2 < \tfrac{4}{3}$ and has no turning points. When $G_0^2 = \tfrac{16}{27}$, these two branches merge and, for $\tfrac{16}{27} < G_0^2 < \tfrac{32}{27}$, there is a single branch which has two turning points in $\kappa > 0$ (and two in $\kappa < 0$) which lie in the region $\tfrac{4}{3} < R_0^2 < 4$ and $0 < \kappa^2 < \tfrac{1}{3}$: one turning point is such that $\tfrac{4}{3} < R_0^2 < \tfrac{8}{3}$ and the other is such that $\tfrac{8}{3} < R_0^2 < 4$. When $G_0^2 = \tfrac{32}{27}$, the two turning points merge and, for $G_0^2 > \tfrac{32}{27}$, there is a single branch with no turning points.

Next we turn to the question of stability, and calculate α_0 and β_0 either from (3.19) using (5.2), or from (3.28) using (5.3). We find that

$$2\alpha_0 = \pi(2 - R_0^2), \qquad \beta_0 = \pi^2\{\kappa^2 + (1 - \tfrac{1}{4}R_0^2)(1 - \tfrac{3}{4}R_0^2)\}. \tag{5.6}$$

Consider first the stability criterion $\epsilon\alpha_0 < 0$ (see theorem 8.2). For $\epsilon > 0$, this implies that the solution is unstable for $R_0^2 < 2$, a result which agrees with the stability criterion (2.28) obtained for non-resonant oscillations. Next, we observe that β_0 has the opposite sign to the right-hand side of (5.5) and, hence, $\beta_0 > 0 \, (< 0)$ when $\kappa \, \mathrm{d}\kappa/\mathrm{d}R_0 < 0 \, (> 0)$. Combining these two results, we see that, for uniform and asymptotic stability, $R_0^2 > 2$ and $\kappa \, \mathrm{d}\kappa/\mathrm{d}R_0 < 0$, with instability otherwise. The stability boundary is shown in figure 8.5. Note that turning points which lie in $R_0^2 > 2$ correspond to a change of stability.

We are now in a position to discuss the phenomenon of **entrainment**. First note that, for any G_0, the response $R_0(\kappa)$ is unstable if $|\kappa|$ is sufficiently large. In this situation the oscillator described by (5.1) will not be able to respond to a forcing of period 2π with a response of the same period. However, as $|\kappa|$ is decreased, eventually a stable response can be found. To be more specific, suppose that the forcing amplitude G_0 is held constant, but the forcing frequency is varied slowly. As in our discussion of the jump phenomenon, this can be modelled by assuming that κ is varied slowly, and the response of the oscillator is described by (3.23), where the response amplitude is $R_0(\kappa)$, provided that the response is stable. Consider the physically interesting case when G_0 is small, so that $0 < G_0^2 < 27$. For $|\kappa|$ sufficiently large, the response amplitude $R_0(\kappa)$ lies on the lower branch, and is unstable. As $|\kappa|$ is decreased, a stable response will occur when $|\kappa|$ reaches the value corresponding to the turning point on the upper branch (the value $|\kappa_1|$ in figure 8.5). For this upper branch, the upper portion of the loop is stable and the lower portion is unstable. Hence, as $|\kappa|$ is decreased further, the oscillator will respond with a stable oscillation. The oscillator is said to be **entrained** to the frequency of the forcing, and $|\kappa| < \kappa_1$ is the entrainment zone. Clearly, this entrainment phenomenon is associated with the existence of the stable limit cycle for the unforced Van der Pol equation, which here corresponds to the point $\kappa = 0$, $R_0 = 2$ when $G_0 = 0$.

Problems

1 For each of the following equations, determine the forced oscillation of period 2π for the non-resonant case (i.e. $\sigma \neq n$, $n = 0, 1, 2, \ldots$), using a power-series expansion in ϵ.

$$\text{(a)} \quad u'' + \sigma^2 u = F_0 \cos t + \epsilon u^2;$$

$$\text{(b)} \quad u'' + \sigma^2 u = F_0 \cos t + \epsilon(u^2 - 1)u';$$

$$\text{(c)} \quad u'' + \sigma^2 u = F_0 \cos t - \epsilon(2u^4 - 5u^2 + 1)u'.$$

In each case determine the stability from the criterion (2.7). Note that, in case (a), a linear damping term, $-\nu u'$, where ν is $O(\epsilon)$, must be added to the right-hand side to resolve the stability (see (2.25) for the analogous situation for the forced Duffing's equation).

2 For the forced simple harmonic oscillator with linear damping:

$$u'' + \nu u' + \sigma^2 u = F_0 \cos t,$$

show that the forced oscillation of period 2π is given by

$$u = R_0 \cos(t - \phi_0),$$

where

$$R_0 \sqrt{(\sigma^2 - 1)^2 + \nu^2} = |F_0|, \qquad \tan \phi_0 = \nu(\sigma^2 - 1)^{-1}.$$

Sketch the response diagram for R_0 as a function of $(\sigma^2 - 1)$.

3 Show that the equation

$$u'' + \nu u' + \sigma^2 u = F_0 \cos t + \epsilon f(u, u', t; \epsilon),$$

where $\nu > 0$ and f satisfies the same conditions specified for (1.1), has, for sufficiently small $|\epsilon|$, a periodic solution of period 2π which is analytic in ϵ and such that, as $\epsilon \to 0$, $u \to u_0$, where u_0 is the solution of the equation if $\epsilon = 0$. Show also that this solution is asymptotically stable for sufficiently small $|\epsilon|$. [*Hint*: Use a procedure analogous to that for the case $\nu = 0$ described in section 7.2. Note that the condition that $\sigma \neq n$ ($n = 0, 1, 2, \ldots$) which holds when $\nu = 0$ is here replaced by the condition that the homogeneous linear equation

$$u'' + \nu u' + \sigma^2 u = 0$$

should have no non-trivial solution of period 2π, which is satisfied for all σ when $\nu > 0$.]

4 For each of the following equations, use the method of section 8.3 to find the forced resonant oscillations of period 2π for $|\epsilon|$ sufficiently small and $|\sigma^2 - 1|$ of $O(\epsilon)$. In each case determine the stability and sketch the response diagram.

(a) $u'' + \sigma^2 u = \epsilon G_0 \cos t + \epsilon(u^3 - u^5)$,

(b) $u'' + \sigma^2 u = \epsilon G_0 \cos t + \epsilon(u^3 + u^5)$,

(c) $u'' + \sigma^2 u = \epsilon G_0 \cos t - \nu u^2 u'$,   where ν is $O(\epsilon)$.

5 For the equation

$$u'' + \sigma^2 u = \epsilon G_0 \cos t + \epsilon u^2,$$

use the method of section 8.3 to show that there are forced resonant oscillations of period 2π, for $|\epsilon|$ sufficiently small and $|\sigma^2 - 1|$ of $O(\epsilon)$, given by

$$u = R_0 \cos t + \epsilon R_0^2(\tfrac{1}{2} - \tfrac{1}{6}\cos 2t) + O(\epsilon^2),$$

where

$$R_0(\kappa - \tfrac{5}{6}\epsilon R_0^2) = G_0 + O(\epsilon^2), \qquad \sigma^2 - 1 = \epsilon\kappa.$$

Determine the stability and sketch the response diagram. [*Hint*: In this case, application of theorem 8.2 only shows that $\kappa R_0 = G_0$, which gives the response curve for the linear oscillator (i.e. the ϵu^2 term is omitted). Hence it is necessary here to calculate u to $O(\epsilon)$, and then use the conditions (3.10), also to $O(\epsilon)$.]

6 Confirm that the resonant oscillations for the forced Duffing's equation with linear damping (i.e. (4.6)) are given by (4.9) and, hence, sketch the response diagram. Show that, for $\hat{\nu}^2 < \tfrac{3}{4}(\tfrac{3}{4}G_0^2)^{2/3}$, the response diagram has the shape shown in figure 8.4 with two turning points and, in this case, confirm that the upper and lower branches are stable, while the middle branch is unstable (i.e. find α_0 and β_0 (4.11)). On the other hand, if $\hat{\nu}^2 > \tfrac{3}{4}(\tfrac{3}{4}G_0^2)^{2/3}$, show that there are no turning points and that the response diagram has a single stable branch. [*Hint*: The value of $\hat{\nu}$ which distinguishes between the two cases can be found by finding the condition that $d\kappa/dR_0$ and $d^2\kappa/dR_0^2$ vanish simultaneously.]

7 For resonant superharmonic oscillations of (1.1):

$$u'' + \sigma^2 u = F_0 \cos t + \epsilon f(u, u', t; \epsilon),$$

let $\sigma^2 = n^2 + \epsilon\kappa$, where $n = 2, 3, \ldots$, and consider the unique solution $u(t; A, B, \epsilon)$ which satisfies the initial conditions

$$u(0) = A + \frac{F_0}{n^2 - 1}, \qquad u'(0) = nB.$$

Show that

$$u = A\cos nt + B\sin nt + \frac{F_0}{n^2 - 1}\cos t + \frac{\epsilon}{n}\int_0^t g(u, u', s; \epsilon)\sin n(t - s)\, ds,$$

$$u' = -nA\sin nt + nB\cos nt - \frac{F_0}{n^2 - 1}\sin t + \epsilon\int_0^t g(u, u', s; \epsilon)\cos n(t - s)\, ds,$$

where

$$g(u, u', t; \epsilon) = -\kappa u + f(u, u', t; \epsilon).$$

Define

$$\epsilon P(A, B; \epsilon) = n\{u(2\pi; A, B, \epsilon) - u(0; A, B, \epsilon)\},$$

$$\epsilon Q(A, B; \epsilon) = u'(2\pi; A, B, \epsilon) - u'(0; A, B, \epsilon),$$

so that periodic solutions of period 2π correspond to $P = Q = 0$. Deduce that, in the limit $\epsilon \to 0$, as $A \to A_0$ and $B \to B_0$, periodic solutions are given by

$$P_0(A_0, B_0) = -\int_0^{2\pi} g(u_0, u_0', s; 0)\sin ns \ ds = 0,$$

$$Q_0(A_0, B_0) = \int_0^{2\pi} g(u_0, u_0', s; 0)\cos ns \ ds = 0,$$

where

$$u_0 = A_0 \cos nt + B_0 \sin nt + \frac{F_0}{n^2 - 1} \cos t,$$

provided that

$$\frac{\partial(P_0, Q_0)}{\partial(A_0, B_0)} \neq 0.$$

Further, show that the periodic solution is uniformly and asymptotically stable if

$$\frac{\partial(P_0, Q_0)}{\partial(A_0, B_0)} > 0 \quad \text{and} \quad \epsilon\left(\frac{\partial P_0}{\partial A_0} + \frac{\partial Q_0}{\partial B_0}\right) < 0,$$

and is unstable if either of these inequalities is reversed. [*Hint*: Follow the procedure used in the proof of theorem 8.2.]

8 Use the method described in problem 7 to obtain resonant superharmonic oscillations of the forced Duffing's equation (1.4):

$$u'' + \sigma^2 u = F_0 \cos t + \epsilon u^3,$$

where $\sigma^2 = 9 + \epsilon\kappa$. Show that

$$u = A_0 \cos 3t + \tfrac{1}{8} F_0 \cos t + O(\epsilon),$$

where

$$A_0\{\kappa - \tfrac{3}{4} A_0^2 - \tfrac{3}{2}(\tfrac{1}{8} F_0)^2\} = \tfrac{1}{4}(\tfrac{1}{8} F_0)^3,$$

and sketch the response diagram. Determine the stability.

9 For resonant subharmonic oscillations of (1.1):

$$u'' + \sigma^2 u = F_0 \cos nt + \epsilon f(u, u', t; \epsilon),$$

where $\sigma^2 = 1 + \epsilon\kappa$ and $n = 2, 3, \ldots$, consider the unique solution $u(t; A, B, \epsilon)$ which satisfies the initial conditions

$$u(0) = A - \frac{F_0}{n^2 - 1}, \qquad u'(0) = B.$$

Show that

$$u = A \cos t + B \sin t - \frac{F_0}{n^2 - 1} \cos nt + \epsilon \int_0^t g(u, u', s; \epsilon)\sin(t - s) \ ds,$$

$$u' = -A\sin t + B\cos t + \frac{nF_0}{n^2-1}\sin nt + \epsilon\int_0^t g(u,u',s;\epsilon)\cos(t-s)\ ds,$$

where

$$g(u,u',s;\epsilon) = -\kappa u + f(u,u',t;\epsilon).$$

Define

$$\epsilon P(A,B;\epsilon) = u(2\pi;A,B,\epsilon) - u(0;A,B,\epsilon),$$

$$\epsilon Q(A,B;\epsilon) = u'(2\pi;A,B,\epsilon) - u'(0;A,B,\epsilon),$$

so that periodic solutions of period 2π correspond to $P = Q = 0$. Deduce that, in the limit $\epsilon \to 0$, as $A \to A_0$ and $B \to B_0$, periodic solutions are given by

$$P_0(A_0,B_0) = -\int_0^{2\pi} g(u_0,u_0',s;0)\sin s\ ds,$$

$$Q_0(A_0,B_0) = \int_0^{2\pi} g(u_0,u_0',s;0)\cos s\ ds,$$

where

$$u_0 = A_0\cos t + B_0\sin t - \frac{F_0}{n^2-1}\cos nt,$$

provided that

$$\frac{\partial(P_0,Q_0)}{\partial(A_0,B_0)} \neq 0.$$

Further, show that the periodic solution is uniformly and asymptotically stable if

$$\frac{\partial(P_0,Q_0)}{\partial(A_0,B_0)} > 0 \quad \text{and} \quad \epsilon\left(\frac{\partial P_0}{\partial A_0} + \frac{\partial Q_0}{\partial B_0}\right) < 0,$$

and is unstable if either of these inequalities is reversed. [*Hint:* Follow the procedure used in the proof of theorem 8.2.]

10 Use the method described in problem 9 to obtain resonant subharmonic oscillations of the forced Duffing's equation (1.4):

$$u'' + \sigma^2 u = F_0\cos 3t + \epsilon u^3,$$

where $\sigma^2 = 1 + \epsilon\kappa$. Show that

$$u = R_0\cos(t-\phi_0) - \tfrac{1}{8}F_0\cos 3t + O(\epsilon),$$

where either

$$\kappa = \tfrac{3}{4}R_0^2 - \tfrac{3}{4}R_0(\tfrac{1}{8}F_0) + \tfrac{3}{2}(\tfrac{1}{8}F_0)^2, \qquad \phi_0 = 0, \pm\tfrac{2}{3}\pi,$$

or

$$\kappa = \tfrac{3}{4}R_0^2 + \tfrac{3}{4}R_0(\tfrac{1}{8}F_0) + \tfrac{3}{2}(\tfrac{1}{8}F_0)^2, \qquad \phi_0 = \pm\tfrac{1}{3}\pi, \pm\pi,$$

and sketch the response diagram. Determine the stability. Note that the transformation $t \to t \pm \tfrac{2}{3}\pi$ leaves the forcing term $F_0\cos 3t$ unchanged, and explains why there are three distinct values of ϕ_0 for each solution branch.

11 For the forced Duffing's equation of problem 10, show that there is an exact subharmonic solution

$$u = A_0 \cos t,$$

provided that

$$\sigma^2 - 1 = \tfrac{3}{4}\epsilon A_0^2, \qquad F_0 = -\tfrac{1}{4}\epsilon A_0^2.$$

What is the relation between this exact solution and the subharmonic solution found in problem 10?

12 Use the method described in problem 9 to show that there are no resonant subharmonic oscillations of the forced Duffing's equation (1.4):

$$u'' + \sigma^2 u = F_0 \cos 2t + \epsilon u^3 - \nu u',$$

where $\sigma^2 = 1 + \epsilon\kappa$, when the damping coefficient $\nu \neq 0$. Assume that ν is $O(\epsilon)$. [*Hint*: Show that when $n = 2$ and $f = u^3 - \hat{\nu}u'$, where $\nu = \epsilon\hat{\nu}$, then A_0 and B_0 are zero if $\hat{\nu} \neq 0$.]

CHAPTER NINE
AVERAGING METHODS

9.1 Averaging methods for autonomous equations

To motivate the form of the equations to be discussed in this section let us first reconsider equation (1.1) of chapter 7,

$$u'' + u = \epsilon f(u, u'; \epsilon), \tag{1.1}$$

which, for $|\epsilon|$ sufficiently small, describes a perturbation of a simple harmonic oscillator. In chapter 7 we described a perturbation method for constructing periodic solutions of (1.1). Here we shall present an alternative procedure which allows us not only to recover the periodic solutions of (1.1), but also to obtain information about all the orbits of (1.1). The first step is to convert (1.1) into a first-order system, by letting

$$x = u, \qquad y = u'. \tag{1.2}$$

Then (1.1) becomes

$$x' = y, \qquad y' = -x + \epsilon f(x, y; \epsilon). \tag{1.3}$$

Now introduce the polar coordinates (r, θ), where

$$x = r \cos \theta, \qquad y = r \sin \theta, \tag{1.4}$$

and hence convert (1.3) into

$$r' = \epsilon \sin \theta f(r \cos \theta, r \sin \theta; \epsilon),$$

$$\theta' = -1 + \frac{\epsilon \cos \theta}{r} f(r \cos \theta, r \sin \theta; \epsilon). \tag{1.5}$$

In deriving this result we use (1.4) of section 6.1. Finally, it is customary in the context of this chapter to replace r by I, where

$$I = \tfrac{1}{2} r^2 = \tfrac{1}{2}(x^2 + y^2). \tag{1.6}$$

Here I is a measure of the energy associated with the oscillator (1.1), at least for sufficiently small values of $|\epsilon|$. Indeed, the variable I was used in our analysis of the Van der Pol limit cycle in section 6.3, where it was denoted by E (see (3.4) of section 6.3) in accord with its interpretation as energy. Here we have preferred the notation I as, in the more general context to be developed in this chapter, I will be a measure of an entity called the action, which in general is

216

distinct from the energy. Using (1.6), equations (1.5) become

$$I' = \epsilon F(I, \theta; \epsilon), \qquad \theta' = -1 + \epsilon G(I, \theta; \epsilon), \qquad (1.7)$$

where here F and G are given by

$$F(I, \theta; \epsilon) = r \sin \theta f(r \cos \theta, r \sin \theta; \epsilon),$$

$$G(I, \theta; \epsilon) = \frac{\cos \theta}{r} f(r \cos \theta, r \sin \theta; \epsilon), \qquad (1.8)$$

where, from (1.6), $r = \sqrt{2I}$.

More generally, we shall show in chapter 11 that the appropriate equations for describing the perturbation of a nonlinear oscillator are

$$I' = \epsilon F(I, \theta; \epsilon), \qquad \theta' = \omega(I) + \epsilon G(I, \theta; \epsilon), \qquad (1.9)$$

where F and G are periodic functions with respect to θ, with period 2π. Also, for simplicity, we shall assume that F, G and ω are analytic functions of their respective arguments. Note that (1.7) has the required form (1.9) with $\omega = -1$ and F and G defined by (1.8). When $\epsilon = 0$, the solutions of (1.9) are $I = I_0$ and $\theta = \theta_0 + \omega(I_0)t$, which describe a nonlinear oscillator of action I_0 and angular frequency $\omega(I_0)$. For each of these periodic solutions, the period is $2\pi/|\omega(I_0)|$ and, in general, is a function of the amplitude, since I_0 is a measure of the amplitude. An exception is the simple harmonic oscillator, for which $\omega = -1$ and the period is 2π. We shall show in chapter 11 that I and θ are action–angle variables for the oscillator when $\epsilon = 0$.

To obtain approximate solutions of (1.9), we shall use the method of averaging. Thus, let

$$\bar{F}(I; \epsilon) = \frac{1}{2\pi} \int_0^{2\pi} F(I, \theta; \epsilon) \, d\theta \qquad (1.10)$$

be the averaged value of F. Let $J(t)$ satisfy the equation

$$J' = \epsilon \bar{F}(J; 0). \qquad (1.11)$$

Note that (1.11) can be regarded as an averaged version of the first equation in (1.9). We shall now show that $J(t)$ can be regarded as an approximation to $I(t)$, obtained by solving (1.9). In essence, the true solution for $I(t)$ contains small oscillations due to the dependence of $F(I, \theta; \epsilon)$ on (1.9) for θ, but these can be averaged out, and the overall trend for $I(t)$ is described by (1.11). The situation is sketched schematically in figure 9.1.

Theorem 9.1: Let $J(t)$ satisfy (1.11) for $0 \le |\epsilon t| \le C_0$, where C_0 is a constant independent of ϵ, and suppose that $|\omega(I)|$ is bounded away from zero for all finite $|I|$. Then, if $I(0) = J(0)$, there exists a constant C_1, independent of ϵ, such that $|I(t) - J(t)| < C_1 \epsilon$ for $0 \le |\epsilon t| \le C_0$, provided that $|\epsilon|$ is sufficiently small.

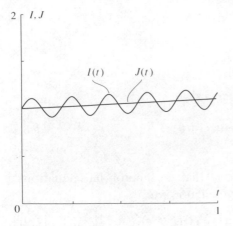

Fig. 9.1 The relation between $I(t)$ and the averaged variable $J(t)$.

Proof: Note that it is easily established that $|J(t) - I(t)|$ is $O(\epsilon)$ for t of $O(1)$. However, to show that the inequality must hold for t of $O(\epsilon^{-1})$ is a more delicate result, and will be achieved by a preliminary change of variables. First, let

$$F(I, \theta; \epsilon) = \bar{F}(I; \epsilon) + \hat{F}(I, \theta; \epsilon). \tag{1.12}$$

By the definition (1.10), $\hat{F}(I, \theta; \epsilon)$ is a periodic function of θ with period 2π, and zero mean, so that

$$\int_0^{2\pi} \hat{F}(I, \theta; \epsilon) \, d\theta = 0.$$

Now let

$$H(I, \theta; \epsilon) = \int_0^{\theta} \hat{F}(I, \hat{\theta}; \epsilon) \, d\theta$$

and we see that $H(I, \theta; \epsilon)$ is also a periodic function of θ, with period 2π. Put

$$\hat{I} = I - \epsilon H \omega^{-1}. \tag{1.13}$$

Then, using (1.9) and (1.12), it follows that

$$\hat{I}' = I' - \epsilon \frac{\partial}{\partial I}\left(\frac{H}{\omega}\right) I' - \epsilon \frac{\partial H}{\partial \theta} \frac{\theta'}{\omega}$$

$$= \epsilon(\bar{F} + \hat{F}) - \epsilon^2 \frac{\partial}{\partial I}\left(\frac{H}{\omega}\right) F - \epsilon \hat{F}\left(1 + \frac{\epsilon G}{\omega}\right)$$

$$= \epsilon \bar{F}(I; \epsilon) + O(\epsilon^2)$$

or

$$\hat{I}' = \epsilon \bar{F}(\hat{I}; \epsilon) + O(\epsilon^2). \tag{1.14}$$

Here the $O(\epsilon^2)$ error term is a periodic function of θ with period 2π. Relative to this error term, we have effectively converted the first equation of (1.9) into the averaged equation (1.11).

To complete the proof, let

$$z = \hat{I} - J. \tag{1.15}$$

Then, from (1.11) and (1.14),

$$z' = \epsilon\{\bar{F}(\hat{I};\epsilon) - \bar{F}(J;0)\} + O(\epsilon^2),$$

or

$$z' = \epsilon\frac{\partial\bar{F}}{\partial J}(J;0)z + O(\epsilon z^2) + O(\epsilon^2). \tag{1.16}$$

Also, since we may assume without loss of generality that $\theta = 0$ at $t = 0$, it follows that $H = 0$ at $t = 0$. Then, recalling that $I(0) = J(0)$, it follows from (1.13) and (1.15) that $z = 0$ at $t = 0$. Hence

$$z(t) = \epsilon\int_0^t \frac{\partial\bar{F}}{\partial J}(J;0)z \ dt' + \int_0^t \{O(\epsilon z^2) + O(\epsilon^2)\} \ dt'. \tag{1.17}$$

But, since $z(0) = 0$, for any $\alpha > 0$ there exists a $t_1 > 0$ such that $|z(t)| \leq \alpha$ for $0 \leq t < t_1$. If also $0 \leq \epsilon t < C_0$, then

$$\left|\frac{\partial\bar{F}}{\partial J}(J;0)\right| \leq K_0,$$

where K_0 is a constant independent of ϵ, while the $O(\epsilon z^2)$ and the $O(\epsilon^2)$ terms in (1.17) are, respectively, bounded by $K_1|\epsilon|\alpha^2$ and $K_2\epsilon^2$, where K_1 and K_2 are also constants independent of ϵ. It now follows from (1.17) that

$$|z(t)| \leq |\epsilon|K_0\int_0^t |z(t')| \ dt' + K_1C_0\alpha^2 + |\epsilon|C_0.$$

But then Gronwall's lemma (see section 1.3 of chapter 1) implies that

$$|z(t)| \leq (K_1C_0\alpha^2 + |\epsilon|C_0)\exp(K_0|\epsilon t|),$$

or

$$|z(t)| \leq (K_1C_0\alpha^2 + |\epsilon|C_0)\exp(K_0C_0).$$

Now choose α so that the right-hand side of this inequality is less than α; the choice $\alpha = 2\epsilon C_0\exp(K_0C_0)$ will suffice for $|\epsilon|$ sufficiently small so that $K_1\alpha^2 < |\epsilon|$. It follows that we may choose t_1 so that $|\epsilon t_1| \geq C_0$, and so $|z(t)| \leq \alpha$ and, hence, $O(\epsilon)$ for $0 \leq |\epsilon t| \leq C_0$. Finally, recalling (1.13) and (1.15), it follows that $|I(t) - J(t)|$ is $O(\epsilon)$, and the theorem is proved. Note that, in establishing this last estimate, it is essential that $|\omega(I)|$ be bounded away from zero.

Theorem 9.1 enables us to use the averaged equation (1.11) as an approximation to (1.9). One of its main uses is to find periodic solutions of (1.9). It is intuitively clear that critical points of the averaged equation (1.11) should correspond to periodic solutions of (1.9), with the stability also corresponding. Indeed, the following result holds.

Theorem 9.2: Under the conditions of theorem 9.1, let J_0 be an isolated critical point of the averaged equation (1.11), so that

$$\bar{F}(J_0;0) = 0, \qquad \frac{\partial \bar{F}}{\partial J_0}(J_0;0) \neq 0.$$

Then there exists a periodic solution of (1.9) for which $I \to J_0$ as $\epsilon \to 0$. Further, this solution is asymptotically orbitally stable if

$$\epsilon \frac{\partial \bar{F}}{\partial J}(J_0;0) < 0$$

and unstable if

$$\epsilon \frac{\partial \bar{F}}{\partial J}(J_0;0) > 0.$$

Proof: The proof is constructed along the lines of that for theorem 7.1 in section 7.2. First, we observe that, since $|\omega(I)|$ is bounded away from zero, it follows from the second equation in (1.9) that θ' is also bounded away from zero for sufficiently small $|\epsilon|$ and, hence, θ is a monotonic function of t. Thus we are entitled to regard I as a function of θ, which satisfies the equation

$$\frac{\mathrm{d}I}{\mathrm{d}\theta} = \epsilon K(I, \theta; \epsilon),$$

where

$$K(I, \theta; \epsilon) = \frac{F(I, \theta; \epsilon)}{\omega(I) + \epsilon G(I, \theta; \epsilon)}. \tag{1.17}$$

Next, let $I = I_0$ at $\theta = 0$, where there is no loss of generality in supposing that $\theta = 0$ coincides with $t = 0$, since the system (1.9) is autonomous, and let us denote the unique solution of this initial-value problem by $I(\theta; I_0, \epsilon)$, which is analytic in θ, I_0 and ϵ. Then put

$$\epsilon N(I_0, \epsilon) = I(2\pi; I_0, \epsilon) - I_0, \tag{1.18}$$

so that periodic solutions of (1.9) correspond to those values of I_0 for which $N(I_0, \epsilon) = 0$.

Now, from (1.17),

$$I = I_0 + \epsilon \int_0^\theta K(I, \hat{\theta}; \epsilon) \, \mathrm{d}\hat{\theta},$$

and so

$$N(I_0, \epsilon) = \int_0^{2\pi} K(I, \theta; \epsilon) \, \mathrm{d}\theta.$$

Thus $I = I_0 + O(\epsilon)$ (for $0 \leqslant \theta \leqslant 2\pi$) and, hence,

$$N(I_0, \epsilon) = \frac{2\pi \bar{F}(I_0; 0)}{\omega(I_0)} + O(\epsilon).$$

Thus, if

$$\bar{F}(J_0; 0) = 0, \qquad \frac{\partial \bar{F}}{\partial J_0}(J_0; 0) \neq 0,$$

the implicit-function theorem shows that there is a unique solution $I_0(\epsilon)$ of $N(I_0, \epsilon) = 0$, which is an analytic function of ϵ and such that $I_0(0) = J_0$. This is sufficient to establish the existence of a periodic solution $I(\theta)$ of (1.17) such that $I \to I_0$ as $\epsilon \to 0$.

To establish the stability, consider a nearby orbit for which $I = I_0^*$ at $\theta = 0$. Then $I(2\pi; I_0^*, \epsilon) = I_0^* + \epsilon N(I_0^*, \epsilon)$, and the periodic solution is asymptotically orbitally stable if $I_0^* \gtrless I_0$ implies that $\epsilon \omega N \leqslant 0$. Note that θ is an increasing (decreasing) function of t when ω is positive (negative). In the limit $\epsilon \to 0$, this condition is equivalent to

$$\epsilon \frac{\partial \bar{F}}{\partial J}(J_0; 0) < 0.$$

Conversely, the periodic solution is unstable if $I_0^* \gtrless I_0$ implies that $\epsilon \omega N \gtrless 0$, which, in the limit $\epsilon \to 0$, is equivalent to

$$\epsilon \frac{\partial \bar{F}}{\partial J}(J_0; 0) > 0.$$

In applying these results, let us first reconsider the perturbed simple harmonic oscillator, described by equation (1.1). In action–angle variables (I, θ), this is described by (1.7) and (1.8), so that $\bar{F}(I; \epsilon)$ (see (1.10)) is given by

$$\bar{F}(I; \epsilon) = \frac{\sqrt{2I}}{2\pi} \int_0^{2\pi} f(\sqrt{2I} \cos \theta, \sqrt{2I} \sin \theta; \epsilon) \sin \theta \ d\theta. \qquad (1.19)$$

To find periodic solutions, we let $I \to J_0$ as $\epsilon \to 0$, and consider the zeros of $\bar{F}(J_0; 0)$, which is here given by

$$\bar{F}(J_0; 0) = \frac{\sqrt{2J_0}}{2\pi} \int_0^{2\pi} f(\sqrt{2J_0} \cos \theta, \sqrt{2J_0} \sin \theta; 0) \sin \theta \ d\theta. \qquad (1.20)$$

Further, in the limit $\epsilon \to 0$, $u = R_0 \cos t + O(\epsilon)$, $\theta = -t + O(\epsilon)$ and $J_0 = \frac{1}{2} R_0^2$, where here $R_0 > 0$. It can then be shown that

$$\bar{F}(J_0; 0) = \frac{R_0 P_0(R_0)}{2\pi}, \qquad (1.21)$$

where $P_0(R_0)$ is defined by (2.12) in section 7.2. It follows that theorem 9.2 is equivalent to theorem 7.1 of section 7.2. Note that, if we put $J = \frac{1}{2} R^2$, so that

$u = R\cos t + O(\epsilon)$ and $\theta = -t + O(\epsilon)$, then the averaged equation (1.11) becomes

$$R' = \frac{\epsilon P_0(R)}{2\pi}.\tag{1.22}$$

The connection between the stability of the critical points R_0 of (1.22) and the stability of the corresponding periodic solutions of (1.1), as prescribed in theorem 7.1, is now immediate.

For instance, consider the Van der Pol equation, for which

$$f(u, u'; \epsilon) = (1 - u^2)u'.\tag{1.23}$$

Then, from (1.19), in this case

$$\bar{F}(I; \epsilon) = I(1 - \tfrac{1}{2}I)$$

and, hence, the averaged equation (1.11) is

$$J' = \epsilon J(1 - \tfrac{1}{2}J).\tag{1.24}$$

This has an asymptotically stable critical point $J = 2$, which corresponds to the Van der Pol limit cycle in the limit $\epsilon \to 0$. The general solution of the averaged equation is

$$J = 2\{1 + C\exp(-\epsilon t)\}^{-1},$$

where C is an arbitrary constant. Using theorem 9.1, we see that, at least for $|\epsilon|$ sufficiently small, all orbits spiral towards the limit cycle as $t \to \infty$.

As a second illustration, consider Duffing's equation with linear damping:

$$u'' + u = \epsilon u^3 - \nu u'.$$

Here the damping coefficient ν is a non-negative constant which we shall suppose is $O(\epsilon)$. When $\nu = 0$, the solutions of Duffing's equation were analyzed in section 6.4, where it was shown that in the x-y phase plane (where $x = u$ and $y = u'$) there is a family of periodic orbits surrounding the origin. In the limit $\epsilon \to 0$ these periodic solutions were analyzed in section 7.1. Here we wish to examine the effect of a small amount of damping on these solutions. Putting $\nu = \epsilon\hat{\nu}$, we see that here

$$f(u, u'; \epsilon) = u^3 - \hat{\nu}u'\tag{1.25}$$

and, hence, from (1.19),

$$\bar{F}(I; \epsilon) = -\hat{\nu}I,$$

and so the averaged equation (1.11) is

$$J' = -\nu J,$$

so that

$$J = J_0 e^{-\nu t},\tag{1.26}$$

where J_0 is an arbitrary constant. Hence $J \to 0$ as $t \to \infty$ and, using theorem 9.1, we can deduce that, at least for $|\epsilon|$ sufficiently small, the orbits spiral towards the origin. Note that the decay rate is just that for the linear oscillator, since here the nonlinear term (i.e. ϵu^3) does not contribute to the averaged equation. It should also be noted that this result, like that in the preceding paragraph for the Van der Pol equation, is restricted to orbits for which I and J are bounded independently of ϵ, since this assumption is implicit in the proof of theorem 9.1. In particular, the above result does not necessarily hold for orbits which pass near the saddle points at $x = \pm \epsilon^{-1/2}$, $y = 0$, since for these orbits I is of order ϵ^{-1}.

Further illustrative examples of the application of theorem 9.1 and 9.2 to (1.9) must wait until chapter 11 when we shall develop the theory of action–angle variables for Hamiltonian systems. It is also clear that the method of averaging has much wider applicability than the particular context, namely equations (1.9), that we have used here. In the next two sections we describe how averaging methods can be used to discuss forced oscillations and a certain class of non-autonomous equations with slowly varying coefficients. More comprehensive accounts of averaging methods can be found in Bogoliubov and Mitropolsky (1951), and Sanders and Verhulst (1985) (see also Arnold, 1983, Ch. 4, Chow and Hale, 1982, Ch. 12 and Guckenheimer and Holmes, 1983, Ch. 4).

9.2 Averaging methods for forced oscillations

To motivate the form of the equations to be discussed in this section, let us first reconsider equation (3.3) of section 8.3 of chapter 8:

$$u'' + u = \epsilon g(u, u', t; \epsilon),$$

where

$$g(u, u', t; \epsilon) = G_0 \cos t - \kappa u + f(u, u', t; \epsilon), \tag{2.1}$$

which for $|\epsilon|$ sufficiently small, describes the forced oscillations near resonance for a perturbation of a simple harmonic oscillator. Here we recall that the natural frequency of the oscillator is σ, where $\sigma^2 - 1 = \epsilon \kappa$ (see (3.1) of section 8.3) and $f(u, u', t; \epsilon)$ is a prescribed function of u, u', t and ϵ, which is an analytic function of all these variables and is a periodic function of t with period 2π. In chapter 8 we described a perturbation method for constructing periodic solutions of (1.1). Here we shall present an alternative procedure which allows us not only to recover the periodic solutions of (2.1) but also to obtain information about all the orbits of (2.1).

The first step is to convert (2.1) into a first-order system by letting

$$x = u, \qquad y = u'. \tag{2.2}$$

Then (2.1) becomes

$$x' = y, \qquad y' = -x + \epsilon g(x, y, t; \epsilon). \tag{2.3}$$

Next we introduce the change of variables

$$\begin{aligned} x &= a\cos t + b\sin t, \\ y &= -a\sin t + b\cos t, \end{aligned} \tag{2.4}$$

where a and b are the new dependent variables. The motivation for this comes from the observation that, when $\epsilon = 0$, the expressions (2.4) with a and b constant provide the general solution of (2.3). Hence, when $|\epsilon|$ is small, we can anticipate that a' and b' will be $O(\epsilon)$. To obtain the equations for a and b we first express (2.4) in the form

$$\begin{bmatrix} x \\ y \end{bmatrix} = X(t) \begin{bmatrix} a \\ b \end{bmatrix},$$

where

$$X(t) = \begin{bmatrix} \cos t & \sin t \\ -\sin t & \cos t \end{bmatrix}.$$

Here $X(t)$ is the fundamental matrix for the system when $\epsilon = 0$, and the change of variables (2.4) is, in effect, an application of the method of 'variation-of-parameters' for obtaining particular solutions for linear equations (see section 2.3). It follows that

$$X(t) \begin{bmatrix} a' \\ b' \end{bmatrix} = \begin{bmatrix} 0 \\ \epsilon g \end{bmatrix}$$

and, hence, the equations for a and b are

$$a' = -\epsilon \sin t\, g(x, y, t; \epsilon), \qquad b' = \epsilon \cos t\, g(x, y, t; \epsilon), \tag{2.5}$$

where, on the right-hand side, x and y are defined in terms of a and b by (2.4).

For an alternative formulation using polar coordinates, let

$$a = r\cos\phi, \qquad b = r\sin\phi, \tag{2.6}$$

so that the transformation (2.4) becomes

$$x = r\cos(t - \phi), \qquad y = -r\sin(t - \phi). \tag{2.7}$$

Then the equations (2.5) are replaced by

$$r' = -\epsilon \sin(t - \phi)\, g(x, y, t; \epsilon), \qquad r\phi' = \epsilon \cos(t - \phi)\, g(x, y, t; \epsilon), \tag{2.8}$$

where, on the right-hand sides, x and y are defined in terms of r and ϕ by (2.7). Note that, in terms of the polar coordinates (r, θ) (see (1.4)), $\theta = -t + \phi$.

Let us now consider the system of equations for u and v given by

$$u' = \epsilon F(u, v, t; \epsilon), \qquad v' = \epsilon G(u, v, t; \epsilon), \tag{2.9}$$

where F and G are periodic functions with respect to t, with period 2π. Also, for simplicity, we shall assume that F and G are analytic functions of u, v, t and ϵ.

Note that both (2.5) and (2.8) have the required form with (u, v) being replaced by (a, b) and (r, ϕ), respectively. To obtain approximate solutions of (2.9), we shall use the method of averaging. Thus, let

$$\begin{bmatrix} \bar{F}(u, v; \epsilon) \\ \bar{G}(u, v; \epsilon) \end{bmatrix} = \frac{1}{2\pi} \int_0^{2\pi} \begin{bmatrix} F(u, v, t; \epsilon) \\ G(u, v, t; \epsilon) \end{bmatrix} dt \tag{2.10}$$

denote the averaged values of (F, G). Then let $U(t)$ and $V(t)$ satisfy the equations

$$U' = \epsilon \bar{F}(U, V; 0), \qquad V' = \epsilon \bar{G}(U, V; 0). \tag{2.11}$$

Note that (2.11) can be regarded as an averaged version of the equations (2.9). We shall now show that $U(t)$ and $V(t)$ can be regarded as an approximation to $u(t)$ and $v(t)$, obtained by solving (2.9). Just as for the autonomous case discussed in section 9.1, the true solution for $u(t)$ and $v(t)$ contains small oscillations which can be averaged out, and the overall trend for $u(t)$ and $v(t)$ is described by (2.11).

Theorem 9.3: Let $U(t)$ and $V(t)$ satisfy (2.11) for $0 \leq |\epsilon t| \leq C_0$, where C_0 is a constant independent of ϵ. Then if $u(0) = U(0)$ and $v(0) = V(0)$, there exists a constant C_1 independent of ϵ, such that $|u(t) - U(t)| + |v(t) - V(t)| < C_1 \epsilon$ for $0 \leq |\epsilon t| \leq C_0$, provided that $|\epsilon|$ is sufficiently small.

Proof: The proof is analogous to that for theorem 9.1. To simplify the notation, let w be the 2-vector with components (u, v) and W the 2-vector with components (U, V). Then (2.9) and (2.11) are replaced by

$$w' = \epsilon H(w, t; \epsilon), \qquad W' = \epsilon \bar{H}(W; 0), \tag{2.12}$$

respectively. Here H is the 2-vector with components (F, G), $\bar{H}$ is the 2-vector with components $(\bar{F}, \bar{G})$, and $\bar{H}$ is clearly the average of H, so that (see (2.10))

$$\bar{H}(w; \epsilon) = \frac{1}{2\pi} \int_0^{2\pi} H(w, t; \epsilon) \, dt. \tag{2.13}$$

In this form the theorem will be seen to hold when w is an n-vector, although we shall use the result only when $n = 2$.

First, let

$$H(w, t; \epsilon) = \bar{H}(w; \epsilon) + \hat{H}(w, t; \epsilon), \tag{2.14}$$

where, from the definition (2.12), $\hat{H}$ is a periodic function of t with period 2π and zero mean. Next, put

$$K(w, t; \epsilon) = \int_0^t \hat{H}(w, \hat{t}; \epsilon) \, d\hat{t},$$

so that $K(w, t; \epsilon)$ is a periodic function of t with period 2π. Let

$$\hat{w} = w - \epsilon K. \tag{2.15}$$

Then, using (2.12), it follows that

$$\hat{w}' = w' - \epsilon \frac{\partial K}{\partial w} w' - \epsilon \frac{\partial K}{\partial t},$$

$$= \epsilon H - \epsilon^2 \frac{\partial K}{\partial w} H - \epsilon \hat{H},$$

or

$$\hat{w}' = \epsilon \bar{H}(\hat{w}; \epsilon) + O(\epsilon^2). \tag{2.16}$$

Here the $O(\epsilon^2)$ error term is a periodic function of t with period 2π. Relative to this error term, we have effectively converted the first equation (2.12) into the averaged equation. To complete the proof, let

$$z = \hat{w} - W, \tag{2.17}$$

so that, using (2.12) and (2.16),

$$z' = \epsilon \{ \bar{H}(\hat{w}; \epsilon) - \bar{H}(W; 0) \} + O(\epsilon^2),$$

or

$$z' = \epsilon \frac{\partial \bar{H}}{\partial W} (W; 0) z + O(\epsilon z^2) + O(\epsilon^2). \tag{2.18}$$

It can now be shown that $|z|$ is $O(\epsilon)$ for $0 \le |\epsilon t| \le C_0$. The proof is analogous to that for theorem 9.1 (see (1.16) and the subsequent discussion), and will be omitted here.

Theorem 9.3 enables us to use the averaged equation (2.11) as an approximation to (2.9). This has the advantage that the non-autonomous system (2.9) is replaced by the autonomous system (2.11), for which the machinery of chapters 5 and 6 is available. One of the main uses is to find solutions of (2.9) which are periodic functions of t with period 2π. It is intuitively clear that the critical points of the averaged equation (2.11) should correspond to periodic solutions of (2.9), with the stability also corresponding. Indeed, the following result holds.

Theorem 9.4: Under the conditions of theorem 9.3, let (U_0, V_0) be an isolated critical point of the averaged equations (2.11), so that

$$\bar{F}(U_0, V_0; 0) = \bar{G}(U_0, V_0; 0) = 0 \quad \text{and} \quad \frac{\partial(\bar{F}, \bar{G})}{\partial(U_0, V_0)} \ne 0. \tag{2.19}$$

Then there exists a periodic solution of (2.9), with period 2π, for which $u \to U_0$ and $v \to V_0$ as $\epsilon \to 0$. Further, this solution is uniformly and asymptotically stable, or unstable, according as the critical point (U_0, V_0) has the corresponding stability properties considered as a solution of (2.11), that is, the solution is uniformly and asymptotically stable if

$$\frac{\partial(\bar{F}, \bar{G})}{\partial(U_0, V_0)} > 0 \quad \text{and} \quad \epsilon\left(\frac{\partial\bar{F}}{\partial V_0} + \frac{\partial\bar{G}}{\partial V_0}\right) < 0, \tag{2.20}$$

and is unstable if either of these inequalities is reversed.

Proof: The proof is constructed along the lines of that for theorem 8.2 of section 8.3. As in the proof of theorem 9.3, we simplify the notation by letting w be the 2-vector with components (u, v) and W the 2-vector with components (U, V), so that (2.9) and (2.11) are replaced by (2.12). Next, let $w = w_0$ at $t = 0$, and denote the unique solution of this initial-value problem by $w(t; w_0, \epsilon)$, which is analytic in t, w_0 and ϵ. Then put

$$\epsilon N(w_0, \epsilon) = w(2\pi; w_0, \epsilon) - w_0, \tag{2.21}$$

so that periodic solutions of (2.9) correspond to those values of w_0 for which $N(w_0, \epsilon) = 0$. Now, from (2.12),

$$w = w_0 + \epsilon \int_0^t H(w, \hat{t}; \epsilon) \, d\hat{t},$$

and so

$$N(w_0, \epsilon) = \int_0^{2\pi} H(w, t; \epsilon) \, dt.$$

Thus, $w = w_0 + O(\epsilon)$ (for $0 \leq t \leq 2\pi$) and, hence,

$$N(w_0, \epsilon) = 2\pi\bar{H}(w_0; 0) + O(\epsilon).$$

Thus if $\bar{H}(W_0; 0) = 0$ and the Jacobian,

$$\frac{\partial(\bar{F}, \bar{G})}{\partial(U_0, V_0)} \neq 0,$$

the implicit-function theorem shows that there is a unique solution $w_0(\epsilon)$ of $N(w_0, \epsilon) = 0$ which is an analytic function of ϵ and such that $w_0(0) = W_0$. This is sufficient to establish the existence of a periodic solution $w(t)$ of (2.9) such that $w \to W_0$ as $\epsilon \to 0$.

To establish the stability, we replace $w(t)$ in (2.9) by $w(t) + \hat{w}$, where $w(t)$ is the periodic solution whose stability is being tested. Using the principle of linearized stability (see section 4.3), we may assume that $\hat{w}$ satisfies the first variational equation:

$$\hat{w}' = \epsilon\hat{A}(t)\hat{w},$$

where

$$\hat{A}(t) = \frac{\partial H}{\partial w}(w(t), t; \epsilon).$$

Here $\hat{A}(t)$ is a periodic matrix function of t with period 2π. From (3.5) of section 4.3, the principal fundamental matrix for this equation is $\partial w / \partial w_0$, and the characteristic multipliers ρ, which determine the stability, are the eigenvalues of

$$\frac{\partial w}{\partial w_0}(t = 2\pi) = E + \epsilon \frac{\partial N}{\partial w_0}.$$

Here the right-hand side has been determined from (2.21), and we recall that E is the unit matrix. Thus, in the limit $\epsilon \to 0$, the characteristic multipliers ρ are the eigenvalues of

$$E + 2\pi\epsilon \frac{\partial \overline{H}}{\partial W_0}(W_0; 0) + O(\epsilon^2)$$

and, hence, are given by

$$\rho = 1 + \epsilon \rho_1 + O(\epsilon^2),$$

where

$$\rho_1 = \alpha_0 \pm \sqrt{\alpha_0^2 - \beta_0}$$

and

$$2\alpha_0 = 2\pi\left(\frac{\partial \overline{F}}{\partial U_0} + \frac{\partial \overline{G}}{\partial V_0}\right), \qquad \beta_0 = (2\pi)^2 \frac{\partial(\overline{F}, \overline{G})}{\partial(U_0, V_0)}.$$

The stability result (2.20) now follows immediately. Note that $\epsilon\rho_1/2\pi$ are just the eigenvalues of the 2×2 constant coefficient matrix obtained when the averaged equations (2.11) are linearized about a critical point (U_0, V_0).

In applying these results, let us reconsider the forced oscillations of the perturbed simple harmonic oscillator (2.1). Using the variables a and b (see (2.4)), the governing equations are (2.5), for which the averaged equations are

$$A' = \frac{\epsilon P_0(A, B)}{2\pi}, \qquad B' = \frac{\epsilon Q_0(A, B)}{2\pi}, \qquad (2.22)$$

where

$$P_0(A, B) = -\int_0^{2\pi} g(x, y, t; 0)\sin t \, dt, \qquad Q_0(A, B) = \int_0^{2\pi} g(x, y, t; 0)\cos t \, dt.$$

Here x and y are obtained from (2.4), and so

$$x = A\cos t + B\sin t + O(\epsilon), \qquad y = -A\sin t + B\cos t + O(\epsilon), \qquad (2.23)$$

where we recall from (2.2) that $x = u$ and $y = u'$. Theorem 9.3 now shows that, with x and y given by (2.23), the solutions of (2.1) may be approximated by the solutions of (2.22). To find periodic solutions, we let $A, B \to A_0, B_0$ as

$\epsilon \to 0$, and consider the zeros of

$$P_0(A_0, B_0) = 0, \qquad Q_0(A_0, B_0) = 0. \qquad (2.24)$$

Thus, (A_0, B_0) are critical points for equations (2.22) and, hence, by theorem 9.4, define periodic solutions of (2.1). Further, it is readily seen that P_0 and Q_0 as defined by (2.22) are precisely the same expressions as defined by (3.11) of section 8.3 and, hence, theorem 9.4 is equivalent to theorem 8.2 of section 8.3. The connection between the stability of the critical points (A_0, B_0) of (2.22), and the stability of the corresponding periodic solutions of (1.1) as prescribed in theorem 8.2, is now immediate; that is, the stability criterion (3.14) of section 8.3 and (2.20) are identical.

In terms of the polar coordinates (r, ϕ) (see (2.6) and (2.7)), the governing equations are (2.8), for which the averaged equations are

$$R' = \frac{\epsilon M_0(R, \Phi)}{2\pi}, \qquad R\Phi' = \frac{\epsilon N_0(R, \Phi)}{2\pi}, \qquad (2.25)$$

where

$$M_0(R, \Phi) = -\int_0^{2\pi} g(x, y, t; 0)\sin(t - \Phi)\, dt,$$

$$N_0(R, \Phi) = \int_0^{2\pi} g(x, y, t; 0)\cos(t - \Phi)\, dt.$$

Here x and y are obtained from (2.7), and so

$$x = R\cos(t - \Phi) + O(\epsilon), \qquad y = -R\sin(t - \Phi) + O(\epsilon). \qquad (2.26)$$

It is readily seen that

$$A = R\cos\Phi, \qquad B = R\sin\Phi,$$

and so

$$M_0(R, \Phi) = P_0\cos\Phi + Q_0\sin\Phi, \qquad N_0(R, \Phi) = -P_0\sin\Phi + Q_0\cos\Phi. \qquad (2.27)$$

It is then a simple exercise to show that the averaged equations (2.25) are equivalent to (2.22). In many instances, the polar form (2.25) is easier to analyze. For periodic solutions, as $A, B \to A_0, B_0$ let $R, \Phi \to R_0, \Phi_0$ as $\epsilon \to 0$ and, in this limit, the critical points of (2.25) are the zeros of

$$M_0(R_0, \Phi_0) = 0, \qquad N_0(R_0, \Phi_0) = 0.$$

It can now be readily established that M_0 and N_0 are precisely the same expressions as defined by (3.25) of section 8.3. In particular, we recall that stability requires that $\alpha_0 < 0$ and $\beta_0 > 0$, where α_0 and β_0 are defined in Cartesian form by (3.19) of section 8.3 and in polar form by (3.28) of section 8.3.

As an illustration, let us consider the resonant oscillations for Duffing's equation, which we discussed in detail in section 8.4. Recalling (4.1) of section 8.4, we see that here

$$f(u, u', t; \epsilon) = u^3, \qquad g(u, u', t; \epsilon) = G_0 \cos t - \kappa u + u^3, \qquad (2.28)$$

where G_0 is the amplitude of the forcing term and κ is the tuning parameter. In section 8.4 we determined the response amplitude, R_0, for resonant oscillations of period 2π, as a function of G_0 and κ. The results are displayed in the response diagram shown in figure 8.3. Here we are able to set that analysis in a wider context by using the averaged equations (2.22). We find that $P_0(A, B)$ and $Q_0(A, B)$ are defined by (4.2) of section 8.4 and, hence, the averaged equations (2.22) are

$$A' = \tfrac{1}{2}\epsilon B\{\kappa - \tfrac{3}{4}(A^2 + B^2)\}, \qquad B' = \tfrac{1}{2}\epsilon[G_0 - A\{\kappa - \tfrac{3}{4}(A^2 + B^2)\}]. \qquad (2.29)$$

In the polar form (2.25), we find that $M_0(R, \Phi)$ and $N_0(R, \Phi)$ are defined by (4.3) of section 8.4 and, hence, the averaged equations (2.25) are

$$R' = \tfrac{1}{2}\epsilon G_0 \sin \Phi, \qquad R\Phi' = \tfrac{1}{2}\epsilon[G_0 \cos \Phi - R(\kappa - \tfrac{3}{4}R^2)]. \qquad (2.30)$$

With the transformation (2.27) this is readily seen to be equivalent to (2.29). The critical points of (2.29), or (2.30), are (R_0, Φ_0), where here $\Phi_0 = 0, \pi$ and

$$R_0(\kappa - \tfrac{3}{4}R_0^2) = \pm G_0,$$

the alternate signs referring to $\Phi_0 = 0$ and π. This agrees, of course, with the result (4.4) obtained in section 8.4 and sketched in figure 8.3. Using a linearized analysis about each critical point, it is readily shown that the upper branch ($\Phi_0 = \pi$) is a centre, while the lower branch ($\Phi_0 = 0$) is a saddle point on its upper unstable portion, and a centre on its lower portion. A typical portrait of the A-B phase plane is shown in figure 9.2 for a case when κ is such that there are three critical points (i.e. $\kappa > \tfrac{9}{4}(\tfrac{2}{3}G_0)^{2/3}$). In obtaining these orbits it is helpful to observe that (2.29) possesses an integral

$$\tfrac{1}{2}\kappa R^2 - \tfrac{3}{16}R^4 - G_0 R \cos \Phi = E,$$

where E is a constant. Note that there is a double saddle connection and three

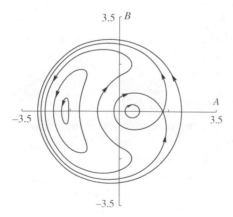

Fig. 9.2 The orbits for equation (2.29). The orbits shown are for $G_0 = 1$ and $\kappa = 2.4$

continuous families of periodic orbits, each separated from the other by the saddle connection. Also note that the existence of a periodic orbit for the averaged equations does not generally imply the existence of a periodic solution for the full equation (2.9), firstly, since the period of the orbit for the averaged equations may not be commensurate with 2π and, secondly, because of the approximate nature of the averaged equations.

For Duffing's equation with linear damping (2.28) is replaced by

$$f(u, u', t; \epsilon) = u^3 - \hat{v}u', \qquad g(u, u', t; \epsilon) = G_0 \cos t - \kappa u - \hat{v}u' + u^3, \quad (2.31)$$

where the damping coefficient $v = \epsilon\hat{v}$. $P_0(A, B)$ and $Q_0(A, B)$ are now defined by (4.7) of section 8.4 and, consequently, (2.29) is replaced by

$$A' = \tfrac{1}{2}\epsilon B\{\kappa - \tfrac{3}{4}(A^2 + B^2)\} - \tfrac{1}{2}\nu A,$$

$$B' = \tfrac{1}{2}\epsilon[G_0 - A\{\kappa - \tfrac{3}{4}(A^2 + B^2)\}] - \tfrac{1}{2}\nu B. \qquad (2.32)$$

For the polar form, $M_0(R, \Phi)$ and $N_0(R, \Phi)$ are now defined by (4.8) of section 8.4 and, consequently, (2.30) is replaced by

$$R' = \tfrac{1}{2}\epsilon G_0 \sin\Phi - \tfrac{1}{2}\nu R, \qquad R\Phi' = \tfrac{1}{2}\epsilon[G_0 \cos\Phi - R(\kappa - \tfrac{3}{4}R^2)]. \qquad (2.33)$$

The critical points (R_0, Φ_0) are now given by $G_0 \sin\Phi_0 = \hat{v}R_0$ and

$$R_0^2\{\hat{v}^2 + (\kappa - \tfrac{3}{4}R_0^2)^2\} = G_0^2,$$

which agrees with (4.9) of section 8.4. The response diagram is plotted in figure 8.4 for a case when the friction coefficient is not too large. The upper stable branch is a stable spiral point for $\kappa < \tfrac{3}{4}R_0^2$; at $\kappa = \tfrac{3}{4}R_0^2$, $dR_0/d\kappa = 0$ (see (4.10) of section 8.4) and, for $\kappa > \tfrac{3}{4}R_0^2$ until the turning point where $d\kappa/dR_0 = 0$ is reached, it is a stable node. Beyond this turning point the middle unstable branch is a saddle point until the second turning point, where $d\kappa/dR_0 = 0$, is reached. The lower stable branch is a stable node for $\kappa < \tfrac{9}{4}R_0^2$ and a stable spiral point for $\kappa > \tfrac{9}{4}R_0^2$. The orbits in the A-B phase plane must now be obtained numerically, and this is left for the exercises (see also Jordan and Smith, 1977, section 7.2, or Guckenheimer and Holmes, 1983, section 4.2). Note, however, that theorem 6.1 of section 6.1 may be used to show that there are no periodic orbits if $v > 0$.

Next we consider the resonant oscillations for the forced Van der Pol equation, which we discussed in detail in section 8.5. Recalling (5.1) of section 8.5, we see that here

$$f(u, u', t; \epsilon) = (1 - u^2)u', \qquad g(u, u', t; \epsilon) = G_0 \cos t - \kappa u + (1 - u^2)u'. \quad (2.34)$$

In section 8.5 the response diagram was obtained and sketched in figure 8.5. To obtain the averaged equations, we follow the procedure described above for the forced Duffing's equation, and find that $P_0(A, B)$ and $Q_0(A, B)$ are defined by (5.2) of section 8.5, so that the averaged equations (2.22) become

$$A' = \tfrac{1}{2}\epsilon\{\kappa B + A - \tfrac{1}{4}A(A^2 + B^2)\},$$

$$B' = \tfrac{1}{2}\epsilon\{G_0 - \kappa A + B - \tfrac{1}{4}B(A^2 + B^2)\}. \tag{2.35}$$

In the polar form, we find that $M_0(R, \Phi)$ and $N_0(R, \Phi)$ are defined by (5.3) of section 8.5, so that the averaged equations (2.25) become

$$R' = \tfrac{1}{2}\epsilon(G_0 \sin\Phi + R - \tfrac{1}{4}R^3), \qquad R\Phi' = \tfrac{1}{2}\epsilon(G_0 \cos\Phi - \kappa R). \tag{2.36}$$

The critical points of (2.35), or (2.36), are (R_0, Φ_0), where $G_0 \cos\Phi_0 = \kappa R_0$ and

$$R_0^2\{\kappa^2 + (1 - \tfrac{1}{4}R_0^2)^2\} = G_0^2.$$

This agrees, of course, with (5.4) of section 8.5, which is sketched in figure 8.5. In determining the stability and type for the critical points we must compute α_0 and β_0 (see (3.19) of section 8.3). For the present case α_0 and β_0 were calculated in (5.6):

$$2\alpha_0 = \pi(2 - R_0^2), \qquad \beta_0 = \pi^2\{\kappa^2 + (1 - \tfrac{1}{4}R_0^2)(1 - \tfrac{3}{4}R_0^2)\}.$$

Now, the eigenvalues of the coefficient matrix obtained when the averaged equations (2.35), or (2.36), are linearized about a critical point (R_0, Φ_0) are

$$\frac{\epsilon}{2\pi}\sqrt{\alpha_0 \pm (\alpha_0^2 - \beta_0)}.$$

Hence, taking the case $\epsilon > 0$, the critical point is a saddle point if $\beta_0 < 0$, a stable (unstable) node if $\alpha_0^2 > \beta_0 > 0$ and $\alpha_0 < 0$ ($\alpha_0 > 0$), and a stable (unstable) spiral point if $\beta_0 > \alpha_0^2$ and $\alpha_0 < 0$ ($\alpha_0 > 0$). The boundaries between these types are given by $\alpha_0 = 0$, $\beta_0 = 0$ and $\alpha_0^2 = \beta_0$; the first two are obtained from the expressions above for α_0 and β_0, and the boundary $\alpha_0^2 = \beta_0$ is just $|\kappa| = \tfrac{1}{4}R_0^2$. These boundaries and the corresponding types of critical points are sketched in figure 9.3. This diagram must now be matched to the response diagram shown in figure 8.5 to determine the type of each critical point. The general picture is clearly quite intricate and the details are left to the exercises (see also Guckenheimer and Holmes, 1983, section 2.1).

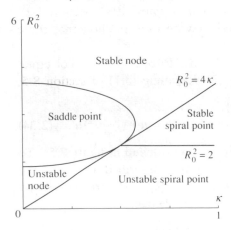

Fig. 9.3 The classification diagram for the critical points of (2.35). The diagram is symmetric in κ and, hence, only $\kappa \geqslant 0$ is shown.

9.3 Adiabatic invariance

In section 9.1 we described an averaging method for autonomous equations. Here we shall present a useful extension of that technique for certain classes of non-autonomous equations. These equations are characterized by the presence of two intrinsic time scales, which are widely separated. To motivate the form of equations to be discussed, we consider

$$u'' + \omega^2(\tau)u = 0, \tag{3.1}$$

which is the equation for a simple harmonic oscillator of frequency ω. Here, however, the frequency is itself a function of time, since we let

$$\tau = \epsilon t, \tag{3.2}$$

where ϵ (> 0) is a small parameter. The simple harmonic oscillator has a local time scale $2\pi\omega^{-1}$, but ω itself varies on a time scale of ϵ^{-1}.

First, let us convert (3.1) into a first-order system by letting

$$x = u, \qquad y = u', \tag{3.3}$$

so that (3.1) becomes

$$x' = y, \qquad y' = -\omega^2(\tau)x. \tag{3.4}$$

Now, assuming without loss of generality that $\omega > 0$, we put

$$x = \sqrt{\frac{2I}{\omega}}\sin\theta, \qquad y = \sqrt{2\omega I}\cos\theta,$$

or

$$I = \frac{1}{2\omega}(y^2 + \omega^2 x^2), \qquad \tan\theta = \frac{\omega x}{y}. \tag{3.5}$$

I and θ can be recognised as a form of polar coordinates. In the form chosen here, I and θ are action–angle variables for the simple harmonic oscillator, a topic which we shall develop in more detail in chapter 11. Note that $\frac{1}{2}\pi - \theta$ is the conventional polar angle, but we have preferred the form (3.5) to agree with our later definition of action–angle variables in chapter 11. In terms of (I, θ), equations (3.4) become

$$I' = -\epsilon\frac{\omega'(\tau)}{\omega(\tau)}I\cos 2\theta, \qquad \theta' = \omega(\tau) + \epsilon\frac{\omega'(\tau)}{2\omega(\tau)}\sin 2\theta. \tag{3.6}$$

Here $\omega'(\tau)$ denotes the derivative of $\omega(\tau)$ with respect to τ.

More generally, we shall show in chapter 11 that the appropriate equations for describing a nonlinear oscillator with two intrinsic time scales are

$$I' = \epsilon F(I, \theta, \tau; \epsilon), \qquad \theta' = \omega(I, \tau) + \epsilon G(I, \theta, \tau; \epsilon), \tag{3.7}$$

where F and G are periodic functions with respect to θ, with period 2π. Also we shall assume that F, G and ω are analytic functions. These equations are the appropriate extension of the autonomous equations (1.9). To obtain approximate solutions of (3.7), we again use the method of averaging. Thus, let

$$\bar{F}(I,\tau;\epsilon) = \frac{1}{2\pi} \int_0^{2\pi} F(I,\theta,\tau;\epsilon)\, d\theta \qquad (3.8)$$

be the averaged value of F, and let J satisfy the averaged equation

$$J' = \epsilon\bar{F}(J,\tau;0). \qquad (3.9)$$

Note that, using the transformation (3.2), this equation is

$$\frac{dJ}{d\tau} = \bar{F}(J,\tau;0) \qquad (3.10)$$

and, hence, we are entitled to assume that $J = J(\tau)$ is a function of the slow time variable τ. We shall now show that $J(\tau)$ can be regarded as an approximation to $I(t)$, obtained by solving (3.7).

Theorem 9.5: Let $J(\tau)$ satisfy (3.10) for $0 \leqslant |\tau| \leqslant C_0$ (i.e. $0 \leqslant |\epsilon t| \leqslant C_0$), where C_0 is a constant independent of ϵ, and suppose that $|\omega(I,\tau)|$ is bounded away from zero for all finite $|I|$ and all τ. Then, if $I(0) = J(0)$, there exists a constant C_1, independent of ϵ, such that $|I(t) - J(\tau)| < C_1\epsilon$ for $0 \leqslant |\tau| \leqslant C_0$, provided that $|\epsilon|$ is sufficiently small.

Proof: The proof is analogous to that for theorem 9.1. Thus, let

$$F(I,\theta,\tau;\epsilon) = \bar{F}(I,\tau;\epsilon) + \hat{F}(I,\theta,\tau;\epsilon),$$

and put

$$H(I,\theta,\tau;\epsilon) = \int_0^\theta \hat{F}(I,\hat{\theta},\tau;\epsilon)\, d\hat{\theta}. \qquad (3.11)$$

Since $\hat{F}$ is periodic with respect to θ with zero mean, H is also a periodic function of θ. Now put

$$\hat{I} = I - \epsilon H\omega^{-1}, \qquad (3.12)$$

and then, using (3.7) and (3.11), it follows that

$$\begin{aligned}
\hat{I}' &= I' - \epsilon\frac{\partial}{\partial I}\left(\frac{H}{\omega}\right)I' - \epsilon\frac{\partial H}{\partial\theta}\frac{\theta'}{\omega} - \epsilon^2\frac{\partial}{\partial\tau}\left(\frac{H}{\omega}\right) \\
&= \epsilon(\bar{F}+\hat{F}) - \epsilon^2\frac{\partial}{\partial I}\left(\frac{H}{\omega}\right)F - \epsilon\hat{F}\left(1+\frac{\epsilon G}{\omega}\right) - \epsilon^2\frac{\partial}{\partial\tau}\left(\frac{H}{\omega}\right) \\
&= \epsilon\bar{F}(I,\tau;\epsilon) + O(\epsilon^2)
\end{aligned}$$

or

$$\hat{I}' = \epsilon\bar{F}(\hat{I},\tau;\epsilon) + O(\epsilon^2). \qquad (3.13)$$

Here the $O(\epsilon^2)$ error term is a periodic function of θ with period 2π. Relative to this error term, we have effectively converted the first equation of (1.9) into the averaged equation (3.9). To complete the proof, let

$$z = \hat{I} - J, \qquad (3.14)$$

so that, using (3.9) and (3.13),

$$z' = \epsilon\{\bar{F}(\hat{I}, \tau; \epsilon) - \bar{F}(J, \tau; 0)\} + O(\epsilon^2),$$

or

$$z' = \epsilon \frac{\partial \bar{F}}{\partial J}(J, \tau; 0)\, z + O(\epsilon z^2) + O(\epsilon^2). \qquad (3.15)$$

It can now be shown that $|z|$ is $O(\epsilon)$ for $0 \le |\epsilon t| \le C_0$. The proof is analogous to that for theorem 9.1 and will be omitted here.

To illustrate the application of this theorem, let us consider equation (3.1) for a simple harmonic oscillator with a variable frequency. In the form appropriate to use the method of averaging, the equation is (3.6). Since the average (3.8) of $\cos 2\theta$ is zero, the averaged equation (3.9) is here

$$J' = 0$$

and, hence, $J = J(0)$ is a constant. It follows from theorem 9.5 that $I = I(0) + O(\epsilon)$ for $0 \le |\tau| \le C_0$, where C_0 is a constant independent of ϵ. In effect the action variable I is approximately a constant, and is said to be an **adiabatic invariant**.

It is interesting to observe here that it is the action I, rather than the energy E, which is the adiabatic invariant. Here, for equation (3.1), the energy E is given by

$$E = \tfrac{1}{2}(y^2 + \omega^2 x^2),$$

and we see that $I = E/\omega$. In terms of E,

$$\omega x = \sqrt{2E} \sin \theta, \qquad y = \sqrt{2E} \cos \theta,$$

and equations (3.6) are replaced by

$$E' = 2\epsilon \frac{\omega'(\tau)}{\omega(\tau)} E \sin^2 \theta, \qquad \theta' = \omega(\tau) + \epsilon \frac{\omega'(\tau)}{2\omega(\tau)} \sin 2\theta.$$

Application of theorem 9.5 now shows that E can be approximated by F, where

$$F' = \epsilon \frac{\omega'(\tau)}{\omega(\tau)} F.$$

It follows that F/ω is a constant, and we have recovered the result that $I = E/\omega$ is the adiabatic invariant. The point to notice here is that the definition of I in (3.5) is designed to ensure that the right-hand side of the equation for I (see (3.6)) has a zero average.

9.4 Multi-scale methods

It is apparent from the analysis of section 9.3 that the approximate action variable J, which satisfies equation (3.9), is a function of τ, where we recall from (3.2) that $\tau = \epsilon t$; indeed, we showed that (3.9) is equivalent to (3.10) and so $J = J(\tau)$. Recalling the full equations (3.7) and the averaging theorem (theorem 9.5) which links the solutions of (3.7) to J, we can deduce that the solutions of (3.7) exhibit two time scales, a rapid variation characterized by a local time scale of $2\pi\omega^{-1}$ and a slow variation of ϵ^{-1} exhibited by a dependence on the slow time variable τ. Similar conclusions can be deduced for the solutions u and v of (2.9) which can be expected from theorem 9.3 to have a dependence on both the fast time variable t and the slow time variable τ, since the averaged variables U and V which satisfy (2.11) depend only on τ.

From these remarks we anticipate that it may be possible to obtain directly approximate solutions which exhibit this multi-scale structure. To illustrate the procedure, let us first reconsider Duffing's equation with linear damping, which we discussed at the end of section 9.1. Thus, let

$$u'' + u = \epsilon u^3 - \nu u', \tag{4.1}$$

where the damping coefficient ν is $O(\epsilon)$, and so we put $\nu = \epsilon\hat{\nu}$. Then seek a solution of the form

$$u = u(t, \tau; \epsilon), \quad \text{where } \tau = \epsilon t. \tag{4.2}$$

In the multi-scale formalism the variables t and τ are regarded as independent, so that

$$u' = u_t + \epsilon u_\tau, \qquad u'' = u_{tt} + 2\epsilon u_{t\tau} + \epsilon^2 u_{\tau\tau}. \tag{4.3}$$

Here the subscripts denote partial derivatives. Next let

$$u = u_0(t, \tau) + \epsilon u_1(t, \tau) + \ldots, \tag{4.4}$$

so that, on substituting (4.3) and (4.4) into (4.1) and equating to zero successive coefficients of ϵ, we find that, at the leading order,

$$u_{0tt} + u_0 = 0 \tag{4.5}$$

and, at the first order in ϵ,

$$u_{1tt} + u_1 = -2u_{0t\tau} + u_0^3 - \hat{\nu}u_{0t}. \tag{4.6}$$

From (3.6), the general solution is

$$u_0 = A(\tau)\cos t + B(\tau)\sin t. \tag{4.7}$$

At this stage, A and B are constants with respect to the fast variable t, but are functions of the slow variable τ, and are yet to be determined. Substituting (4.7) into the right-hand side of (4.6), we obtain

$$u_{1tt} + u_1 = f_1, \tag{4.8}$$

where

$$f_1 = 2A_\tau \sin t - 2B_\tau \cos t + (A \cos t + B \sin t)^3 + \hat{\nu}(A \sin t - B \cos t).$$

Since the left-hand side of this equation for u_1 contains only derivatives with respect to t, it is an ordinary differential equation with independent variable t whose general solution is readily obtained. Using standard trigonometric identities, f_1 is seen to be a linear expression in $\cos t$, $\sin t$, $\cos 3t$ and $\sin 3t$. The corresponding particular solutions for u_1 are $\frac{1}{2}t\sin t$, $-\frac{1}{2}t\cos t$, $-\frac{1}{8}\cos 3t$ and $-\frac{1}{8}\sin 3t$, respectively. The first two of these particular solutions are secular, since they describe a resonant growth in u_1 on the fast timescale. Such solutions are incompatible with the philosophy of a multi-scale expansion, and must be removed by equating to zero the coefficients of the secular-producing, or resonant, terms $\cos t$ and $\sin t$ in f_1. The result is

$$2A_\tau = -\hat{\nu}A - \tfrac{3}{4}B(A^2 + B^2), \qquad 2B_\tau = -\hat{\nu}B + \tfrac{3}{4}A(A^2 + B^2). \tag{4.9}$$

These form a pair of ordinary differential equations in τ from which $A(\tau)$ and $B(\tau)$ are determined. Thus, at this stage, the leading-order term u_0 (4.7) has been constructed. With the condition (4.9) we next solve (4.8) for u_1 to obtain

$$u_1 = A_1(\tau)\cos t + B_1(\tau)\sin t - \tfrac{1}{32}(A^3 - 3AB^2)\cos 3t + \tfrac{1}{32}(B^3 - 3A^2B)\sin 3t, \tag{4.10}$$

where A_1 and B_1 are, at this stage, undetermined functions of τ. They are found by considering the equation for u_2, and again removing the secular-producing terms. The procedure can, clearly, be repeated indefinitely.

To solve (4.9) for A and B we introduce the polar coordinates (R, Φ), where

$$A = R\cos\Phi, \qquad y = R\sin\Phi, \tag{4.11}$$

and note that then

$$u_0 = R\cos(t - \Phi). \tag{4.12}$$

It is readily shown that

$$R_\tau = -\tfrac{1}{2}\hat{\nu}R,$$

so that

$$R = R_0 \exp\{-\tfrac{1}{2}\hat{\nu}\tau\}$$

and

$$\Phi_\tau = \tfrac{3}{8}R^2. \tag{4.13}$$

Note that, since $\nu = \epsilon\hat{\nu}$ and $\tau = \epsilon t$,

$$R = R_0 \exp\{-\tfrac{1}{2}\nu t\},$$

and this agrees with the result (1.26) obtained from the averaging method of section 9.1, since here the approximate action variable J is just $\frac{1}{2}R^2$, to leading order in ϵ. In the absence of any damping (i.e. $\hat{\nu} = 0$), $R = R_0$ and

$$\Phi = \Phi_0 + \tfrac{3}{8}R_0^2\tau.$$

In this instance, the expression (4.12) for u_0 becomes

$$u_0 = R_0 \cos\{(1 - \tfrac{3}{8}\epsilon R_0^2)t - \Phi_0\}.$$

Hence we can deduce the presence of periodic orbits of period $2\pi\omega^{-1}$, where, to leading order in ϵ,

$$\omega = 1 - \tfrac{3}{8}\epsilon R_0^2.$$

This result agrees with that obtained from the Poincaré–Lindstedt method described in section 7.1 (see (1.23) of section 7.1).

At least in this particular case, we see that the results obtained from the multi-scale method agree with those obtained using other approximation procedures. Indeed, more can be said, since, if the multi-scale hypothesis (4.4) is applied to equation (2.1), which includes (1.1) as a special case, then it is readily shown that

$$A_\tau = (2\pi)^{-1}P_0(A,B), \qquad B_\tau = (2\pi)^{-1}Q_0(A,B), \qquad (4.14)$$

where $P_0(A,B)$ and $Q_0(A,B)$ are defined by (2.22). Indeed, these equations are precisely the same equations (2.22) obtained by the averaging method described in section 8.2 and, hence, to leading order in ϵ, the multi-scale method gives precisely the same results as the averaging method. Hence, as anticipated, the averaging theorems (theorems 9.3 and 9.4) provide an indirect justification of the multi-scale method.

In applying the multi-scale method, we have assumed so far that the fast time variable is simply t. However, this is not appropriate for equations of the form (3.1), since in that case the local time scale is $2\pi\omega^{-1}$, which itself varies on the slow time scale τ. Hence we must determine an appropriate fast time scale as part of the multi-scale procedure. Thus we replace the hypothesis (4.2) by

$$u = u(s,\tau;\epsilon),$$

where

$$s = \frac{1}{\epsilon}\int_0^\tau S(\tau')\,\mathrm{d}\tau'. \qquad (4.15)$$

Here s is the fast time variable, constructed so that its derivative with respect to t is

$$s_t = S(\tau),$$

which varies only on the slow time scale. The function $S(\tau)$ has yet to be determined. As an illustration, let us reconsider equation (3.1). With u given by (4.15),

$$u' = S(\tau) u_s + \epsilon u_\tau, \qquad u'' = S^2(\tau) u_{ss} + 2\epsilon S(\tau) u_{\tau s} + \epsilon S_\tau(\tau) u_s + \epsilon^2 u_{\tau\tau}, \quad (4.16)$$

and we let

$$u = u_0(s, \tau) + \epsilon u_1(s, \tau) + \dots, \quad (4.17)$$

so that, on substituting (4.16) and (4.17) into (3.1), we find that

$$S^2(\tau) u_{0ss} + \omega^2(\tau) u_0 = 0, \qquad S^2(\tau) u_{1ss} + \omega^2(\tau) u_1 = -2S(\tau) u_{0\tau s} - S_\tau(\tau) u_{0s}. \quad (4.18)$$

The general solution for u_0 is

$$u_0 = R(\tau)\cos\{\lambda(\tau) s - \Phi(\tau)\},$$

where

$$\lambda(\tau) = \frac{\omega(\tau)}{S(\tau)}, \quad (4.19)$$

and R and Φ are constants with respect to the fast variable s, but are functions of the slow variable τ, and are yet to be determined. We now observe that we must choose λ equal to a constant, as otherwise $u_{0\tau}$ will contain secular terms which are proportional to s. We choose $\lambda = 1$ so that $S(\tau) = \omega(\tau)$, and the fast variable s (4.15) is

$$s = \frac{1}{\epsilon} \int_0^\tau \omega(\tau') \, d\tau'. \quad (4.20)$$

Thus u_0 is periodic in s, with period 2π, and we shall impose this condition on all terms in the expansion (4.17) for u.

The equation (4.18) for u_1 is now

$$u_{1ss} + u_1 = f_1,$$

where

$$f_1 = \left(\frac{2R_\tau}{\omega} + \frac{R\omega_\tau}{\omega^2}\right)\sin(s - \Phi) - \frac{2R\Phi_\tau}{\omega}\cos(s - \Phi). \quad (4.21)$$

Again we must equate to zero the coefficients of the secular-producing, or resonant, terms in f_1. Here these are $\sin(s - \Phi)$ and $\cos(s - \Phi)$ and, hence, we obtain

$$\frac{2R_\tau}{R} = -\frac{\omega_\tau}{\omega}, \qquad \Phi_\tau = 0. \quad (4.22)$$

This pair of equations now determines R and Φ. We see that

$$\tfrac{1}{2}R^2\omega = J_0, \qquad \Phi = \Phi_0, \quad (4.23)$$

where J_0 and Φ_0 are constants. The first of these results agrees with the result obtained in section 9.3 using averaging methods, since here the action variable I (see (3.5)) is given by

$$I = \frac{1}{2\omega}(u'^2 + \omega^2 u^2),$$

and the result obtained above shows that $I = J_0 + O(\epsilon)$. It is also pertinent to notice here that the expression (4.20) for s and the result in (4.23) for Φ agree, to leading order in ϵ, with the determination of the phase variable θ from (3.6). The details are left to the exercises.

Once again, we see that the averaging theorem, in this case theorem 9.5, can be used as an indirect justification of the multi-scale method. Some justification of this kind is necessary as the multi-scale formalism alone gives no indication of the region of validity of the approximation. That some caution is needed can be seen by considering the case when $\omega(\tau)$ in (3.1) is a periodic function with respect to τ of period T. If the mean value of $\omega(\tau)$ is $\bar{\omega}$, then the Floquet theory for linear equations with periodic coefficients described in chapter 3 shows that unstable solutions are to be expected whenever $\bar{\omega}T$ is approximately $\epsilon n\pi$, where n is an integer (see (2.21) of section 3.2). Note that the period of ω with respect to t is $T\epsilon^{-1}$. On the other hand, the multi-scale formalism, for which the leading-order result is (4.23), gives no indication of unstable behaviour nor, indeed, of the necessity to modify the approximation procedure when the conditions for parametric resonance are met (i.e. $\bar{\omega}T \approx \epsilon n\pi$). Of course, the averaging method remains valid and, in particular, theorem 9.5 holds even when $\omega(\tau)$ is a periodic function. However, we note that theorem 9.5 validates the averaging procedure only for $0 \leqslant |\tau| \leqslant C_0$, where C_0 is a constant independent of ϵ, and it is clear that, in the presence of parametric resonance, this is, in general, the best that can be achieved.

Problems

1 For each of the following equations, use the averaging method of section 9.1 and theorems 9.1 and 9.2 to obtain the limit cycles as $\epsilon \to 0$, and determine the stability. Compare with the Poincaré–Linstedt method (problem 2 of chapter 7) and numerical integration (problem 3 of chapter 7).

   (a) $u'' + \epsilon(5u^4 - 1)u' + u = 0$;

   (b) $u'' + \epsilon(u^2 - 1)u' + u - \epsilon u^3 = 0$;

   (c) $u'' + \epsilon(u^4 - 10u^2 + 8)u' + u = 0$.

2 For the equation

$$u'' + u = \epsilon f(u) - \nu u',$$

where $\nu \, (> 0)$ is $O(\epsilon)$, use the averaging method of section 9.1 to describe the orbits in the u-u' phase plane. Note that, when $\nu = 0$, the equation is that for a conservative nonlinear oscillator and, for sufficiently small $|\epsilon|$, there is a family of periodic orbits enclosing the origin (see section 6.4).

3 Use the transformations (1.4) and (1.6) to put the equations

$$x' = y + \epsilon F(x, y; \epsilon), \qquad y' = -x + \epsilon G(x, y; \epsilon),$$

into the form (1.9). Then use theorems 9.1 and 9.2 to obtain the limit cycles as $\epsilon \to 0$, and determine the stability. Compare with the Poincaré–Lindstedt method described in problem 6 of chapter 7. Illustrate when

$$F(x, y; \epsilon) = x - \tfrac{1}{3}x^3, \qquad G = -y^3,$$

and compare with problem 7(b) of chapter 7.

4 Use the transformations (1.4) and (1.6) to put the equations

$$x' = y(1 + r^2) + \epsilon(x - \tfrac{1}{3}x^3), \qquad y' = -x(1 + r^2) + \epsilon y,$$

into the form (1.9). Then use the averaging method of section 9.1 to describe the orbits.

5 For each of the following equations, use the averaging method of section 9.1 and theorems 9.3 and 9.4 to obtain the forced resonant oscillations of period 2π as $\epsilon \to 0$, and determine the stability. Compare with the perturbation method of chapter 8 (problem 4 of chapter 8).

(a) $u'' + \sigma^2 u = \epsilon G_0 \cos t + \epsilon(u^3 - u^5)$;

(b) $u'' + \sigma^2 u = \epsilon G_0 \cos t + \epsilon(u^3 + u^5)$;

(c) $u'' + \sigma^2 u = \epsilon G_0 \cos t - \nu u^2 u'$, where ν is $O(\epsilon)$.

Here $|\sigma^2 - 1|$ is $O(\epsilon)$.

6 For the forced Duffing's equation, with linear damping:

$$u'' + \nu u' + \sigma^2 u - \epsilon u^3 = \epsilon G_0 \cos t,$$

where ν is $O(\epsilon)$ and $\sigma^2 = 1 + \epsilon\kappa$, use the averaging method of section 9.2 to show that forced resonant oscillations are described by (2.32), where

$$u = A \cos t + B \sin t + O(\epsilon).$$

For these equations (2.32) determine the critical points and use numerical integration to determine the orbits in the A-B phase plane. Compare with the results obtained from numerical integration of the equation for u.

7 For the forced Van der Pol equation:

$$u'' + \epsilon(u^2 - 1)u' + \sigma^2 u = \epsilon G_0 \cos t,$$

where $\sigma^2 = 1 + \epsilon\kappa$, use the averaging method of section 9.2 to show that forced resonant oscillations are described by (2.35), where

$$u = A \cos t + B \sin t + O(\epsilon).$$

For these equations (2.35) determine the critical points and verify the classification shown in figure 9.5. Use numerical integration to determine the orbits in the A-B phase plane, and compare with the results obtained from numerical integration of the equation for u.

8 For the forced Duffing's equation:

$$u'' + \sigma^2 u = F_0 \cos 3t + \epsilon u^3,$$

where $\sigma^2 = 1 + \epsilon \kappa$, use the transformation

$$u = a\cos t + b\sin t - \tfrac{1}{8}F_0\cos 3t, \qquad u' = -a\sin t + b\cos t + \tfrac{3}{8}F_0\sin 3t,$$

to show that forced subharmonic oscillations are described by the equations

$$a' = -\epsilon \sin t(u^3 - \kappa u), \qquad b' = \epsilon \cos t(u^3 - \kappa u).$$

Here the right-hand sides are functions of a, b and t which are periodic in t with period 2π, by virtue of the above expression for u. Since these equations have the required form (2.9), use the averaging method of section 9.2 and theorems 9.3 and 9.4 to describe the subharmonic oscillations. Compare with the perturbation method used in chapter 8 (see problem 10 of chapter 8).

9 Use the averaging method of section 9.2 to discuss parametric resonance for Mathieu's equation (see section 3.3):

$$u'' + (\delta + \epsilon \cos 2t)u = 0,$$

when $\delta = 1 + O(\epsilon)$ and $|\epsilon| \ll 1$. Thus, put

$$x = u, \qquad y = u',$$

and

$$x = a\cos t + b\sin t, \qquad y = -a\sin t + b\cos t,$$

and then show that

$$a' = \epsilon \sin t(\kappa + \cos 2t)(a\cos t + b\sin t), \qquad b' = -\epsilon \cos t(\kappa + \cos 2t)(a\cos t + b\sin t),$$

where

$$\delta = 1 + \epsilon \kappa.$$

Now derive the averaged equations

$$A' = \tfrac{1}{2}\epsilon(\kappa - \tfrac{1}{2})B, \qquad B' = -\tfrac{1}{2}\epsilon(\kappa + \tfrac{1}{2})A,$$

and use the solutions to obtain approximate solutions for a and b and, hence, for u. Compare the result with the perturbation method used in section 3.3.

10 For the linear oscillator:

$$u'' + \nu(\tau)u' + \omega^2(\tau)u = 0, \qquad \tau = \epsilon t,$$

with a variable damping coefficient $\nu(\tau)$ as well as a variable frequency $\omega(\tau)$, use the transformations (3.3) and (3.5) to show that the equation analogous to (3.6) is

$$I' = -\epsilon \frac{\omega'(\tau)}{\omega(\tau)} I \cos 2\theta - 2\nu(\tau) I \sin^2 \theta, \qquad \theta' = \omega(\tau) + \tfrac{1}{2}\left(\epsilon \frac{\omega'(\tau)}{\omega(\tau)} + \nu(\tau)\right)\sin 2\theta.$$

Then use theorem 9.5 to obtain an approximate solution from the averaged equation

$$J' = -\nu(\tau)J,$$

assuming that $\nu(\tau)$ is $O(\epsilon)$. Deduce that, if

$$\omega(\tau) = \omega_0 \exp\left\{\int_0^\tau \nu(s) \, ds\right\},$$

where ω_0 is a constant, then the energy $E \, (= \omega I)$ of the oscillator is constant to this order of approximation.

11 For the linear oscillator of problem 9, suppose that the damping coefficient ν and the frequency ω are constants and, hence, show that, for $\nu^2 < 4\omega^2$,

$$u = r_0 \exp(-\tfrac{1}{2}\nu t)\cos\sigma(t - t_0),$$

where

$$\sigma^2 = \omega^2 - \tfrac{1}{4}\nu^2,$$

and r_0 and t_0 are arbitrary constants. Then use the transformations (3.3) and (3.5) to show that

$$2\omega I = r_0^2 \exp(-\nu t)\{\omega^2 + \tfrac{1}{2}\sigma\nu \sin 2\sigma(t - t_0) + \tfrac{1}{4}\nu^2 \cos 2\sigma(t - t_0)\}.$$

Compare this exact result with the approximate solution obtained using the averaging theorem (theorem 9.1 or 9.5).

12 For the linear oscillator (3.1), we have shown that the transformations (3.3) and (3.5) lead to (3.6) for the action–angle variables (I, θ). Use the method described in the proof of theorem 9.5 to show that the transformation (see (3.12))

$$\hat{I} = I + \epsilon\frac{\omega'(\tau)}{2\omega^2}I\sin 2\theta,$$

leads to the following equation for $\hat{I}$:

$$\hat{I}' = \epsilon^2\left(\frac{\omega'(\tau)}{2\omega^2}\right)'\hat{I}\sin 2\theta + O(\epsilon^3).$$

Adapt the proof of theorem 9.5 to deduce that $\hat{I} = \hat{I}(0) + O(\epsilon^2)$ for $0 \leqslant |\epsilon t| \leqslant C_0$, where C_0 is a constant independent of ϵ. This is an example of second-order averaging.

13 Use the multi-scale hypothesis (4.2) to obtain approximate solutions for each of the equations in problem 1, and compare with the approximate solutions obtained using the averaging method of section 9.1.

14 For Duffing's equation:

$$u'' + u = \epsilon u^3,$$

use the multi-scale hypothesis (4.2) to show that

$$u = u_0(t, \tau) + \epsilon u_1(t, \tau) + \epsilon^2 u_2(t, \tau) + O(\epsilon^3),$$

where

$$u_0 = A(\tau)\cos t + B(\tau)\sin t, \qquad u_1 = -\tfrac{1}{32}(A^3 - 3AB^2)\cos 3t + \tfrac{1}{32}(B^3 - 3A^2B)\sin 3t.$$

Here $\tau = \epsilon t$ (see 14.2)), and $A(\tau)$ and $B(\tau)$ satisfy the equations

$$2A_\tau - \epsilon B_{\tau\tau} = -\tfrac{3}{4}B(A^2 + B^2) + \tfrac{3}{128}\epsilon B(A^2 + B^2)^2 + O(\epsilon^2),$$

$$2B_\tau - \epsilon A_{\tau\tau} = \tfrac{3}{4}A(A^2 + B^2) + \tfrac{3}{128}\epsilon A(A^2 + B^2)^2 + O(\epsilon^2).$$

Use the polar-coordinate transformation (4.11) to obtain (approximate) solutions of these equations. [*Hint*: Proceed as in the text example to find the first approximate equation (4.9) for A and B, and equation (4.10) for u_1. Then, from the equation for u_2, remove the secular-producing terms to obtain equations for A_1 and B_1. Finally, combine the equations for A, B and A_1, B_1 by substituting $A + \epsilon A_1$ and $B + \epsilon B_1$ for A and B, respectively.]

15 For the damped linear oscillator:

$$u'' + \nu u' + \omega^2 u = 0,$$

suppose that ν is $O(\epsilon)$ and use the multi-scale hypothesis (4.2) to construct an approximate solution correct to $O(\epsilon^2)$. Compare your result with the exact solution constructed in problem 9.

16 For the damped linear oscillator with variable coefficients of problem 10, use the multi-scale hypothesis (4.15) to construct an approximate solution, assuming that $\nu(\tau)$ is $O(\epsilon)$ and $|\omega(\tau)|$ is bounded away from zero. Compare your answer with the results obtained from the averaging method in problem 10.

17 For the linear oscillator with a variable frequency (see (3.1)):

$$u'' + \omega(\tau) u = 0,$$

use the multi-scale hypothesis (4.15) to construct an approximate solution correct to $O(\epsilon^2)$. Compare your result with that obtained in problem 12 using second-order averaging.

CHAPTER TEN
ELEMENTARY BIFURCATION THEORY

10.1 Preliminary notions

In general, bifurcation theory is concerned with dynamical systems which contain one, or more, external parameters, and with the manner in which the solution set may undergo structural changes as the parameter, or parameters, are varied. As we shall see, such behaviour is essentially determined by the stability of solutions and the manner in which this may change as the parameter, or parameters, vary. In this chapter we shall give an elementary and introductory account of bifurcation theory. For more advanced accounts the reader is referred to the texts by Chow and Hale (1982), Guckenheimer and Holmes (1983), and Iooss and Joseph (1980).

For simplicity, we shall consider only first-order autonomous systems which depend on a single parameter μ. Thus we let

$$x' = f(x;\mu) \tag{1.1}$$

where we shall suppose, again for simplicity, that f is an analytic function of x and μ. Let $x_0(\mu)$ be a critical point of (1.1), so that

$$f(x_0(\mu);\mu) = 0.$$

To determine the stability of $x_0(\mu)$ we let

$$z = x - x_0(\mu)$$

so that

$$z' = A(\mu)z + O(z^2),$$

where

$$A(\mu) = \frac{\partial f}{\partial x}(x_0(\mu);\mu). \tag{1.2}$$

Now we have shown in chapter 4 (see also section 5.1) that, using the principle of linearized stability, the stability of the critical point $x_0(\mu)$ is essentially determined by the eigenvalues of the matrix $A(\mu)$. Let us denote these by $\lambda_1(\mu), \lambda_2(\mu), \ldots, \lambda_n(\mu)$. Then, from theorem 4.4, if all the eigenvalues have a negative real part, then the critical point is uniformly and asymptotically stable. On the other hand, from theorem 4.5, if one or more eigenvalues have a positive real part, then the critical point is unstable.

The new feature here is that, since f depends on the parameter μ, the matrix A and, hence, its eigenvalues, also depend on μ. Thus the stability may change as the parameter μ varies. Indeed, it is clear from the preceding paragraph, that a change of stability can be expected to occur whenever, for $\mu = \mu_0$, say,

$$\operatorname{Re}\lambda_i(\mu_0) = 0 \quad \text{(for some } i), \qquad \operatorname{Re}\lambda_j(\mu_0) < 0 \quad \text{(for all } j \neq i). \qquad (1.3)$$

The values of μ_0 for which (1.3) holds define the **bifurcation points**. Assuming that $\operatorname{Re}\lambda_i(\mu)$ changes sign as μ passes through μ_0, the bifurcation points correspond to a change of stability. The nomenclature indicates that, associated with the change of stability, we can anticipate that there will be a bifurcation, that is, the character of the solution will undergo a structural change as μ passes through μ_0.

Since we are assuming that $f(x;\mu)$ is an analytic function of x and μ, it follows from the implicit-function theorem that $x_0(\mu)$ will be an analytic function of μ, provided that

$$\det\left[\frac{\partial f}{\partial x}(x_0(\mu);\mu)\right] \neq 0.$$

From (1.2), this is equivalent to the requirement that $\det A(\mu) \neq 0$ or, equivalently, that $A(\mu)$ has no zero eigenvalues. Thus, apart possibly from the bifurcation points, we can assume that $x_0(\mu)$ is an analytic function of μ and, hence, $A(\mu)$ is also an analytic function of μ. Note, however, that this analyticity may fail at a bifurcation point. Next, since the eigenvalues are determined by

$$\det[\lambda E - A(\mu)] = 0,$$

which is an nth-degree polynomial equation in λ whose coefficients are analytic functions of μ, it follows, again from the implicit-function theorem, that distinct eigenvalues are also analytic functions of μ, except possibly at the bifurcation points μ_0. For simplicity, in this text we shall assume the generic case in which all the eigenvalues of $A(\mu)$ are distinct (i.e. $\lambda_i(\mu) \neq \lambda_j(\mu)$ for $i \neq j$ and all μ near μ_0). Also we note that, since $A(\mu)$ is a real-valued matrix, the eigenvalues of $A(\mu)$ are either also real-valued or occur as complex-conjugate pairs. It follows that a bifurcation point can arise in one of two ways:

either (a) $\lambda_1(\mu)$ is real-valued, $\lambda_1(\mu_0) = 0$ and $\operatorname{Re}\lambda_i(\mu_0) < 0$ (for $i = 2,\ldots,n$);   (1.4)

or (b) $\lambda_1(\mu) = \overline{\lambda_2(\mu)} = \alpha(\mu) + i\beta(\mu)$ and $\lambda_2(\mu)$ form a complex-conjugate pair such that $\alpha(\mu_0) = 0$, $\beta(\mu_0) \neq 0$ and $\operatorname{Re}\lambda_i(\mu_0) < 0$ (for $i = 3,\ldots,n$).   (1.5)

We shall call case (a) **one-dimensional bifurcation**, since we shall show that, in essence, only the eigenspace corresponding to the single eigenvalue $\lambda_1(\mu)$ is active in the bifurcation process. Note that for a one-dimensional system (i.e.

$n = 1$) only case (a) can occur. We shall follow the current convention and call case (b) **Hopf bifurcation** in acknowledgement of the work of Hopf in the 1940s, although the essentials of the theory had been previously developed by Andronov in the 1930s. In figure 10.1 we show a schematic diagram of the variation of the eigenvalues as μ varies through μ_0 for a two-dimensional system (i.e. $n = 2$).

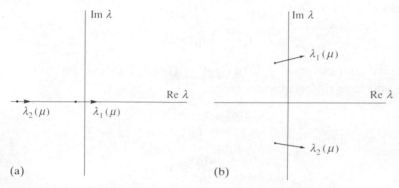

Fig. 10.1 Schematic diagrams for (a) one-dimensional bifurcation and (b) Hopf bifurcation.

In this situation, for case (a) both eigenvalues are real-valued and the bifurcation corresponds to the transition of the critical point $x_0(\mu)$ from a stable node into a saddle point. In case (b) both eigenvalues are complex-valued and the bifurcation corresponds to the transition of the critical point $x_0(\mu)$ from a stable spiral point into an unstable spiral point.

10.2 One-dimensional bifurcations

In this section we shall suppose that x is a scalar variable, and so (1.1) is a one-dimensional system (i.e. $n = 1$). In this situation, only case (a) (1.4) can occur. Let $x = \epsilon$ be a critical point, so that

$$f(\epsilon ; \mu) = 0. \tag{2.1}$$

Here, $\epsilon = \epsilon(\mu)$ is a single-valued analytic function of μ, provided that $\lambda_1(\mu) \neq 0$, where

$$\lambda_1(\mu) = f_x(\epsilon ; \mu). \tag{2.2}$$

Note that, in this one-dimensional situation, the matrix $A(\mu)$ (1.2) is 1×1 and equal to its single eigenvalue $\lambda_1(\mu)$. Hence the critical point is uniformly and asymptotically stable if $\lambda_1(\mu) < 0$ and unstable if $\lambda_1(\mu) > 0$. The bifurcation point μ_0 (see (1.3)) is here simply defined by $\lambda_1(\mu_0) = 0$. Using (2.1) and (2.2), we see that the bifurcation points are given by

$$f(\epsilon ; \mu) = f_x(\epsilon ; \mu) = 0. \tag{2.3}$$

The solution of these simultaneous equations locates the bifurcation points in the μ-ϵ plane.

Now, when $\lambda_1(\mu) \neq 0$, (2.1) can be solved to give $\epsilon = \epsilon(\mu)$, and then differentiation of

$$f(\epsilon(\mu); \mu) = 0$$

with respect to μ shows that

$$f_x \frac{d\epsilon}{d\mu} + f_\mu = 0.$$

Alternatively, suppose that $f_\mu(\epsilon; \mu) \neq 0$. Then (2.1) can be solved to give $\mu = \mu(\epsilon)$, and then differentiation of

$$f(\epsilon; \mu(\epsilon)) = 0$$

with respect to ϵ shows that

$$f_x + f_\mu \frac{d\mu}{d\epsilon} = 0,$$

or

$$\lambda_1(\mu) = -f_\mu \frac{d\mu}{d\epsilon}. \tag{2.4}$$

Hence, if $f_\mu \neq 0$ at the bifurcation point where $\lambda_1(\mu) = 0$, it follows that $d\mu/d\epsilon = 0$ there. A typical situation is sketched in figure 10.2(a). This kind of configuration is called a saddle–node bifurcation since, for a two-dimensional system, it corresponds to a situation where a critical point changes from a stable node to a saddle point.

Our aim now is to describe a local analysis of the system (1.1) in the vicinity of a bifurcation point. Without loss of generality, we shall suppose that the bifurcation point is $\epsilon = \mu = 0$, so that the condition (2.3) becomes

$$f(0; 0) = f_x(0; 0) = 0. \tag{2.5}$$

Before developing the general theory, it is instructive to consider the following examples which describe the four kinds of bifurcation usually encountered. Note that in each example the condition (2.5) is satisfied.

First consider

$$x' = \mu - x^2. \tag{2.6}$$

Here the critical points determined by (2.1) are $\epsilon^2 = \mu$ and this is plotted in figure 10.2(a). For (2.6),

$$f_x = -2x,$$

and so $\lambda_1 = -2\epsilon$, showing that the critical point is stable or unstable, according as $\epsilon > 0$ or $\epsilon < 0$. This information is also contained in figure 10.2(a) which is called a **bifurcation diagram**. This example is a typical illustration of a

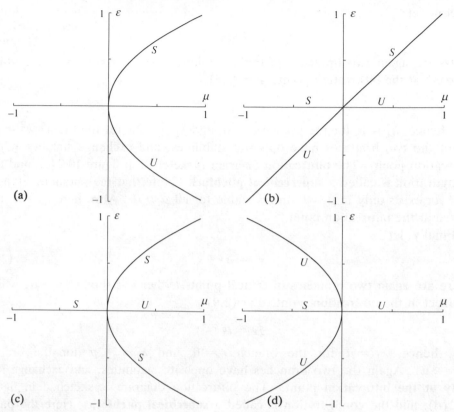

Fig. 10.2 Bifurcation diagrams. (a) Saddle–node (equation 2.6), (b) transcritical (equation 2.7), (c) supercritical pitchfork (equation 2.8) and (d) subcritical pitchfork (equation 2.9).

saddle–node bifurcation, in which a single branch of critical points undergoes a transition from a stable to an unstable state. Note that, for (2.6), $f_\mu = 1$ and, hence, does not vanish at the bifurcation point. Also note that, in the remainder of this chapter, we shall use the term stable in place of the more accurate description, uniformly and asymptotically stable.

Next consider the logistic equation

$$x' = \mu x - x^2. \tag{2.7}$$

There are now two branches of critical points, either $\epsilon = 0$ or $\epsilon = \mu$, which intersect at the bifurcation point. For (2.7),

$$f_x = \mu - 2x$$

and, hence, $\lambda_1 = \mu$ for the branch $\epsilon = 0$, and $\lambda_1 = -\mu$ for the branch $\epsilon = \mu$. Thus the two branches have opposite stabilities, and exchange stability at the bifurcation point. The bifurcation diagram is sketched in figure 10.2(b), and the configuration is called a **transcritical bifurcation**. Note that here $f_\mu = x$ and is zero at the bifurcation point.

Now let

$$x' = \mu x - x^3. \tag{2.8}$$

There are again two branches of critical points, either $\epsilon = 0$ or $\epsilon^2 = \mu$, which intersect at the bifurcation point. For (2.8),

$$f_x = \mu - 3x^2$$

and, hence, $\lambda_1 = \mu$ for the branch $\epsilon = 0$, and $\lambda_1 = -2\mu$ for the branch $\epsilon^2 = \mu$. Again the two branches have opposite stabilities, and exchange stability at the bifurcation point. The bifurcation diagram is sketched in figure 10.2 (c), and the configuration is called a **supercritical pitchfork**. Note that the parabolic branch, $\epsilon^2 = \mu$, exists only in $\mu \geqslant 0$ and is stable for all $\mu > 0$. Also, here $f_\mu = x$ and is zero at the bifurcation point.

Finally, let

$$x' = \mu x + x^3. \tag{2.9}$$

There are again two branches of critical points, either $\epsilon = 0$ or $\epsilon^2 = -\mu$, which intersect at the bifurcation point. For (2.9)

$$f_x = \mu + 3x^2$$

and, hence, $\lambda_1 = \mu$ for the branch $\epsilon = 0$, and $\lambda_1 = -2\mu$ for the branch $\epsilon^2 = -\mu$. Again the two branches have opposite stabilities, and exchange stability at the bifurcation point. The bifurcation diagram is sketched in figure 10.2 (d), and the configuration is called a **subcritical pitchfork**. Here the parabolic branch $\epsilon^2 = -\mu$ exists only in $\mu \leqslant 0$, and is unstable for all $\mu < 0$. Also, here $f_\mu = x$ and is zero at the bifurcation point.

Let us now return to a consideration of the general case (2.1), where $\epsilon = \mu = 0$ is a bifurcation point, so that (2.5) holds. The following theorem can be established.

Theorem 10.1: Let $f(\epsilon; \mu)$ be an analytic function of ϵ and μ in a neighbourhood of $\epsilon = \mu = 0$, and suppose that $\epsilon = \mu = 0$ is a bifurcation point so that (2.5) holds.
 (i) If $f_\mu(0; 0) \neq 0$, there exists, in some neighbourhood of the bifurcation point, a single branch of critical points which undergoes a saddle–node bifurcation at the bifurcation point.
 (ii) Suppose that $f_\mu(0; 0) = 0$, and let

$$D = f_{\mu\mu}(0; 0)f_{xx}(0; 0) - f_{x\mu}^2(0; 0). \tag{2.10}$$

Then, if $D > 0$, the bifurcation point is the only critical point in a neighbourhood of the origin, and is said to be an isolated point. On the other hand, if $D < 0$ there are two branches of critical points which intersect and exchange stability at the bifurcation point. The bifurcation is either transcritical or a pitchfork.

Proof: The classification as a saddle–node, transcritical or pitchfork bifurcation is based on the examples (2.6–9) discussed above and will be defined more precisely in the course of the proof.

Since $f(\epsilon;\mu)$ is analytic, we may write, in a neighbourhood of the bifurcation point,

$$f(\epsilon;\mu) = \alpha\mu + \tfrac{1}{2}a\mu^2 + b\mu\epsilon + \tfrac{1}{2}c\epsilon^2 + O(\epsilon^3, \epsilon^2\mu, \epsilon\mu^2, \mu^3),$$

where

$$\alpha = f_\mu(0;0), \qquad a = f_{\mu\mu}(0;0), \qquad b = f_{\mu x}(0;0), \qquad c = f_{xx}(0;0). \quad (2.11)$$

Here we have used the conditions (2.5), and we note from (2.10) that

$$D = ac - b^2. \tag{2.12}$$

Also, we note that

$$f_x(\epsilon;\mu) = b\mu + c\epsilon + O(\epsilon^2, \epsilon\mu, \mu^2),$$

$$f_\mu(\epsilon;\mu) = \alpha + a\mu + b\epsilon + O(\epsilon^2, \epsilon\mu, \mu^2). \tag{2.13}$$

(i) In this case $\alpha \neq 0$, and the implicit-function theorem establishes the existence of a solution $\mu = \mu(\epsilon)$ of (2.2) which is an analytic function of ϵ and such that $\mu(0) = 0$. From (2.11) it follows that

$$\mu = -\frac{c}{2\alpha}\epsilon^2 + O(\epsilon^3)$$

and, from (2.2) and (2.13),

$$\lambda_1 = c\epsilon + O(\epsilon^2).$$

Note that this last result could also be deduced from (2.4). It follows that $d\mu/d\epsilon = 0$ at the bifurcation point, where $\lambda_1 = 0$. In the generic case when $c \neq 0$, the eigenvalue λ_1 changes sign at the bifurcation point, near which the branch has the shape of a parabola. These are the characteristic features of a saddle–node bifurcation, for which a typical bifurcation diagram is shown in figure 10.2 (a) for the example (2.6).

(ii) For the case $\alpha = 0$, let us first suppose that $D > 0$, where D is defined by (2.12). Then the quadratic form

$$\tfrac{1}{2}a\mu^2 + b\mu\epsilon + \tfrac{1}{2}c\epsilon^2$$

cannot vanish for any real values of μ and ϵ except $\mu = \epsilon = 0$. It follows that the only solution of (2.1) in a neighbourhood of the bifurcation point is the bifurcation point itself.

Next suppose that $D < 0$, and at first assume that $c \neq 0$. In this case, put

$$\epsilon = \mu v,$$

and define

$$g(v;\mu) = \frac{f(\mu v;\mu)}{\mu^2}.$$

Note that, from (2.11), $\mu = 0$, $\epsilon \neq 0$ cannot be a solution of (2.1) near the bifurcation point and, hence, the solutions of

$$g(v;\mu) = 0$$

will define all the critical points, that is, all the solutions of (2.1). Now, from (2.11),

$$g(v;\mu) = \tfrac{1}{2}a + bv + \tfrac{1}{2}cv^2 + O(\mu).$$

Hence

$$g(v;0) = 0$$

has the two distinct solutions, $v = \gamma_1$ and $v = \gamma_2$, where

$$\gamma_1, \gamma_2 = \frac{-b \pm \sqrt{-D}}{c}.$$

Since $D < 0$, both these solutions are real-valued. Also,

$$g_v(\gamma_{1,2};0) = \pm\sqrt{-D} \neq 0$$

and, hence, the implicit-function theorem establishes the existence of two distinct solutions, $v = v_1(\mu)$ and $v = v_2(\mu)$ of $g(v;\mu) = 0$, which are analytic functions of μ and such that $v_1(0) = \gamma_1$ and $v_2(0) = \gamma_2$. Thus we have established the existence of two distinct branches of critical points $\epsilon = \mu v_1(\mu)$ and $\epsilon = \mu v_2(\mu)$ which intersect at the bifurcation point. Note that γ_1 and γ_2 are the slopes of these branches at the bifurcation point. Thus we have a transcritical bifurcation, for which a typical bifurcation diagram is shown in figure 10.2 (b) for the example (2.7).

For the stability of these branches we observe that, from (2.2) and (2.13),

$$\lambda_1 = b\mu + c\mu v + O(\mu^2) = \pm\mu\sqrt{-D} + O(\mu^2).$$

Hence the two branches have opposite stabilities, and exchange stability at the bifurcation point.

When $D < 0$ and $c = 0$, then $b \neq 0$ (see (2.12)). In this case we first proceed as above (for the case $c \neq 0$), putting $\epsilon = \mu v$ and defining $g(v;\mu) = f(\mu v;\mu)/\mu^2$, where now

$$g(v;\mu) = \tfrac{1}{2}a + bv + O(\mu).$$

Hence $g(v;0) = 0$ has just one solution $v = -a/2b$. Since

$$g_v\left(-\frac{1}{2b};0\right) = b \neq 0,$$

the implicit-function theorem establishes the existence of a solution $v(\mu)$ which is an analytic function of μ and such that $v(0) = -a/2b$. Thus we have established the existence of a branch of critical points $\epsilon = \mu v(\mu)$. For the stability we see that

$$\lambda_1 = b\mu + O(\mu^2)$$

and, hence, there is an exchange of stability at the bifurcation point.

To obtain a second branch, we put

$$\mu = \epsilon w$$

and define

$$h(\epsilon; w) = \frac{f(\epsilon; \epsilon w)}{\epsilon^2}.$$

From (2.11),

$$h(\epsilon; w) = \tfrac{1}{2}aw^2 + bw + O(\epsilon).$$

Now, $h(0; w) = 0$ has a solution $w = 0$ and, since

$$h_w(0; 0) = b \neq 0,$$

the implicit-function theorem establishes the existence of a solution $w(\epsilon)$ which is an analytic function of ϵ and such that $w(0) = 0$. Thus we have established the existence of a second branch of critical points $\mu = \epsilon w(\epsilon)$. Note that, since $w(0) = 0$, μ is $O(\epsilon^2)$. In the generic case $\mu = \beta\epsilon^2 + O(\epsilon^3)$. For the stability, we find from (2.2) and (2.13) that

$$\lambda_1 = -f_\mu(\epsilon; \epsilon w)\frac{d\mu}{d\epsilon} = -\{b\epsilon + O(\epsilon^2)\}\frac{d\mu}{d\epsilon}.$$

In the generic case $\lambda_1 = -2b\mu$ to leading order and, hence, this branch has the opposite stability to the first branch, where $\lambda_1 = b\mu$ to leading order (see the preceding paragraph). Note that, in the generic case, the second branch is approximately given by the parabola $\mu = \beta\epsilon^2$ near the bifurcation point, and is either always stable (if $b\beta > 0$) or always unstable (if $b\beta < 0$) provided, of course, that $\mu \neq 0$. Hence we have a supercritical pitchfork if $b\beta > 0$, or a subcritical pitchfork if $b\beta < 0$. Typical bifurcation diagrams are shown in figure 10.2(c) and 10.2(d) for the examples (2.8) and (2.9). Finally, note that, if $a \neq 0$, $h(0; w) = 0$ has a second solution $w = -2b/a$; however, this is clearly just the first branch in a different formulation.

To illustrate how these results are used in analyzing the solutions of (1.1), consider the example

$$x' = x(\mu - x^2)(x - \mu + 2). \tag{2.14}$$

Here there are three branches of critical points given by $\epsilon = 0$, $\epsilon^2 = \mu$ and

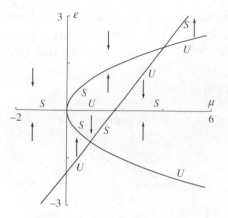

Fig. 10.3 The bifurcation diagram for equation (2.14).

$\epsilon = \mu - 2$, where we are again denoting x by ϵ for the critical points of (2.14). They are sketched in the μ-ϵ plane in figure 10.3.

Theorem 10.1 implies that the bifurcation points for transcritical and pitchfork bifurcations are determined by locating the places where the branches intersect, while saddle–node bifurcations are found by locating the points where $d\mu/d\epsilon = 0$. Hence for (2.14) the bifurcation points are

$$(1) \ \mu = 0, \ \ \epsilon = 0; \qquad (2) \ \mu = 2, \ \ \epsilon = 0;$$

$$(3) \ \mu = 1, \ \ \epsilon = -1; \qquad (4) \ \mu = 4, \ \ \epsilon = 2. \qquad (2.15)$$

Referring to figure 10.3, we see that (1) is the intersection of the branches $\epsilon = 0$ and $\epsilon^2 = \mu$ and is a pitchfork bifurcation, (2) is the intersection of $\epsilon = 0$ and $\epsilon = \mu - 2$ and is a transcritical bifurcation, while (3) and (4) are the intersections of $\epsilon^2 = \mu$ and $\epsilon = \mu - 2$ and are also transcritical bifurcations. To determine the stability, we must calculate λ_1 from (2.2), where here

$$f_x(x;\mu) = (\mu - x^2)(x - \mu + 2) - 2x^2(x - \mu + 2) + x(\mu - x^2). \qquad (2.16)$$

Hence λ_1 for each branch is given by

$$\lambda_1 = \begin{cases} \mu(2-\mu) & \text{(for } \epsilon = 0), \\ -2\epsilon^2(2+\epsilon-\epsilon^2) & \text{(for } \epsilon^2 = \mu), \\ \epsilon(2+\epsilon-\epsilon^2) & \text{(for } \epsilon = \mu - 2). \end{cases} \qquad (2.17)$$

Hence the branch $\epsilon = 0$ is stable for $\mu < 0$ and $\mu > 2$, but is unstable for $0 < \mu < 2$. The branch $\epsilon^2 = \mu$ is stable for $-1 < \epsilon < 2$, but is unstable for $\epsilon < -1$ and $\epsilon > 2$. The branch $\epsilon = \mu - 2$ is stable for $-1 < \epsilon < 0$ and $\epsilon > 2$, but is unstable for $\epsilon < -1$ and $0 < \epsilon < 2$. In agreement with the predictions of theorem 10.1, changes of stability occur at the bifurcation points (2.15). The complete bifurcation diagram is shown in figure 10.3, where we see, in particular, that the bifurcation point (1) ($\mu = 0$, $\epsilon = 0$) is a supercritical pitchfork bifurcation. Note also that there is, as required by theorem 10.1, an exchange of stability at each bifurcation point. Finally, we note that, while the bifurcation diagram shows only the critical points and their stability, in this one-dimensional case it can be used to infer the behaviour of the solutions of (2.14) as functions of t, for each fixed μ. Thus, for each fixed $\mu < 0$, there are two critical points, $x = 0$ and $x = \mu - 2$, which are, respectively, stable and unstable. Hence solutions which commence in $x > 0$, or in $\mu - 2 < x < 0$, are attracted to $x = 0$ as $t \to \infty$, while solutions which commence in $x < \mu - 2$ will pass to infinity. The arrows in figure 10.3 indicate this. Next, for each fixed μ $(0 < \mu < 1)$, there are four critical points, $x = \sqrt{\mu}$, 0, $-\sqrt{\mu}$ and $\mu - 2$, which are, respectively, stable, unstable, stable and unstable. Hence solutions which commence in the region $x > 0$ are attracted to the critical point $x = \sqrt{\mu}$, while solutions which commence in $\mu - 2 < x < 0$ are attracted to the critical point $x = -\sqrt{\mu}$; solutions which commence in $x < \mu - 2$ will pass to infinity. Again the arrows in figure 10.3 indicate this. A similar interpretation holds for the zones $1 < \mu < 2$, $2 < \mu < 4$ and $\mu > 4$, and the results are shown in figure 10.3.

10.3 One-dimensional bifurcations (continued)

Here we shall continue our discussion of one-dimensional bifurcations (i.e. case (a), see (1.4)), but will now consider first-order systems of the form (1.1), where $n \geqslant 2$. In fact, it will be sufficient to consider the planar case when $n = 2$, as the method we shall describe for dealing with this case can be used for the general case ($n \geqslant 2$) as well. For this planar case we shall suppose that the dependent variables are the scalars x and y and (1.1) is replaced by

$$x' = f(x, y; \mu), \qquad y' = g(x, y; \mu), \tag{3.1}$$

where f and g are analytic functions of x, y and μ. Without loss of generality, we may suppose that $x = y = 0$ and $\mu = 0$ is a bifurcation point. Then, on expanding f and g in a series expansion in powers of x and y, we see that (3.1) takes the form

$$\begin{bmatrix} x' \\ y' \end{bmatrix} = \begin{bmatrix} a(\mu) \\ b(\mu) \end{bmatrix} + A_0(\mu) \begin{bmatrix} x \\ y \end{bmatrix} + \begin{bmatrix} F(x, y; \mu) \\ G(x, y; \mu) \end{bmatrix}. \tag{3.2}$$

Here F and G are $O(r^2)$ as $r \to 0$, where $r^2 = x^2 + y^2$, and are analytic functions of x, y and μ. Also $a(\mu)$, $b(\mu)$ and the 2×2 matrix $A_0(\mu)$ are analytic functions of μ. Note that

$$A_0(\mu) = \begin{bmatrix} f_x & f_y \\ g_x & g_y \end{bmatrix} \quad (\text{at } x = y = 0), \tag{3.3}$$

and differs from the matrix $A(\mu)$ defined by (1.2), although, of course, they agree at the bifurcation point $\mu = 0$. Since $x = y = 0$ and $\mu = 0$ is a bifurcation point, it is a critical point for (3.2) and, hence,

$$a(0) = b(0) = 0. \tag{3.4}$$

Further, let $\gamma_1(\mu)$ and $\gamma_2(\mu)$ be the eigenvalues of $A_0(\mu)$. Since, at $\mu = 0$, these are also the eigenvalues of $A(\mu)$, it follows that, for a one-dimensional bifurcation (see case (a), (1.4)),

$$\gamma_1(0) = 0, \qquad \gamma_2(0) < 0. \tag{3.5}$$

Consequently, in a neighbourhood of $\mu = 0$, $\gamma_1(\mu)$ and $\gamma_2(\mu)$ are distinct real-valued eigenvalues of $A_0(\mu)$ and, hence, are analytic functions of μ. Further, we may assume that $\gamma_2(\mu) < 0$ for μ in a neighbourhood of zero.

In the preceding paragraph we have given the general formulation of a one-dimensional bifurcation for the planar system (3.2). However, in the subsequent theoretical development, it is useful to suppose that the matrix $A_0(\mu)$ has the canonical form (see section 5.2):

$$A_0(\mu) = \begin{bmatrix} \gamma_1(\mu) & 0 \\ 0 & \gamma_2(\mu) \end{bmatrix}. \tag{3.6}$$

This can always be achieved, at least in a neighbourhood of $\mu = 0$, by a linear transformation in x and y, which is analytic in μ and, being linear, preserves the property that F and G are $O(r^2)$ as $r \to 0$. But note that, in practical examples, this may not necessarily be a useful way to proceed. When $A_0(\mu)$ has the form (3.6) the planar system (3.2) becomes

$$x' = a(\mu) + \gamma_1(\mu)x + F(x, y; \mu), \qquad y' = b(\mu) + \gamma_2(\mu)y + G(x, y; \mu), \tag{3.7}$$

where we recall the bifurcation conditions (3.4) and (3.5). Note that now

$$f(x, y; \mu) = a(\mu) + \gamma_1(\mu)x + F(x, y; \mu), \qquad g(x, y; \mu) = b(\mu) + \gamma_2(\mu)y + G(x, y; \mu), \tag{3.8}$$

and we recall that the nonlinear terms F and G are $O(r^2)$ as $r \to 0$. We are now in a position to state the following theorem.

Theorem 10.2: Let $a(\mu)$, $b(\mu)$, $\gamma_1(\mu)$, $\gamma_2(\mu)$, $F(x, y; \mu)$ and $G(x, y; \mu)$ be analytic functions of their respective arguments in a neighbourhood of $x = y = 0$ and $\mu = 0$. Suppose that $x = y = 0$ and $\mu = 0$ is a bifurcation point for a one-dimensional bifurcation point, so that (3.4) and (3.5) hold. Then, for the planar system (3.7), if $a'(0) \neq 0$, there exists, in some neighbourhood of the bifurcation point, a single branch of critical points which undergoes a saddle–node bifurcation. If $a'(0) = 0$ and $D > 0$, the bifurcation point is an isolated point, where D is defined by (3.12). Finally, if $a'(0) = 0$ and $D < 0$, then there are two branches of critical points which intersect and exchange stability at the bifurcation point; the bifurcation is either transcritical or a pitchfork.

Proof: Comparing the statement of this theorem with theorem 10.1 for the one-dimensional case discussed in the preceding section, we see that it is completely analogous. Indeed, our proof of theorem 10.2 will be to show that the determination of the critical points of (3.7) can be reduced to the analysis of an equivalent one-dimensional, or scalar, problem. Indeed, this is our reason for the nomenclature, one-dimensional bifurcation.

The critical points of (3.7) are given by

$$f(x,y;\mu) = g(x,y;\mu) = 0,$$

where f and g are defined by (3.8). Now first consider $g(x,y;\mu) = 0$, which we shall regard as an equation for y as a function of x and μ. Since, from (3.8), using (3.4) and (3.5),

$$g(0,0;0) = 0, \qquad g_y(0,0;0) = \gamma_2(0) \neq 0,$$

it follows from the implicit-function theorem that $g(x,y;\mu) = 0$ has a unique solution,

$$y = Y(x;\mu), \tag{3.9}$$

which is an analytic function of x and μ in a neighbourhood of the bifurcation point $x = y = 0$ and $\mu = 0$. Further, it can be shown that

$$Y(x;\mu) = -\frac{b'(0)\mu}{\gamma_2(0)} + O(x^2, \mu x, \mu^2). \tag{3.10}$$

In particular, note that $Y_x(0;0) = 0$. Next, we define

$$\hat{f}(x;\mu) = f(x, Y(x;\mu);\mu) = a(\mu) + \gamma_1(\mu)x + F(x, Y(x;\mu);\mu), \tag{3.11}$$

so that the determination of the critical points of (3.8) is equivalent to finding the zeros of $\hat{f}(x;\mu)$. This is the equivalent one-dimensional problem, and the proof is completed by appealing to theorem 10.1. Note that

$$\hat{f}_x = f_x + f_y Y_x = \gamma_1(\mu) + F_x + F_y Y_x,$$

$$\hat{f}_\mu = f_\mu + f_y Y_\mu = a'(\mu) + \gamma_1'(\mu)x + F_\mu + F_y Y_\mu.$$

In particular, we see that

$$\hat{f}(0;0) = \hat{f}_x(0;0) = 0,$$

so that the conditions (2.5) for a one-dimensional bifurcation are met. Here we should recall that F and G are $O(r^2)$ as $r \to 0$ and, hence, their leading-order terms are quadratic in x and y. Further,

$$\hat{f}_\mu(0;0) = a'(0)$$

and, if we define

$$D = \hat{f}_{\mu\mu}(0;0)\hat{f}_{xx}(0;0) - \hat{f}_{x\mu}^2(0;0), \tag{3.12}$$

then we are in a position to use theorem 10.1.

To complete the proof it must be shown that the stability of the equivalent one-dimensional problem can be used to infer the stability of the corresponding planar problem. Now, the stability of the critical points of (3.7) is determined by the eigenvalues $\lambda_1(\mu)$ and $\lambda_2(\mu)$ of the matrix $A(\mu)$ (1.2), where here

$$A(\mu) = \begin{bmatrix} f_x & f_y \\ g_x & g_y \end{bmatrix}, \quad (\text{at } f = g = 0). \tag{3.13}$$

From (3.8), we see that, at the bifurcation point,

$$A = \begin{bmatrix} 0 & 0 \\ 0 & \gamma_2(0) \end{bmatrix} \quad (\text{for } x = y = 0) \quad \text{and} \quad \mu = 0.$$

Hence the eigenvalues are 0 and $\gamma_2(0)$, and we can deduce that $\lambda_1(0) = 0$ and $\lambda_2(0) = \gamma_2(0)$. Since $\gamma_2(0) < 0$ by hypothesis (see (3.5)), it follows that $\lambda_2(\mu) < 0$ in a neighbourhood of $\mu = 0$. Hence the stability is determined by the eigenvalue $\lambda_1(\mu)$. Note, in particular, that both eigenvalues $\lambda_1(\mu)$ and $\lambda_2(\mu)$ are real-valued in a neighbourhood of $\mu = 0$.

To find $\lambda_1(\mu)$, we first note that, since the product of the eigenvalues is equal to the determinant of $A(\mu)$,

$$\lambda_1 \lambda_2 = f_x g_y - f_y g_x.$$

But

$$g(x, Y(x, \mu); \mu) = 0,$$

and so

$$g_x + g_y Y_x = 0.$$

Thus,

$$\lambda_1 \lambda_2 = g_y(f_x + f_y Y_x).$$

But, from (3.11),

$$\hat{f}_x = f_x + f_y Y_x,$$

and so,

$$\lambda_1 \lambda_2 = g_y \hat{f}_x.$$

But $g_y(0, 0; 0) = \gamma_2(0)$, and we have shown above that $\lambda_2(0) = \gamma_2(0)$. Hence

$$\lambda_1 = \hat{f}_x\{1 + O(\mu, x)\}.$$

It follows that, in a neighbourhood of the bifurcation point, the sign of λ_1 is exactly that of $\hat{f}_x$. Hence the stability of the critical points of the planar problem is exactly the same as the stability of the equivalent one-dimensional problem. This completes the proof of theorem 10.2.

To illustrate how bifurcation diagrams are constructed for planar systems, we consider the example:

$$x' = x(\mu - x) - (x + 1)y^2, \qquad y' = y(x - 1). \tag{3.14}$$

This system can be regarded as a model of a Lotka–Volterra food chain, where

x and y represent the resource and predator, respectively. There are three critical points,

$$(1)\ \ x = 0,\ \ y = 0, \qquad (2)\ \ x = \mu,\ \ y = 0, \qquad (3)\ \ x = 1,\ \ 2y^2 = \mu - 1.$$

$$(3.15)$$

These are plotted in figure 10.4 as functions of the parameter μ.

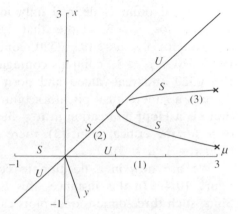

Fig. 10.4 The bifurcation diagram for equation (3.14). $\times$ denotes a Hopf bifurcation and, for $\mu > 3$, this branch is unstable. The three branches correspond to the classification in (3.15).

It can be seen that there are two bifurcation points. The first at $\mu = 0$ is the intersection of (1) and (2) and is a transcritical bifurcation; the second at $\mu = 1$ is the intersection of (2) and (3) and is a pitchfork bifurcation.

To determine the stability of each critical point, we must find the eigenvalues of the matrix $A(\mu)$, given by (3.13). Here

$$\begin{bmatrix} f_x & f_y \\ g_x & g_y \end{bmatrix} = \begin{bmatrix} \mu - 2x - y^2 & -2(x+1)y \\ y & x - 1 \end{bmatrix}.$$

Thus, for the critical point (1),

$$A(\mu) = \begin{bmatrix} \mu & 0 \\ 0 & -1 \end{bmatrix},$$

and so the eigenvalues are μ and -1. There is a bifurcation point at $\mu = 0$, and the critical point is stable for $\mu < 0$ and unstable for $\mu > 0$. Next, for the critical point (2),

$$A(\mu) = \begin{bmatrix} -\mu & 0 \\ 0 & \mu - 1 \end{bmatrix},$$

and so the eigenvalues are now $-\mu$ and $\mu - 1$. Hence this critical point is stable for $0 < \mu < 1$, and is unstable for $\mu < 0$ and $\mu > 1$. As we have already anticipated, there are bifurcation points at $\mu = 0$ and $\mu = 1$. Finally, for the critical point (3),

$$A(\mu) = \begin{bmatrix} y^2 - 1 & 4y \\ y & 0 \end{bmatrix},$$

where

$$2y^2 = \mu - 1.$$

Note that it is more convenient here to use y, rather than μ, as the argument of the matrix. The eigenvalues are now given by

$$\tfrac{1}{2}\{y^2 - 1 \pm \sqrt{(y^2 - 9)^2 - 80}\} \quad \text{or} \quad \tfrac{1}{4}\{\mu - 3 \pm \sqrt{(\mu - 19)^2 - 320}\}.$$

The critical point is defined only in $\mu \geq 1$, and is stable for $1 < \mu < 3$ and unstable for $\mu > 3$. Note that here the eigenvalues are real-valued and negative for $1 < \mu < 19 - \sqrt{320}$, complex-conjugates with negative real part for $19 - \sqrt{320} < \mu < 3$, complex-conjugates with positive real part for $3 < \mu < 19 + \sqrt{320}$ and real-valued and positive for $\mu > 19 + \sqrt{320}$. As already noted, there is a supercritical pitchfork bifurcation at $\mu = 1$. Further, we now see that there is a Hopf bifurcation at $\mu = 3$; indeed, since y is a double-valued function of μ for the critical point (3), there are two Hopf bifurcation points at $y = \pm 1$ (and $x = 1$, $\mu = 3$).

We are now in a position to construct the bifurcation diagram shown in figure 10.4. In this instance, this is a three-dimensional plot of x, y and μ. Since such three-dimensional plots can be relatively complicated, it is sometimes useful to plot two-dimensional cross-sections. Thus, here, let us examine the bifurcation diagram in the x-μ plane, shown in figure 10.5(a); this shows very clearly the transcritical bifurcation at $\mu = 0$, but does not reveal the correct structure of the pitchfork bifurcation at $\mu = 1$, although we note that its type can be inferred since the branch $x = 1$ exists only in $\mu > 1$.

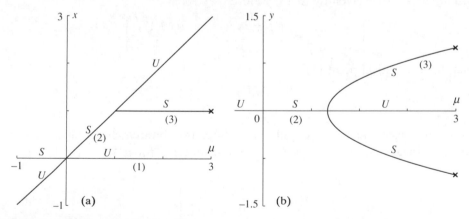

Fig. 10.5 Projection plots of the bifurcation diagram shown in figure 10.4 for equation (3.14). (a) The x-μ plane and (b) the y-μ plane.

Next consider the y-μ plane, shown in figure 10.5(b); note that here only the critical points (2) and (3) can be shown. The structure of the pitchfork bifurcation at $\mu = 1$ is now clearly seen. However, the transcritical bifurcation at $\mu = 0$ cannot be seen. This demonstrates that it is often necessary to use different two-dimensional cross-sections to reveal different aspects of the bifurcation diagram. It should also be noted that the bifurcation diagram cannot yet be considered complete as we have yet to determine the consequences of the Hopf bifurcation at $\mu = 3$. This aspect will be taken up in the next section.

10.5 Hopf bifurcation

This is case (b) of section 10.1 (see (1.5)) and occurs when a pair of complex-conjugate eigenvalues describing the stability of a critical point are such that, at the bifurcation point, they have zero real parts but non-zero imaginary parts. Clearly, the first-order system (1.1) must have dimension $n \geq 2$. We will consider only the planar case $n = 2$, when (1.1) is replaced by the system (3.1) for the scalar variables x and y. Systems of dimension $n \geq 3$ can be treated by methods similar to that which we shall describe here; for further details see Chow and Hale (1982), Guckenheimer and Holmes (1983), Iooss and Joseph (1980). As in section 10.3, we shall suppose that f and g are analytic functions of x, y and μ. Let $(x_0(\mu), y_0(\mu))$ be a critical point, so that

$$f(x_0(\mu), y_0(\mu); \mu) = g(x_0(\mu), y_0(\mu); \mu) = 0.$$

Its stability is determined by the eigenvalues $\lambda_1(\mu)$ and $\lambda_2(\mu)$ of the 2×2 matrix $A(\mu)$, defined by (3.13). Assuming, without loss of generality, that the bifurcation point is $\mu = 0$, we recall from (1.5) that, for a Hopf bifurcation,

$$\lambda_1(\mu) = \overline{\lambda_2(\mu)} = \alpha(\mu) + i\beta(\mu),$$

and

$$\alpha(0) = 0, \qquad \beta(0) \neq 0. \tag{4.1}$$

It follows that, in a neighbourhood of $\mu = 0$, $\det A(\mu) \neq 0$ since $A(\mu)$ has no zero eigenvalues. Hence the implicit-function theorem guarantees that $x_0(\mu)$ and $y_0(\mu)$ are analytic functions of μ in a neighbourhood of $\mu = 0$.

Next we use the transformation (1.2) which, in our present notation, means that we replace the variables x and y by $x + x_0(\mu)$ and $y + y_0(\mu)$, respectively. This is an analytic transformation, which preserves the analyticity of f and g. Its effect is that we may assume that, in the transformed variables, the critical point is the identically zero solution $x = y = 0$, so that

$$f(0, 0; \mu) = g(0, 0; \mu) = 0. \tag{4.2}$$

Next, we expand f and g in a series expansion in powers of x and y, and then (3.1) takes the form

$$\begin{bmatrix} x' \\ y' \end{bmatrix} = A(\mu) \begin{bmatrix} x \\ y \end{bmatrix} + \begin{bmatrix} F(x, y; \mu) \\ G(x, y; \mu) \end{bmatrix}. \tag{4.3}$$

Here F and G are $O(r^2)$ as $r \to 0$, and are analytic functions of x, y and μ.

Also the 2×2 matrix $A(\mu)$ is an analytic function of μ, with eigenvalues $\lambda_1(\mu)$ and $\lambda_2(\mu)$ given by (4.1).

Next, in the subsequent theoretical development, it is useful to suppose that the matrix $A(\mu)$ has the canonical form (see section 5.2):

$$A(\mu) = \begin{bmatrix} \alpha(\mu) & \beta(\mu) \\ -\beta(\mu) & \alpha(\mu) \end{bmatrix}. \tag{4.4}$$

This can always be achieved, at least in a neighbourhood of $\mu = 0$, by a linear transformation in x and y, which is analytic in μ and, being linear, preserves the property that F and G are $O(r^2)$ as $r \to 0$. With the canonical form (4.4), the planar system (4.3) becomes

$$x' = \alpha(\mu)x + \beta(\mu)y + F(x,y;\mu), \qquad y' = -\beta(\mu)x + \alpha(\mu)y + G(x,y;\mu). \tag{4.5}$$

Recalling our analysis of critical points in chapter 5, we see that, at least in the vicinity of the bifurcation point, the origin $x = y = 0$ is a spiral point whose stability is governed by the sign of $\alpha(\mu)$ and, in general, since $\alpha(0) = 0$, can be expected to change stability as μ passes through zero. In analyzing spiral points, we recall that it was found useful to use the polar co-ordinates (r, θ), where

$$x = r\cos\theta, \qquad y = r\sin\theta. \tag{4.6}$$

We shall find these polar co-ordinates useful here too, but, before introducing them, we shall first analyze (4.5) in terms of the complex variable z and its complex conjugate $\bar{z}$, where

$$z = x + iy = e^{i\theta}, \qquad \bar{z} = x - iy = e^{-i\theta}. \tag{4.7}$$

In terms of z and $\bar{z}$ the system (4.5) becomes

$$z' = \{\alpha(\mu) - i\beta(\mu)\}z + N(z, \bar{z}; \mu), \tag{4.8}$$

where

$$N(z, \bar{z}; \mu) = F(x, y; \mu) + iG(x, y; \mu).$$

Here, in the arguments of F and G, we put

$$x = \tfrac{1}{2}(z + \bar{z}), \qquad y = -\tfrac{1}{2}i(z - \bar{z}). \tag{4.9}$$

In effect, we have replaced the pair x and y by the pair z and $\bar{z}$. Note that the equation for $\bar{z}$ is simply the complex conjugate of (4.8). Here $N(z, \bar{z}; \mu)$ is $O(|z|^2)$ as $|z| \to 0$, and so is quadratic to leading order in z and $\bar{z}$.

The analysis of (4.8) now proceeds in two stages. In the first stage the system (4.8) is further transformed using normal-form analysis. In the second stage it will be shown that Hopf bifurcation leads to the existence of a periodic solution of (4.8) as the zero solution changes stability. The first stage is described in the following theorem.

Theorem 10.3: There exists a near-identity analytic transformation

$$\zeta = z + S(z, \bar{z}; \mu), \tag{4.10}$$

where S is $O(|z|^2)$ as $|z| \to 0$, such that the system (4.8) adopts the **normal form**

$$\zeta' = \{\alpha(\mu) - i\beta(\mu)\}\zeta + \{\gamma(\mu) + i\delta(\mu)\}|\zeta|^2\zeta + O(|\zeta|^4), \tag{4.11}$$

where $\gamma(\mu)$ and $\delta(\mu)$ are analytic functions of μ.

Proof: The near-identity transformation (4.10) will be constructed in two iterative steps. First, since $N(z, \bar{z}; \mu)$ is quadratic to leading order in z and $\bar{z}$, we let

$$N(z, \bar{z}; \mu) = \tfrac{1}{2}n_1 z^2 + n_2 z\bar{z} + \tfrac{1}{2}n_3 \bar{z}^2 + O(|z|^3),$$

where n_1, n_2 and n_3 are (analytic) functions of μ. Then we introduce the transformation

$$w = z + Q(z, \bar{z}; \mu), \tag{4.12}$$

where

$$Q(z, \bar{z}; \mu) = \tfrac{1}{2}q_1 z^2 + q_2 z\bar{z} + \tfrac{1}{2}q_3 \bar{z}^2.$$

This is a near-identity transformation, since Q is $O(|z|^2)$ as $|z| \to 0$. At this point the coefficients q_1, q_2 and q_3 are undetermined, and our aim is to show they may be chosen to eliminate all the quadratic terms in the system (4.8) when the system is formulated in terms of the variable w. Note that the inverse of (4.12) is

$$z = w - Q(w, \bar{w}; \mu) + O(|w|^3).$$

Now, from (4.12),

$$w' = z' + (q_1 z + q_2 \bar{z}) z' + (q_2 z + q_3 \bar{z}) \bar{z}'.$$

Then, using (4.8) and the inverse of (4.12), we see that

$$w' = (\alpha - i\beta)\{w - \tfrac{1}{2}q_1 w^2 - q_2 w\bar{w} - \tfrac{1}{2}q_3 \bar{w}^2 + O(|w|^3)\}$$
$$+ \tfrac{1}{2}n_1 w^2 + n_2 w\bar{w} + \tfrac{1}{2}n_3 \bar{w}^2 + O(|w|^3)$$
$$+ (q_1 w + q_2 \bar{w})(\alpha - i\beta) w + (q_2 w + q_3 \bar{w})(\alpha + i\beta) \bar{w} + O(|w|^3),$$

or

$$w' = (\alpha - i\beta) w + \tfrac{1}{2}\hat{n}_1 w^2 + \hat{n}_2 w\bar{w} + \tfrac{1}{2}\hat{n}_3 \hat{w}^2 + O(|w|^3),$$

where

$$\hat{n}_1 = n_1 + (\alpha - i\beta) q_1, \qquad \hat{n}_2 = n_2 + (\alpha + i\beta) q_2, \qquad \hat{n}_3 = n_3 + (\alpha + 3i\beta) q_3.$$

Since, at the bifurcation point $\mu = 0$, $\alpha(0) = 0$ but $\beta(0) \neq 0$, we can, clearly, choose q_1, q_2 and q_3 so that $\hat{n}_1$, $\hat{n}_2$ and $\hat{n}_3$ are all zero, since the coefficients of q_1, q_2 and q_3 are all non-zero for μ in a neighbourhood of $\mu = 0$. Note that q_1, q_2 and q_3 are analytic functions of μ and, hence, (4.12) is an analytic transformation. The equation for w is now

$$w' = (\alpha - i\beta)w + M(w, \bar{w}; \mu), \qquad (4.13)$$

where M is $O(|w|^3)$ as $|w| \to 0$, and is analytic in w, $\bar{w}$ and μ.

Next, let

$$M(w, \bar{w}; \mu) = \tfrac{1}{3}m_1 w^3 + m_2 w^2 \bar{w} + m_3 w \bar{w}^2 + \tfrac{1}{3}m_4 \bar{w}^3 + O(|w|^4),$$

where m_1, m_2, m_3 and m_4 are (analytic) functions of μ. Then we introduce a second near-identity transformation:

$$\zeta = w + R(w, \bar{w}; \mu), \qquad (4.14)$$

where

$$R(w, \bar{w}; \mu) = \tfrac{1}{3}r_1 w^3 + r_2 w^2 \bar{w} + r_3 w \bar{w}^2 + \tfrac{1}{3}r_4 \bar{w}^3.$$

Thus R is $O(|w|)^3)$ as $|w| \to 0$. At this point the coefficients r_1, r_2, r_3 and r_4 are undetermined, and our aim is to choose them to eliminate as many of the cubic terms in (4.13) as possible. Note that the inverse of (4.14) is

$$w = \zeta - R(\zeta, \bar{\zeta}; \mu) + O(|\zeta|^4).$$

Now from (4.14)

$$\zeta' = w' + (r_1 w^2 + 2r_2 w\bar{w} + r_3 \bar{w}^2)w' + (r_2 w^2 + 2r_3 w\bar{w} + r_4 \bar{w}^2)\bar{w}'.$$

Then, using (4.13) and the inverse of (4.14), we find that

$$\zeta' = (\alpha - i\beta)\{\zeta - \tfrac{1}{3}r_1 \zeta^3 - r_2 \zeta^2\bar{\zeta} - r_3 \zeta\bar{\zeta}^2 - \tfrac{1}{3}r_4 \bar{\zeta}^3 + O(|\zeta|^4)\}$$
$$+ \tfrac{1}{3}m_1 \zeta^3 + m_2 \zeta^2\bar{\zeta} + m_3 \zeta\bar{\zeta}^2 + \tfrac{1}{3}m_4 \bar{\zeta}^3 + O(|\zeta|^4)$$
$$+ (r_1 \zeta^2 + 2r_2 \zeta\bar{\zeta} + r_3 \bar{\zeta}^2)(\alpha - i\beta)\zeta$$
$$+ (r_2 \zeta^2 + 2r_3 \zeta\bar{\zeta} + r_4 \bar{\zeta}^2)(\alpha + i\beta)\bar{\zeta} + O(|\zeta|^4)\},$$

or

$$\zeta' = (\alpha - i\beta)\zeta + \tfrac{1}{3}\hat{m}_1 \zeta^3 + \hat{m}_2 \zeta^2\bar{\zeta} + \hat{m}_3 \zeta\bar{\zeta}^2 + \tfrac{1}{3}\hat{m}_4 \bar{\zeta}^3 + O(|\zeta|^4),$$

where

$$\hat{m}_1 = m_1 + 2(\alpha - i\beta)r_1, \qquad \hat{m}_2 = m_2 + 2\alpha r_2,$$
$$\hat{m}_3 = m_3 + 2(\alpha + i\beta)r_3, \qquad \hat{m}_4 = m_4 + 2(\alpha + 2i\beta)r_4.$$

Now, at the bifurcation point $\mu = 0$, $\alpha(0) = 0$ but $\beta(0) \neq 0$. Hence we can, clearly, choose r_1, r_3 and r_4 so that $\hat{m}_1$, $\hat{m}_3$ and $\hat{m}_4$ are all zero, since the coefficients of r_1, r_3 and r_4 are all non-zero in a neighbourhood of $\mu = 0$.

However, r_2 cannot be chosen to eliminate $\hat{m}_2$, since its coefficient is $\alpha(\mu)$, which is zero at the bifurcation point. Instead we choose $r_2 = 0$. Note that r_1, r_3 and r_4 are analytic functions of μ and, hence, (4.14) is an analytic transformation. The equation for ζ is now

$$\zeta' = (\alpha - i\beta)\zeta + m_2|\zeta|^2\zeta + O(|\zeta|^4), \tag{4.15}$$

where m_2 is an analytic function of μ. Setting

$$m_2 = \gamma + i\delta,$$

we see that (4.15) has the required form (4.11). This completes the proof, since the composition of the near-identity transformations (4.12) and (4.14) will produce the transformation (4.10).

There are a number of features to note about the proof of theorem 10.3. First, the theorem shows that the nonlinear term $N(z, \bar{z}; \mu)$ in the original system (4.8) can be transformed to remove all quadratic terms and all cubic terms except one, namely, the term $|\zeta|^2\zeta$. The reason is that this is the lowest-order nonlinear term which has the same phase as ζ and, hence, is the lowest-order term which can produce a resonance. Note that, at the bifurcation point $\mu = 0$, the linearized version of (4.8) has the solution

$$z = z_0 \exp(i\omega_0 t), \quad \text{where } \omega_0 = -\beta(0).$$

The quadratic terms in $N(z, \bar{z}; \mu)$ are clearly not resonant with this and, of the cubic terms, only $|z|^2z$ (or $|\zeta|^2\zeta$ after carrying out the near-identity transformation) is resonant. Also, note that the composite near-identity transformation is, with $r_2 = 0$,

$$\zeta = z + \tfrac{1}{2}q_1z^2 + q_2z\bar{z} + \tfrac{1}{2}q_3\bar{z}^2 + \tfrac{1}{3}r_1z^3 + r_3z\bar{z}^2 + \tfrac{1}{3}r_4\bar{z}^3 + O(|z|^4), \tag{4.16}$$

and contains only non-resonant terms. It is useful to observe here that, in the final step from (4.13) to (4.15), the coefficient m_2 of the resonant term is unchanged by the near-identity transformation (4.14). Hence, in the usual situation where we are mainly interested only in the final form (4.15), it is not necessary to construct the transformation (4.14) explicitly (i.e. there is no need to calculate r_1, r_3 and r_4 explicitly).

Since the coefficient $m_2 = \gamma + i\delta$ in the normal-form equation (4.15) contains all the information about the nonlinear terms in the original equation (4.8), it is worthwhile to develop an explicit formula for m_2. First, suppose that the original equation (4.8) is given by

$$z' = (\alpha - i\beta)z + \tfrac{1}{2}n_1z^2 + n_2z\bar{z} + \tfrac{1}{2}n_3\bar{z}^2 + \tfrac{1}{3}n_4z^3 + n_5z^2\bar{z}$$

$$+ n_6z\bar{z}^2 + \tfrac{1}{3}n_7\bar{z}^3 + O(|z|^4). \tag{4.17}$$

Now, we have shown in the course of theorem 10.3 that the transformation (4.12) takes (4.17) into (4.13), or

$$w' = (\alpha - i\beta)w + \tfrac{1}{3}m_1 w^3 + m_2 w^2 \bar{w} + m_3 w\bar{w}^2 + \tfrac{1}{3}m_4 \bar{w}^3 + O(|w|^4). \qquad (4.18)$$

A further transformation (4.14) takes (4.18) into (4.15), but, as we have already noticed above, this leaves the coefficient m_2 unchanged. Hence we need only consider (4.17) and (4.18) to find m_2. Also we recall from the proof of theorem 10.3 that the coefficients q_1, q_2 and q_3 in (4.12) are given by

$$(\alpha - i\beta)q_1 + n_1 = 0, \qquad (\alpha + i\beta)q_2 + n_2 = 0, \qquad (\alpha + 3i\beta)q_3 + n_3 = 0. \quad (4.19)$$

Now, to find m_2, we substitute (4.12) into (4.18) to get

$$z' + (q_1 z + q_2 \bar{z})z' + (q_2 z + q_3 \bar{z})\bar{z}' = (\alpha - i\beta)(z + \tfrac{1}{2}q_1 z^2 + q_2 z\bar{z} + \tfrac{1}{2}q_3 \bar{z}^2)$$

$$+ \tfrac{1}{3}m_1 z^3 + m_2 z^2 \bar{z} + m_3 z\bar{z}^2$$

$$+ \tfrac{1}{3}m_4 \bar{z}^3 + O(|z|^4). \qquad (4.20)$$

Then, using (4.17), equating coefficients of like powers of z and $\bar{z}$ and using (4.19), we eventually find that

$$m_2 = n_5 - \frac{n_1 n_2}{\alpha - i\beta} - \frac{n_1 n_2}{2(\alpha + i\beta)} - \frac{|n_2|^2}{\alpha + i\beta} - \frac{|n_3|^2}{2(\alpha + 3i\beta)}. \qquad (4.21)$$

Similar expressions can, of course, be obtained for m_1, m_3 and m_4, but are not usually needed and, hence, are left as an exercise for the reader.

Next, it is clear that there is a pattern to the determination of the successive near-identity transformations, and that the process can be repeated to an arbitrarily high order (see, for instance, Guckenheimer and Holmes, 1983, pp. 138–45). Thus (4.8) can be transformed into the form

$$\zeta' = (\alpha - i\beta)\zeta + m_2 |\zeta|^2 \zeta + \ldots + m_{2n} |\zeta|^{2n} \zeta + O(|\zeta|^{2n+3}),$$

where $n \geqslant 1$, and can, in principle, be chosen arbitrarily large. In practice, however, it is usually sufficient to truncate at the lowest-order (i.e. $n = 1$), except in the rare event that $\mathrm{Re}\, m_2$ is identically zero at the bifurcation point, $\mu = 0$.

The second stage in the description of a Hopf bifurcation is the analysis of the normal form (4.11). First, we omit the error term $O(|\zeta|^4)$, and consider the model equation

$$\zeta' = \{\alpha(\mu) - i\beta(\mu)\}\zeta + \{\gamma(\mu) + i\delta(\mu)\}|\zeta|^2 \zeta. \qquad (4.22)$$

To analyze this, we use the polar co-ordinates (R, ϕ), where

$$\zeta = R e^{i\phi}. \qquad (4.23)$$

Note that, from the near-identity transformation (4.10), (R, ϕ) are related to the original polar co-ordinates (r, θ) (see (4.6) and (4.7)) by

$$R = r + O(r^2), \qquad \phi = \theta + O(r), \qquad (4.24)$$

where both error terms are periodic functions of θ with period 2π. Hence a description of the solution in the (R, ϕ) co-ordinates provides an approximate description in terms of the original (r, θ) co-ordinates. In terms of the polar co-ordinates (R, ϕ), (4.22) becomes

$$R' = \alpha(\mu) R + \gamma(\mu) R^3, \qquad \phi' = -\beta(\mu) + \delta(\mu) R^2. \tag{4.25}$$

Since these equations are now uncoupled, we may construct the general solution explicitly by solving the first equation for R, and then solving the second equation for ϕ. Here it is not necessary to find the general solution, since it will be sufficient to examine the critical points of the first equation in (4.25), namely, the equation for R.

Assuming the generic case when $\gamma(0) \neq 0$, and so $\gamma(\mu)$ is non-zero in a neighbourhood of $\mu = 0$, we see that there are two critical points,

$$\text{either} \quad R = 0 \quad \text{or} \quad R^2 = -\frac{\alpha(\mu)}{\gamma(\mu)}. \tag{4.26}$$

The first of these corresponds to the zero solution of (4.8), whose change of stability at $\mu = 0$ leads to the Hopf bifurcation. The second critical point is given by

$$R^2 = -\frac{\alpha'(0)}{\gamma(0)}\mu + O(\mu^2) \tag{4.27}$$

and, for the generic condition that $\alpha'(0) \neq 0$, is a pitchfork bifurcation. The bifurcation diagram is sketched in figure 10.6, where we note that here the pitchfork is one-sided since the radial variable R is intrinsically non-negative.

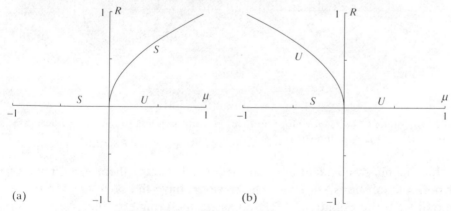

Fig. 10.6 The bifurcation diagram for a Hopf bifurcation (see (4.27)). (a) Supercritical and (b) subcritical.

The stability of each of these solutions is determined by the sign of $\lambda_1(\mu)$, where

$$\lambda_1 = \frac{\partial}{\partial R}(\alpha R + \gamma R^3) \quad (\text{at } \alpha R + \gamma R^3 = 0),$$

or

268 *Chapter 10*

$$\lambda_1 = \begin{cases} \alpha & \text{(for } R = 0), \\ -2\alpha & \text{(for } R^2 = -\alpha/\gamma). \end{cases}$$

Hence the two branches have, as expected, opposite stabilities.

Although, for the R-variable, the bifurcation appears to be one-dimensional, we must now recall the angular variable ϕ. For the second critical point, it follows from (4.25) and (4.26) that

$$\phi = \phi_0 + \omega(\mu)t,$$

where

$$\omega = -\beta - \frac{\delta\alpha}{\gamma}.$$

Hence the solution is periodic, with a period $2\pi/|\omega(\mu)|$. As $\mu \to 0$, $\omega \to \omega_0$, where $\omega_0 = -\beta(0)$. In effect, as μ varies through the bifurcation point $\mu = 0$, the spiral point given by $R = 0$ gives birth to a periodic solution whose period at the bifurcation point is precisely that of the centre to which the spiral point reduces at $\mu = 0$. The situation is sketched in figure 10.7.

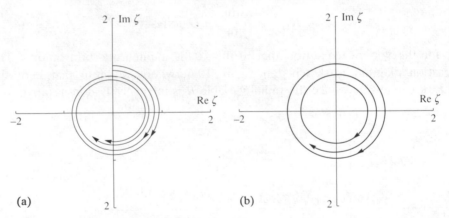

Fig. 10.7 The orbits of (4.25) for Hopf bifurcation, corresponding to the supercritical case (see figure 10.6 (a)). (a) $\mu < 0$ and (b) $\mu > 0$.

Note that the phase constant ϕ_0 is arbitrary and, hence, there is an infinite family of periodic solutions, all of which, however, have the same orbit in the phase plane. Indeed, the equation (4.21) possesses the symmetry that the transformation ζ into $\zeta \exp(i\phi_0)$ leaves the equation invariant, and it is precisely this property that causes Hopf bifurcation to be generically a pitchfork bifurcation.

Let us now return to the full equation (4.11). In the generic case, when $\alpha'(0)$ and $\gamma(0)$ are not zero, the following theorem holds.

Theorem 10.4: For μ in a neighbourhood of zero, (4.11) has a family of periodic solutions, given approximately by (4.27), whose period is $2\pi/|\omega_0|$ as $\mu \to 0$, which are stable (unstable) when $\alpha(\mu) > 0$ ($\alpha(\mu) < 0$). Here $\omega_0 = -\beta(0)$.

Proof: In terms of the polar co-ordinates (4.23), equation (4.11) becomes

$$R' = \alpha(\mu)R + \gamma(\mu)R^3 + O(R^4), \qquad \phi' = -\beta(\mu) + \delta(\mu)R^2 + O(R^3).$$

Next we put $R = \epsilon\rho$ and $\mu = \epsilon^2\nu$, where ϵ is a small parameter. Then these equations are equivalent to

$$\frac{d\rho}{d\phi} = \epsilon^2 H(\rho, \phi; \epsilon), \qquad \omega_0 H(\rho, \phi; \epsilon) = \alpha'(0)\nu\rho + \gamma(0)\rho^3 + O(\epsilon),$$

and we recall that $\omega_0 = -\beta(0)$. Hence, for the unique solution $\rho(\phi; \rho_0, \epsilon)$ such that $\rho = \rho_0$ at $\phi = 0$,

$$\rho = \rho_0 + \epsilon^2 \int_0^\phi H(\rho(\phi'), \phi'; \epsilon)\, d\phi'.$$

This solution is periodic if and only if $\rho(2\pi; \rho_0, \epsilon) = \rho_0$, or

$$P(\rho_0, \epsilon) = \int_0^{2\pi} H(\rho(\phi'), \phi'; \epsilon)\, d\phi' = 0.$$

In the limit $\epsilon \to 0$, $P(\rho_0; \epsilon) \to P_0(\rho_0)$, where

$$P_0(\rho_0) = \frac{2\pi}{\omega_0}\{\alpha'(0)\nu\rho_0 + \gamma(0)\rho_0^3\}.$$

This has the solutions $\rho_0 = 0$ and $\rho_0^2 = -\alpha'(0)\nu/\gamma(0)$. Only the latter needs to be considered, as $\rho_0 = 0$ clearly corresponds to the zero solution of (4.11). Further, for this latter solution,

$$\frac{\partial P}{\partial \rho_0}(\rho_0, 0) = \frac{2\pi}{\omega_0}\{-2\alpha'(0)\nu\}$$

and, since this is not zero for $\nu \neq 0$, the implicit-function theorem establishes the existence of a unique solution $\rho_0 = \rho_0(\epsilon)$ which is analytic in ϵ and such that $\rho_0(0)$ is given by $\rho_0(0)^2 = -\alpha'(0)\nu/\gamma(0)$. The period is found by considering the equation for ϕ; the stated result follows immediately, but note that it is essential for $\omega_0 \neq 0$.

The stability of this solution can be determined either by appealing to theorem 6.4, or, more directly, by using the same kind of argument employed in theorem 7.1. Thus, consider an orbit which commences at $\rho = \rho_0^*$ when $\phi = 0$. After one circuit of the origin, it will reach a point $\rho = \rho^*$, where

$$\rho^* = \rho_0^* + \frac{\epsilon^2}{\omega_0} P(\rho_0^*; \epsilon).$$

Assuming that $\omega_0 > 0$, the periodic solution is clearly stable if $\rho^* \lessgtr \rho_0^*$ according as $\rho_0^* \gtrless \rho_0$. In the limit $\epsilon \to 0$, this is equivalent to

$$\frac{\partial P}{\partial \rho_0}(\rho_0, 0) < 0 \quad \text{or} \quad \omega_0\alpha'(0)\nu > 0,$$

where we have used the expression stated above for $\partial P/\partial\rho_0$. Similarly the periodic solution is unstable if $\omega_0\alpha'(0)\nu < 0$. For $\omega_0 < 0$ an analogous argument produces the same result, where we note that ϕ increases (decreases) with t when $\omega_0 > 0$ ($\omega_0 < 0$). Since $\mu = \epsilon^2\nu$ and $\alpha(\mu) \approx \epsilon^2\alpha'(0)\nu$, the stated result for stability is proven.

We shall conclude by discussing two illustrative examples. First, consider

$$x' = \mu x + y, \qquad y' = -x + \mu y - x^2 y. \tag{4.28}$$

This is a 'Van der Pol'-type of equation, since, with $x = u$, the system is equivalent to

$$u'' + (u^2 - 2\mu)u' + (1 + \mu^2)u - \mu u^3 = 0$$

and, hence, describes a nonlinear oscillator with a nonlinear damping term $(u^2 - 2\mu)u'$, whose coefficient is positive for $u^2 > 2\mu$ but is negative for $u^2 < 2\mu$. Since $x = y = 0$ is a critical point for (4.27), this system already has the structure (4.3). Further, the matrix $A(\mu)$ for the linear part of the system (4.27) is in the canonical form (4.4) with, here, $\alpha = \mu$ and $\beta = 1$. Thus (4.28) is already in the required form (4.5), and the critical point $x = y = 0$ is a stable spiral point for $\mu < 0$ and an unstable spiral point for $\mu > 0$. Hence there is a Hopf bifurcation at $\mu = 0$, and we can deduce from theorems 10.3 and 10.4 that there exists a family of periodic solutions bifurcating from $\mu = 0$. Our main aim now is to determine whether this bifurcation is supercritical and stable, or subcritical and unstable.

To achieve this, we first use the transformation (4.7) to the complex variables z and $\bar{z}$, and obtain the complex form (4.8) for the system (4.28). We find that

$$z' = (\mu - i)z - ix^2 y,$$

or

$$z' = (\mu - i)z - \tfrac{1}{8}(z^3 + z^2\bar{z} - z\bar{z}^2 - \bar{z}^3). \tag{4.29}$$

In this instance the nonlinear terms are already cubic and, hence, there is no need for the quadratic transformation (4.12). Instead, theorem 10.3 implies that there is a near-identity transformation

$$\zeta = z + R(z, \bar{z}; \mu),$$

where R is $O(|z|^3)$ as $|z| \to 0$, such that the system (4.29) adopts the normal form

$$\zeta' = (\mu - i)\zeta - \tfrac{1}{8}|\zeta|^2\zeta + O(|\zeta|^4). \tag{4.30}$$

Note that the coefficient m_2 in (4.15), given in general by the expression (4.20), is here exactly given by the coefficient of the resonant term $|z|^2 z$ in (4.29), due to the absence of the quadratic terms in (4.29) (i.e. n_1, n_2 and n_3 are all zero in (4.20). Also note that there is no need to construct the cubic term $R(z, \bar{z}; \mu)$ explicitly.

Now that the normal form (4.30) has been obtained, theorem 10.4 shows that, to leading order in μ, the error term $O(|\zeta|^4)$ in (4.30) may be omitted, and the periodic solutions of (4.30) may be constructed approximately by solving the truncated equation which has the form (4.21). Using the polar co-ordinates (R, ϕ) (4.23), we see that (4.30) becomes

$$R' = \mu R - \tfrac{1}{8}R^3 + O(R^4), \qquad \phi' = -1 + O(R^2).$$

The periodic solution is given by

$$R^2 = 8\mu + O(\mu^2)$$

and, hence, is stable, being a supercritical pitchfork bifurcation (see figure 10.6(a)).

Next, we reconsider the example (3.14). In section 10.3 we showed that this had three critical points given by (3.15), and the corresponding bifurcation diagram is sketched in figure 10.4. Here our attention is focussed on the third critical point $x = 1$, $2y^2 = \mu - 1$, which exists only for $\mu \geqslant 1$, and is stable for $1 < \mu < 3$ but is unstable for $\mu > 3$. There is a Hopf bifurcation at $\mu = 3$. Our aim now is to analyze this Hopf bifurcation and in particular determine whether it is supercritical and stable, or subcritical and unstable.

The first task is to put the equation (3.14) into the canonical form (4.5). Hence we put

$$X = x - 1, \qquad Y = y - \epsilon,$$

where

$$\mu = 1 + 2\epsilon^2. \tag{4.31}$$

The critical point $x = 1$, $2y^2 = \mu - 1$ is now given by $X = 0$, $Y = 0$, and the parameter μ has been replaced by ϵ. The Hopf bifurcations now occur at $\epsilon = \pm 1$. With the transformation (4.31), the system (3.14) becomes

$$X' = (\epsilon^2 - 1)X - 4\epsilon Y - X^2 - 2\epsilon XY - 2Y^2 - XY^2, \qquad Y' = \epsilon X + XY. \tag{4.32}$$

This has the form (4.3), where the coefficient matrix for the linear terms is

$$A(\epsilon) = \begin{bmatrix} \epsilon^2 - 1 & -4\epsilon \\ \epsilon & 0 \end{bmatrix}.$$

The eigenvalues are $\alpha(\epsilon) \pm i\beta(\epsilon)$, where

$$\alpha(\epsilon) = \tfrac{1}{2}(\epsilon^2 - 1), \qquad \beta(\epsilon) = \tfrac{1}{2}\sqrt{80 - (\epsilon^2 - 9)^2}. \tag{4.33}$$

As anticipated, there are Hopf bifurcations at $\epsilon = \pm 1$. The critical point $X = 0$, $Y = 0$ is stable for $\epsilon^2 < 1$ and unstable for $\epsilon^2 > 1$. Note that, since the Hopf bifurcations occur at $\epsilon^2 = 1$, it is sufficient to restrict the parameter ϵ^2 to a neighbourhood of this value, within which $\alpha(\epsilon)$ and $\beta(\epsilon)$ are both real-valued. Indeed, $\beta(\epsilon)$ is real-valued for $9 - \sqrt{80} < \epsilon^2 < 9 + \sqrt{80}$.

The next step is to convert (4.32) into the canonical form (4.5). This is achieved by a linear transformation of the form

$$\begin{bmatrix} X \\ Y \end{bmatrix} = S \begin{bmatrix} \hat{x} \\ \hat{y} \end{bmatrix},$$

where the 2×2 matrix S is constructed by putting the real and imaginary parts of the eigenvector of A corresponding to the eigenvalue $\alpha + i\beta$ into the columns of S (see (5.12) of section 2.5 and the subsequent discussion). Hence here

$$S = \begin{bmatrix} 4\epsilon & 0 \\ \frac{1}{2}(\epsilon^2 - 1) & -\beta(\epsilon) \end{bmatrix}.$$

We then find that (4.31) is transformed into

$$\hat{x}' = \alpha\hat{x} + \beta\hat{y} + F(\hat{x}, \hat{y}; \epsilon), \qquad \hat{y}' = -\beta\hat{x} + \alpha\hat{y} + G(\hat{x}, \hat{y}; \epsilon),$$

where

$$F = -4\epsilon\hat{x}^2 - 2\epsilon\hat{x}(\alpha\hat{x} - \beta\hat{y}) - \left(\frac{1}{2\epsilon} + \hat{x}\right)(\alpha\hat{x} - \beta\hat{y})^2, \qquad G = \frac{\alpha}{\beta}F - \frac{4\epsilon\hat{x}}{\beta}(\alpha\hat{x} - \beta\hat{y}).$$

$$\text{(4.34)}$$

This now has the required form (4.5). Putting

$$z = \hat{x} + i\hat{y}, \tag{4.35}$$

we finally obtain the required complex form (4.8), or (4.17), which here is given by

$$z' = (\alpha - i\beta)z + N(z, \bar{z}; \epsilon),$$

where

$$N = F + iG = \tfrac{1}{2}n_1 z^2 + n_2 z\bar{z} + \tfrac{1}{2}n_3 z^2 + \tfrac{1}{3}n_4 z^3 + n_5 z^2\bar{z} + n_6 z\bar{z}^2 + \tfrac{1}{3}n_7 \bar{z}^3. \tag{4.36}$$

In this expression $\alpha(\epsilon)$ and $\beta(\epsilon)$ are defined by (4.33), and F and G are given in (4.34), from which the coefficients $n_1, \ldots, n_7$ can be determined. In the sequel, these coefficients are used only to determine m_2 from (4.21) at the bifurcation points $\epsilon = \pm 1$. Hence it is sufficient to evaluate them only at these values, where $\alpha = 0$ and $\beta = 2$. We find that, for $\epsilon = \pm 1$,

$$n_1 = \pm(1 - 2i), \qquad n_2 = \mp 3, \qquad n_3 = \pm(-3 + 2i), \qquad n_5 = -\tfrac{1}{2}. \tag{4.37}$$

Note from (4.21) that it is not necessary to calculate n_4, n_6 or n_7.

We are now in a position to use theorem 10.3, which shows that there is a near-identity transformation (4.10) (or (4.16)) which transforms (4.36) into the normal form (see (4.11) or (4.15)):

$$\zeta' = \{\alpha(\epsilon) - i\beta(\epsilon)\}\zeta + \{\gamma(\epsilon) + i\delta(\epsilon)\}|\zeta|^2\zeta + O(|\zeta|^4). \tag{4.38}$$

Here the coefficient $m_2 = \gamma + i\delta$ is given by (4.21), where here n_1, n_2, n_3 and n_5 are given by (4.37). Hence, evaluating m_2 at the bifurcation points $\epsilon = \pm 1$, we find that

$$m_2 = 1 + \tfrac{19}{3}i + O(\epsilon \mp 1). \tag{4.39}$$

Also, from (4.33),

$$\alpha = \pm(\epsilon \mp 1) + \tfrac{1}{2}(\epsilon \mp 1)^2.$$

Hence, using the polar co-ordinates (4.23) the periodic solution is given by (4.27), where here μ is replaced by $\pm(\epsilon \mp 1)$. Thus we find that

$$R^2 = \mp(\epsilon \mp 1) + O((\epsilon \mp 1)^2).$$

For both cases the bifurcation is subcritical and unstable.

Problems

1 Discuss the stability and bifurcation phenomena for the critical points of the following one-dimensional equations. In each case sketch a bifurcation diagram:

(a) $x' = 2\mu + x^2$;

(b) $x' = \mu x - 3x^2$;

(c) $x' = -\mu x + ax^3$, for both $\alpha > 0$ and $\alpha < 0$;

(d) $x' = x(\mu - 2x + x^2)$;

(e) $x' = x(\mu + x - x^2)(1 - \mu + x^2)$;

(f) $x' = x^2 - \mu^2 x$;

(g) $x' = x(9 - \mu x)(\mu + 2x - x^2)\{(\mu - 10)^2 + (x - 3)^2 - 1\}$;

(h) $x' = \mu^2 - x^4$;

(i) $x' = \mu^2 x - x^3$;

(j) $x' = x(x - \mu^2)(\mu - x)$.

Note that in (f) and (h) the origin ($x = 0$, $\mu = 0$) is a cusp point, and in (i) and (j) the origin is a triple point.

2 Discuss the stability and bifurcation phenomena for the critical points of the following one-dimensional equations, and sketch the bifurcation diagrams in the μ-x plane:

(a) $x' = \mu x - x^2 + \delta$;   (b) $x' = \mu x - 2x^2 - x^3 + \delta$.

Note that here there are *two* parameters μ and δ; δ is called the imperfection parameter. Consider the cases $\delta = 0$, and δ small but either $\delta > 0$ or $\delta < 0$.

3 Discuss the stability and bifurcation phenomena for the critical points of the following planar systems, and sketch the bifurcation diagrams:

(a) $x' = x(\mu - 2x) - xy$, $y' = y(x - 1)$;

(b) $x' = x(\mu - 2x) - xy$, $y' = y(x - 1) + y^2$;

(c) $x' = x(\mu - 2x) - xy$, $y' = y(x - 1) - y^2$;

[(a), (b) and (c) are examples of Lotka–Volterra systems; see, for instance, (3.31) of section 5.3;]

(d) $x' = 2(1 - \mu)x + x^2 - xy$, $y' = x + \tfrac{1}{2}x^2 - y$;

[a model for a biochemical interaction between an autocatalytic activator x and an inhibitor y;]

(e) $x' = \mu x - 2x(x^2 + y^2)$, $y' = (\mu - 1)y - y(x^2 + y^2)$;

(f) $x' = y + \mu x - xy + y^2$, $y' = \mu x + xy - x^2$.

[Here the origin is a bifurcation point where the eigenvalue changing stability has algebraic multiplicity two but geometric multiplicity one. Show that the zero solution exchanges stability at the bifurcation point, but the bifurcating solution is unstable on both sides of the bifurcation point.]

$$(g) \quad x' = \mu x - xy - x^2 + y^2, \qquad y' = \mu y + xy.$$

[Here the origin is a bifurcation point where the eigenvalue changing stability has algebraic and geometric multiplicity two. In general, for this situation, there may be zero, one or three bifurcating solutions. In this case there are three.]

4 For each of the following planar systems show that there is a Hopf bifurcation from the zero solution $x = y = 0$ at $\mu = 0$, and use the method of normal forms to determine the bifurcating solution for small amplitudes:

(a) $x' = \mu x + y - xy^2$, $y' = -x + \mu y$;

(b) $x' = \mu x + y - x^3$, $y' = -x + \mu y + 2y^3$;

(c) $x' = \mu x + y - x^2$, $y' = -x + \mu y + x^2$;

(d) $x' = \mu x + y - x^2 + xy^2$, $y' = -x + \mu y - y^2$.

5 For each of the following planar systems show that there is a Hopf bifurcation from the zero solution $x = y = 0$ at $\mu = 0$, and use the method of normal forms to determine the bifurcating solution for small amplitudes. Note that, unlike problem 4, it is first necessary to apply a linear transformation before using the normal-form analysis.

(a) $x' = y + \mu x - \frac{1}{3}x^3$, $y' = -x$.

[This is the Van der Pol system. The transformation $x, y \rightarrow \sqrt{\epsilon}x, \sqrt{\epsilon}y$, where $\mu = \epsilon$ produces the standard form. See (3.3) of section 6.3, and compare the results obtained here with the results obtained in section 6.3 for the Van der Pol limit cycle as $\epsilon \rightarrow 0$.]

(b) $x' = y + \mu x - y^2$, $y' = -4x + \mu y + y^2$.

6 Show that the planar system

$$x' = r^2(\mu - r^2)x + \omega(r^2)y, \qquad y' = r^2(\mu - r^2)y - \omega(r^2)x,$$

where $r^2 = x^2 + y^2$ and $\omega(0) = 1$ has a Hopf bifurcation at $x = y = 0$ when $\mu = 0$. [*Hint*: Transform to polar co-ordinates.]

7 For the planar system

$$x' = 2(1 - \mu)x + x^2 - xy, \qquad y' = x + \frac{1}{2}x^2 - y,$$

show that the critical point for which $y = x + \frac{1}{2}x^2$ and $x^2 = 4(1 - \mu)$ has a Hopf bifurcation at $x = 1$, $y = \frac{3}{2}$ and $\mu = \frac{3}{4}$ (see problem 3 (d) for the bifurcation diagram of the critical points of this system). Determine the form of the bifurcating periodic solution near the bifurcation point. [*Hint*: Use $x = \epsilon$ as the parameter for this critical point, and first make the transformation $x = \epsilon + \hat{x}$, $y = \epsilon + \frac{1}{2}\epsilon^2 + \hat{y}$ and $4(1 - \mu) = \epsilon^2$. Then use normal-form analysis similar to that requested in problem 5.]

8 For the normal-form equation (4.11), let $\zeta = X + iY$, and show that in these Cartesian coordinates, it becomes

$$X' = F(X, Y; \mu) = \alpha X + \beta Y + R^2(\gamma X - \delta Y) + O(R),$$

$$Y' = G(X, Y; \mu) = -\beta X + \alpha Y + R^2(\delta X + \gamma Y) + O(R),$$

where $R^2 = X^2 + Y^2$. Deduce that

$$F_X + G_Y = 2\alpha + 4\gamma R^2 + O(R^3).$$

Then use theorem 6.4 to prove the stability result of theorem 10.4. [*Hint*: Note that, for the bifurcating periodic solution, $\gamma R^2 = -\alpha + O(R^3)$ and, hence, $F_X + G_Y = -2\alpha + O(R)$.]

CHAPTER ELEVEN

HAMILTONIAN SYSTEMS

11.1 Hamiltonian and Lagrangian dynamics

A Hamiltonian system is one for which the equations can be obtained from a single scalar function, $H(p,q,t)$, called the **Hamiltonian**, in terms of which the equations are

$$p' = -\frac{\partial H}{\partial q}, \qquad q' = \frac{\partial H}{\partial p}. \tag{1.1}$$

In this chapter we shall follow the traditional custom for Hamiltonian systems that the dependent variables are denoted by p and q. Historically the interest in Hamiltonian systems is because frictionless, mechanical systems can be cast into the Hamiltonian form (1.1). In this context there is a beautiful and profound theory for Hamiltonian systems (see, for instance, Landau and Lifshitz, 1960, or Arnold, 1978). In general p and q are m-vectors, but here, in keeping with the general spirit of this text, we shall consider only the planar case when $m = 1$ and p and q are scalar variables. However, many, but not all, of the results that we shall describe have their counterparts for the general case.

Before proceeding to discuss the system (1.1), we shall digress to show how a frictionless mechanical system can be cast into the Hamiltonian form. We have already given a brief account of this in section 6.4 for the autonomous case, and the more general non-autonomous case proceeds similarly. Thus, let $q(t)$ represent the displacement at time t of a particle moving in a straight line subject to a potential-energy field $U(q,t)$ per unit mass. Then the equation of motion is

$$q'' + \frac{\partial U}{\partial q} = 0. \tag{1.2}$$

To put this into the Hamiltonian form, let $p = q'$. Then (1.2) becomes

$$p' = -\frac{\partial U}{\partial q}, \qquad q' = p. \tag{1.3}$$

This has the Hamiltonian form (1.1) if we define the Hamiltonian to be

$$H(p,q,t) = \tfrac{1}{2}p^2 + U(q,t). \tag{1.4}$$

In this context p is the momentum per unit mass, and the Hamiltonian can be identified as the energy of the system, being the sum of the kinetic and potential energies.

More generally, it can be shown that a frictionless mechanical system with a single degree of freedom represented by the scalar dependent variable $q(t)$ can be described by a **Lagrangian**, $L(q', q, t)$, in terms of which the equation of motion is

$$\frac{\mathrm{d}}{\mathrm{d}t}\left(\frac{\partial L}{\partial q'}\right) - \frac{\partial L}{\partial q} = 0. \tag{1.5}$$

This equation is called Lagrange's equation of motion, and is a single equation of the second order for $q(t)$. To convert this to a planar, first-order Hamiltonian system, let

$$p = \frac{\partial L}{\partial q'}. \tag{1.6}$$

This relation defines p as a function of q', q and t. It is customary to call q a generalized coordinate, and then p is the corresponding generalized momentum. We assume that the relation between p and q' is one-to-one, and so (1.6) can be inverted to give q' as a function of p, q and t. Then define

$$H(p, q, t) = pq' - L(q', q, t). \tag{1.7}$$

Hence,

$$\frac{\partial H}{\partial p} = q' + p\frac{\partial q'}{\partial p} - \frac{\partial L}{\partial q'}\frac{\partial q'}{\partial p} = q', \qquad \frac{\partial H}{\partial q} = p\frac{\partial q'}{\partial q} - \frac{\partial L}{\partial q'}\frac{\partial q'}{\partial q} - \frac{\partial L}{\partial q} = -\frac{\partial L}{\partial q},$$

where, in each case, the last result follows on using the definition (1.6). But Lagrange's equation of motion (1.5) is just

$$p' = \frac{\partial L}{\partial q},$$

where we have again used (1.6). Hence we have shown that (1.5) has the Hamiltonian form (1.1).

The converse is also true, and there is a complete equivalence between the Hamiltonian form (1.1) and the Lagrangian form (1.5). Thus, given Hamilton's equations of motion (1.1), let the equation

$$q' = \frac{\partial H}{\partial p}$$

define a one-to-one relation between q' and p which can be inverted to give p as a function of q', q and t. Then define

$$L(q', q, t) = pq' - H(p, q, t).$$

This is, of course, equivalent to (1.7). Hence,

$$\frac{\partial L}{\partial q'} = p + q'\frac{\partial p}{\partial q'} - \frac{\partial H}{\partial p}\frac{\partial p}{\partial q'} = p, \qquad \frac{\partial L}{\partial q} = q'\frac{\partial p}{\partial q} - \frac{\partial H}{\partial p}\frac{\partial p}{\partial q} - \frac{\partial H}{\partial q} = -\frac{\partial H}{\partial q},$$

where, in each case, the last result follows from the second of Hamilton's equations. The first of these results is equivalent to (1.6), and the second is equivalent to the equation of motion (1.5), since, from (1.1), the first of Hamilton's equations is

$$p' = -\frac{\partial H}{\partial q}.$$

In frictionless mechanical systems the Lagrangian is the difference between the kinetic energy $T(x', x, t)$ and the potential energy $U(x, t)$, so that $L = T - U$. To show this, consider a mechanical system consisting of a single particle which is constrained to move on a smooth curve subject to a potential-energy field $U(x, t)$. Here $x(t)$ is the position vector of the particle, and its equation of motion is

$$x'' + \nabla_x U = F, \tag{1.8}$$

where F is the constraint force per unit mass required to keep the particle on the curve. Since the particle has a single degree of freedom, we may put

$$x = x(q, t), \tag{1.9}$$

where q is a parameter on the curve. The explicit dependence on t is included to allow for the case when the curve itself has some prescribed motion. Since the curve is smooth, the constraint force is assumed to do no work, so that

$$F \cdot dx = 0$$

for arbitrary displacements on the curve. But, from (1.9),

$$dx = \frac{\partial x}{\partial q} dq,$$

and so

$$F \cdot \frac{\partial x}{\partial q} = 0,$$

since dq is arbitrary in the first of these expressions. But then (1.8) becomes, on taking the scalar product with $\partial x / \partial q$,

$$x'' \cdot \frac{\partial x}{\partial q} + \nabla_x U \cdot \frac{\partial x}{\partial q} = 0,$$

or

$$x'' \cdot \frac{\partial x}{\partial q} + \frac{\partial U}{\partial q} = 0. \tag{1.10}$$

Here, we have used (1.9) to let $U(x, t)$ become a function $U(q, t)$ of q and t alone. Using (1.9), we can now let $q(t)$ be the generalized coordinate, and then (1.10) is the equation of motion. It remains to show that (1.10) has the Lagrangian form (1.5). Let the Lagrangian be defined by

$$L = \tfrac{1}{2}|\mathbf{x}'|^2 - U. \tag{1.11}$$

Here, from (1.9),

$$\mathbf{x}' = \frac{\partial \mathbf{x}}{\partial q}q' + \frac{\partial \mathbf{x}}{\partial t}, \tag{1.12}$$

and so $\tfrac{1}{2}|\mathbf{x}'|^2$, which is the kinetic energy T per unit mass, is a function of q', q and t. Note that T is positive definite, and is quadratic in q'. Next we regard (1.12) as defining $\mathbf{x}'$ as a function of q', q and t, and so

$$\frac{\partial \mathbf{x}'}{\partial q'} = \frac{\partial \mathbf{x}}{\partial q}.$$

Hence, from (1.11),

$$\frac{\partial L}{\partial q'} = \mathbf{x}' \cdot \frac{\partial \mathbf{x}'}{\partial q'} = \mathbf{x}' \cdot \frac{\partial \mathbf{x}}{\partial q}, \qquad \frac{\partial L}{\partial q} = \mathbf{x}' \cdot \frac{\partial \mathbf{x}'}{\partial q} - \frac{\partial U}{\partial q}.$$

Also,

$$\frac{d}{dt}\left(\frac{\partial \mathbf{x}}{\partial q}\right) = \frac{\partial^2 \mathbf{x}}{\partial q^2}q' + \frac{\partial^2 \mathbf{x}}{\partial q \partial t} = \frac{\partial \mathbf{x}'}{\partial q},$$

where the last result follows from (1.12). Finally, using these results, we see that

$$\begin{aligned}
\frac{d}{dt}\left(\frac{\partial L}{\partial q'}\right) &= \mathbf{x}'' \cdot \frac{\partial \mathbf{x}}{\partial q} + \mathbf{x}' \cdot \frac{d}{dt}\left(\frac{\partial \mathbf{x}}{\partial q}\right) \\
&= \mathbf{x}'' \cdot \frac{\partial \mathbf{x}}{\partial q} + \mathbf{x}' \cdot \frac{\partial \mathbf{x}'}{\partial q} \\
&= \mathbf{x}'' \cdot \frac{\partial \mathbf{x}}{\partial q} + \frac{\partial}{\partial q}(\tfrac{1}{2}|\mathbf{x}'|^2) \\
&= -\frac{\partial U}{\partial q} + \frac{\partial}{\partial q}(L+U) \\
&= \frac{\partial L}{\partial q}.
\end{aligned}$$

The penultimate line follows from the equation of motion (1.10) and the definition of the Lagrangian (1.11). The last line is the required result that the equation of motion has the Lagrangian form (1.5).

Before proceeding to consider the application of these general results, it is useful to consider the relationship between the Hamiltonian H, defined by (1.7), and the energy per unit mass of the system, given by

$$E = T + U, \quad \text{where } T = \tfrac{1}{2}|\mathbf{x}'|^2. \tag{1.13}$$

Now, from (1.12), $\mathbf{x}'$ is a linear expression in q' and, hence, the kinetic energy T is a quadratic expression in q'. We may write

$$T = \tfrac{1}{2}T_2(q,t)\,q'^2 + T_1(q,t)\,q' + \tfrac{1}{2}T_0(q,t).$$

But then, from (1.11),

$$L(q',q,t) = \tfrac{1}{2}T_2(q,t)\,q'^2 + T_1(q,t)\,q' + \tfrac{1}{2}T_0(q,t) - U(q,t).$$

Next, the generalized momentum p is defined by (1.6), and so

$$p = \frac{\partial L}{\partial q'},$$

or

$$p = \frac{\partial T}{\partial q'} = T_2(q,t)\,q' + T_1(q,t).$$

Note that, since the kinetic energy T is a positive-definite expression in q', $T_2 > 0$ and the relation between p and q' is one-to-one as required. Now, the Hamiltonian is defined by (1.7) and, hence, using these expressions for L and p, we find that

$$H(p,q,t) = \tfrac{1}{2}T_2(q,t)\,q'^2 - \tfrac{1}{2}T_0(q,t) + U(q,t),$$

where here, of course, q' must be expressed as a function of p, q and t. Hence,

$$H = E - T_1(q,t)\,q' - T_0(q,t),$$

and, in general, differs from the energy E. Note that (see (1.12))

$$T_1(q,t) = \frac{\partial \boldsymbol{x}}{\partial q} \cdot \frac{\partial \boldsymbol{x}}{\partial t}, \qquad T_0(q,t) = \left|\frac{\partial \boldsymbol{x}}{\partial t}\right|^2.$$

Hence the terms which cause H to differ from E are both due to the possible explicit dependence of $\boldsymbol{x}$ on t in the respresentation (1.9), included to allow for the case when the curve, on which the particle moves, has itself a prescribed motion.

To illustrate how the Hamiltonian form (1.1) is constructed for a mechanical system, consider a simple pendulum. This is a single particle, attached to a fixed point by an inextensible rod, or string, of length l whose mass, relative to that of the particle, can be ignored. We suppose that the pendulum is constrained to swing in a vertical plane and, hence, its motion can be described entirely by the angle θ between the vertical and the pendulum (see figure 11.1). Choosing the x-axis horizontal, the y-axis vertical and the z-axis normal to the plane of the motion, the particle position is given by

$$\boldsymbol{x} = (l\sin\theta, l(1-\cos\theta), 0). \tag{1.14}$$

Here we have chosen the origin to be at the stable equilibrium position when the pendulum is vertical and pointing downwards. This is the required relation (1.9), and the generalized coordinate $q = \theta$. Using (1.12), it is readily shown that

$$T = \tfrac{1}{2}|\boldsymbol{x}'|^2 = \tfrac{1}{2}l^2 q'^2.$$

Fig. 11.1 The simple pendulum

The potential energy per unit mass is

$$U = gl(1 - \cos q)$$

and, hence, the Lagrangian L (1.11) is given by

$$L(q', q, t) = \tfrac{1}{2}l^2 q'^2 - gl(1 - \cos q). \tag{1.15}$$

Hence Lagrange's equation of motion is

$$q'' + \Omega^2 \sin q = 0, \quad \text{where } \Omega^2 = \frac{g}{l}. \tag{1.16}$$

To find the Hamiltonian form, we first find p from (1.6) and (1.15). Here

$$p = l^2 q',$$

and then the Hamiltonian is found from (1.7). On using (1.15), we find that

$$H(p, q, t) = \frac{p^2}{2l^2} + gl(1 - \cos q). \tag{1.17}$$

In this instance the Hamiltonian is just the energy E. Hamilton's equations of motion (1.1) are

$$p' = -gl \sin q, \qquad q' = \frac{p}{l^2}, \tag{1.18}$$

which are clearly equivalent to (1.16).

Next consider the situation when the pendulum is made to rotate around the vertical axis at a constant angular rate ω. In place of (1.14) the particle position is now given by

$$x = (l \sin \theta \cos \omega t, l(1 - \cos \theta), l \sin \theta \sin \omega t). \tag{1.19}$$

Again, putting $q = \theta$ and using (1.12), we now find that

$$T = \tfrac{1}{2}|x'|^2 = \tfrac{1}{2}l^2(q'^2 + \omega^2 \sin^2 q).$$

The potential energy per unit mass is again

$$U = gl(1 - \cos q)$$

and, hence, the Lagrangian L (1.11) is now given by

$$L(q',q,t) = \tfrac{1}{2}l^2(q'^2 + \omega^2 \sin^2 q) - gl(1 - \cos q). \tag{1.20}$$

Lagrange's equation of motion is

$$q'' + \Omega^2 \sin q - \omega^2 \sin q \cos q = 0. \tag{1.21}$$

From (1.6) and (1.20), $p = l^2 q'$, and the Hamiltonian is found from (1.7) and (1.20). We find that

$$H(p,q,t) = \frac{p^2}{2l^2} + gl(1 - \cos q) - \tfrac{1}{2}\omega^2 l^2 \sin^2 q. \tag{1.22}$$

Here the Hamiltonian differs from the energy E (1.13) by the term $T_0 = \omega^2 l^2 \sin^2 q$ (for this example $T_1 = 0$). Hamilton's equations of motion (1.1) are

$$p' = -gl \sin q + \omega^2 l^2 \sin q \cos q, \qquad q' = \frac{p}{l^2}, \tag{1.23}$$

which are equivalent to (1.21). Note that, although the expression (1.19) for the particle position depends explicitly on t, the Hamiltonian (1.22) is explicitly independent of t, and so the resulting Hamiltonian system is autonomous.

Although Hamiltonian systems arise most naturally for frictionless mechanical systems, their occurrence is not confined to this particular context. For instance, consider the Lotka–Volterra equations:

$$x' = x(a - cy), \qquad y' = y(-b + dx),$$

where a, b, c and d are positive constants. This system was discussed in section 5.1 and the orbits are sketched in figure 5.3. Confining attention to the case most relevant for applications when $x, y > 0$, we put

$$p = \ln x, \qquad q = \ln y.$$

In these variables the equations become

$$p' = a - ce^q, \qquad q' = -b + de^p.$$

It is readily seen by inspection that these are in Hamiltonian form with Hamiltonian

$$H = -aq + ce^q - bp + de^p.$$

This last example is an illustration of the general result that the planar system

$$p' = f(p,q,t), \qquad q' = g(p,q,t),$$

is Hamiltonian if and only if

$$\frac{\partial f}{\partial p} + \frac{\partial g}{\partial q} = 0.$$

Here we require f and g to have continuous first derivatives, and the condition should hold in a simply-connected region. The result is established by standard methods of vector analysis and will be proved in section 11.2 (see theorem 11.3).

Let us now return to a preliminary discussion of Hamilton's equations (1.1). First we establish the following theorem.

Theorem 11.1: If the Hamiltonian is autonomous, that is, H is explicitly independent of t $(\partial H/\partial t = 0)$, then H is an integral of Hamilton's equations (1.1).

Proof: Using Hamilton's equations (1.1), we see that

$$\frac{dH}{dt} = \frac{\partial H}{\partial p}p' + \frac{\partial H}{\partial q}q' + \frac{\partial H}{\partial t}$$

$$= \frac{\partial H}{\partial p}\left(-\frac{\partial H}{\partial q}\right) + \frac{\partial H}{\partial q}\left(\frac{\partial H}{\partial p}\right) + \frac{\partial H}{\partial t},$$

or

$$\frac{dH}{dt} = \frac{\partial H}{\partial t}. \tag{1.24}$$

Hence, if $\partial H/\partial t = 0$, the result follows. Thus, for autonomous Hamiltonians, $H(p,q)$ is a constant on each orbit in the p-q phase plane and, hence, in principle, p can be determined as a function of q (or vice-versa) on each orbit. We have already noted this result for autonomous Hamiltonian systems in section 6.4.

For the remainder of this section we shall consider the case when the Hamiltonian system (1.1) is autonomous, and consequently $H(p,q)$ is an integral. The discussion to follow parallels that of section 6.4, which can be regarded as a special case of (1.1) when the Hamiltonian has the form (see (1.4))

$$\tfrac{1}{2}p^2 + U(q),$$

where $U(q)$ is the potential energy. From (1.1) we see that the critical points are associated with the stationary values of $H(p,q)$. Thus, if (p_0,q_0) is a critical point,

$$\frac{\partial H}{\partial p} = \frac{\partial H}{\partial q} = 0 \quad \text{(for } p = p_0, q = q_0\text{)}. \tag{1.25}$$

To analyze the orbits near the critical point, we put

$$x = p - p_0, \qquad y = q - q_0$$

and linearize the system (1.1) about the critical point. Thus, to leading order in x and y,

$$x' = -bx - cy,$$
$$y' = ax + by,$$

where

$$a = \frac{\partial^2 H}{\partial p^2}, \qquad b = \frac{\partial^2 H}{\partial p \partial q}, \qquad c = \frac{\partial^2 H}{\partial q^2} \qquad \text{(for } p = p_0, q = q_0\text{)}. \quad (1.26)$$

The classification of the critical point now depends on the eigenvalues λ of the matrix

$$\begin{bmatrix} -b & -c \\ a & b \end{bmatrix}.$$

They are given by

$$\lambda^2 = b^2 - ac.$$

Thus, for $b^2 < ac$, the eigenvalues are purely imaginary and the critical point is a centre, while, for $b^2 > ac$, the eigenvalues are real-valued and of opposite sign, so that the critical point is a saddle point (see section 5.2). Thus the critical points of a Hamiltonian system are either centres, corresponding to a maximum or minimum of the Hamiltonian, or saddle points, corresponding to a saddle for the Hamiltonian. We shall give an alternative geometrical proof of this result in section 11.2 (see corollary 11.1). Further, it is clear that, whenever the critical point is a centre, the fact that $H(p,q)$ is a constant on each orbit ensures that there will be a family of closed orbits surrounding the centre.

Armed with these results, we return to a consideration of the simple pendulum, for which the Hamiltonian is given by (1.17). It is autonomous, and the integral $H(p,q)$ is equivalent in this instance to the conservation of energy, since the Hamiltonian is the energy in this case. With the normalization $g = l = 1$, this system was discussed in section 6.4 and the orbits are sketched in figure 6.15. As can be anticipated by the preceding discussion, there is a sequence of alternating centres and saddle points on the q-axis. There is a family of periodic orbits surrounding each centre, bounded by saddle connections connecting the saddle points.

Next, consider the rotating pendulum for which the Hamiltonian is given by (1.22). It is also autonomous, but the integral $H(p,q)$ is not now equivalent to the conservation of energy, since, for this case, the Hamiltonian differs from the energy. For $\omega^2 < \Omega^2$, where ω is the angular frequency of the rotation and Ω is the natural frequency of the pendulum (see (1.16)), the critical points are given by

$$p = 0, \qquad q = 0, \pm\pi, \pm 2\pi, \ldots . \qquad (1.27)$$

These are centres for $q = 0, \pm 2\pi, \ldots$, and saddle points for $q = \pm\pi, \pm 3\pi, \ldots$. The situation is analogous to that for the non-rotating pendulum sketched in figure 6.15. However, for $\omega^2 > \Omega^2$ the situation is reversed and all the critical points (1.27) are saddle points. In addition, there is now another set of critical points given by

$$p = 0, \qquad \cos q = \frac{\Omega^2}{\omega^2},$$ (1.28)

These are all centres and, hence, are surrounded by a family of periodic orbits. The phase plane is sketched in figure 11.2.

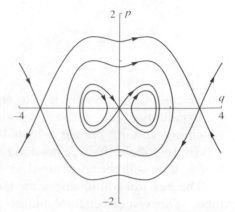

Fig. 11.2 The orbits for the rotating pendulum (1.22) when $\omega^2 > \Omega^2$ ($\Omega^2 = g/l$).

Thus, for $\omega^2 < \Omega^2$ the stable equilibrium position is that with the pendulum vertical and pointing downwards; the pendulum oscillates periodically about this position. However, for $\omega^2 > \Omega^2$, this position becomes unstable and, instead, a stable equilibrium position appears at an angle $\cos^{-1}(\Omega^2/\omega^2)$ to the downward vertical; the pendulum now oscillates periodically about this position. Note that the periodic orbits surrounding the centres (1.28) are themselves enclosed within a double saddle connection through the saddle point at the origin (see figure 11.2), which is itself enclosed by saddle connections from the saddle point at $p = 0$, $q = -\pi$ to that at $p = 0$, $q = \pi$. Between these two saddle connections is a third family of periodic orbits.

11.2 Liouville's theorem

One of the remarkable properties of Hamiltonian systems (1.1) is the existence of a number of integral invariants. These are conserved quantities which relate not to a single solution, but, instead, to a family of solutions. To set the scene for our discussion, we allow a family of solutions $(p(t), q(t))$ of (1.1) to generate a **phase flow** in the (p, q, t)-space. For instance, consider the family of solutions which have initial conditions at $t = t_0$ for which (p, q) lies within a region S_0 with boundary C_0. At subsequent times t these solutions will generate a tube whose cross-section at time t is the region S with boundary C (see figure 11.3). We shall suppose that C_0 is a smooth curve (i.e. a typical parametric representation, $p = p_0(u)$, $q = q_0(u)$, say, is continuously differentiable), and that the Hamiltonian $H(p, q, t)$ has continuous second derivatives for all p, q and t.

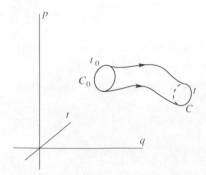

Fig. 11.3 A phase flow.

These conditions are sufficient to ensure that solutions to (1.1) exist for any initial conditions, and define a smooth phase flow for all finite times $t \geq t_0$ for which $p(t)$ and $q(t)$ remain bounded (see chapter 1, sections 1.3 and 1.4). In the results to follow, it is understood that these smoothness conditions hold and, hence, they will not be referred to explicitly again.

The key integral invariant for Hamiltonian systems is the area S of the phase tube. This is a direct consequence of the anti-symmetric structure of Hamilton's equations (1.1). The result is established in the following theorem.

Theorem 11.2 (Liouville's theorem): For the Hamiltonian system (1.1), the phase flow $(p(t), q(t))$ preserves the area

$$\mathcal{S} = \iint_S \mathrm{d}p \, \mathrm{d}q. \tag{2.1}$$

Proof: Let (p_0, q_0) be the initial conditions at $t = t_0$ for the solution $(p(t), q(t))$. Then there is a one-to-one correspondence between $(p(t), q(t))$ and (p_0, q_0). Hence the area $\mathcal{S}$ given by (2.1) can be expressed in the form

$$\mathcal{S} = \iint_{S_0} J \, \mathrm{d}p_0 \, \mathrm{d}q_0,$$

where

$$J = \frac{\partial(p, q)}{\partial(p_0, q_0)}. \tag{2.2}$$

Then

$$\frac{\mathrm{d}\mathcal{S}}{\mathrm{d}t} = \iint_{S_0} \frac{\mathrm{d}J}{\mathrm{d}t} \, \mathrm{d}p_0 \, \mathrm{d}q_0. \tag{2.3}$$

Thus, to prove the result, we must show that $\mathrm{d}J/\mathrm{d}t = 0$ and that the Jacobian J is a constant for each solution.

This result, in turn, is a direct consequence of (3.6) of section 4.3. For convenience, we repeat the argument here. Thus, consider the planar system

$$p' = f(p,q,t), \qquad q' = g(p,q,t) \tag{2.4}$$

and let $p = p(t;p_0,q_0)$, $q = q(t;p_0,q_0)$ be the solution with the initial conditions p_0 and q_0, respectively, at $t = t_0$. Then let

$$X = \begin{bmatrix} \dfrac{\partial p}{\partial p_0} & \dfrac{\partial p}{\partial q_0} \\[2mm] \dfrac{\partial q}{\partial p_0} & \dfrac{\partial q}{\partial q_0} \end{bmatrix}.$$

Since the system itself does not contain any explicit dependence on p_0 and q_0, differentiation of the equations with respect to p_0 and q_0 shows that

$$X' = AX,$$

where

$$A = \begin{bmatrix} \dfrac{\partial f}{\partial p} & \dfrac{\partial f}{\partial q} \\[2mm] \dfrac{\partial g}{\partial p} & \dfrac{\partial g}{\partial q} \end{bmatrix}. \tag{2.5}$$

This is a linear equation, for which X is a fundamental matrix. Hence, from theorem 2.3 of section 2.2 (see (2.9))

$$\frac{dJ}{dt} = (\operatorname{tr} A)J,$$

where

$$J = \det X = \frac{\partial(p,q)}{\partial(p_0,q_0)}. \tag{2.6}$$

Here,

$$\operatorname{tr} A = \frac{\partial f}{\partial p} + \frac{\partial g}{\partial q}. \tag{2.7}$$

But, for a Hamiltonian system (1.1),

$$f = -\frac{\partial H}{\partial q}, \qquad g = \frac{\partial H}{\partial p},$$

and so

$$\frac{\partial f}{\partial p} + \frac{\partial g}{\partial q} = -\frac{\partial^2 H}{\partial p \partial q} + \frac{\partial^2 H}{\partial q \partial p} = 0.$$

Thus $dJ/dt = 0$, and the result is proven.

Corollary 11.1: For autonomous Hamiltonian systems, critical points are either saddle points or centres.

Proof: We have already proven this result in section 11.1 (see (1.20) and the following discussion). Here, however, we see that it is a direct consequence of Liouville's theorem, since it is clear from our analysis of critical points in section 5.2 and section 5.3 that, for a node or a spiral point, stable or unstable, the phase flow is contracting, or expanding, a result which contradicts Liouville's theorem. There is an analogous result for periodic solutions.

Corollary 11.1: For Hamiltonian systems, the characteristic exponents which determine the stability of a periodic solution are either both pure imaginary, or both real and of opposite signs.

Proof: The result is again an immediate consequence of Liouville's theorem. Alternatively, and more directly, the linear system which governs the stability of a periodic solution is given by (1.26), where now the coefficients a, b and c are periodic functions of t. Since, for this system, the trace of the coefficient matrix is zero, it follows that (see (1.4) of section 3.1)

$$\rho_1 \rho_2 = 1,$$

where ρ_1 and ρ_2 are the characteristic multipliers. Since the characteristic exponents μ_1 and μ_2 are given by $\rho_{1,2} = \exp(\mu_{1,2} T)$, where T is the period, it follows that

$$\mu_1 + \mu_2 = 0.$$

The stated result now follows immediately.

The central role of Liouville's theorem in Hamiltonian systems can be seen in the following theorem, which is the converse of theorem 11.2.

Theorem 11.3: Consider the planar system

$$p' = f(p, q, t), \qquad q' = g(p, q, t). \tag{2.8}$$

Then, if the phase flow is area preserving, the system is Hamiltonian, so that there exists a Hamiltonian $H(p, q, t)$ such that

$$f = -\frac{\partial H}{\partial q}, \qquad g = \frac{\partial H}{\partial p}. \tag{2.9}$$

Proof: It is now given that the area $\mathcal{S}$ (2.1) is constant. Thus, using (2.3), it follows that

$$\frac{dJ}{dt} = 0,$$

where the Jacobian J is defined by (2.2). But we showed in the course of proving theorem 11.2 that J satisfies the equation (2.6) and, hence, from (2.7) we deduce that

$$\frac{\partial f}{\partial p} + \frac{\partial g}{\partial q} = 0.$$

It remains to show that this condition leads to the required result (2.9). Now, for any area D in the p-q plane with boundary ∂D,

$$\oint_{\partial D} f \, \mathrm{d}q - g \, \mathrm{d}p = \iint_D \left(\frac{\partial f}{\partial p} + \frac{\partial g}{\partial q} \right) \mathrm{d}p \, \mathrm{d}q = 0,$$

where the first equality is a consequence of Green's theorem in the plane. Hence the line integral of $f \, \mathrm{d}q - g \, \mathrm{d}p$ is path-independent, and defines (up to an arbitrary function of t alone) a function

$$H(p, q, t) = - \int^{(p,q)} f \, \mathrm{d}q - g \, \mathrm{d}p.$$

This is the required Hamiltonian and differentiation leads to (2.9).

Before proceeding to consider another class of integral invariants, we shall describe how Liouville's theorem leads to the phenomenon of Poincaré recurrence. First we prove the following theorem.

Theorem 11.4 (Poincaré recurrence theorem): Let M be any area-preserving, continuous one-to-one mapping defined within a bounded region D. Then, in any neighbourhood N which lies in D, there exists a point $\boldsymbol{p} = (p, q)$ in N which returns to N, that is, $M^s(\boldsymbol{p})$ is in N for some integer $s > 0$. Further, this recurrence occurs infinitely often. Here M^s denotes the map M iterated s times (i.e. $M^2(\boldsymbol{p}) = M(M(\boldsymbol{p}))$, etc.).

Proof: Consider the images of N (see figure 11.4),

$$M, \qquad M(N), \qquad \ldots, \qquad M^r(N), \qquad \ldots.$$

All of these have the same area, and all lie in D. If there were no intersections then D would have an infinite area, which is a contradiction as D is bounded. Hence, there exist integers $k, l \geqslant 0$ with $k > l$ such that the intersection of $M^k(N)$ with $M^l(N)$ is not empty. But then, applying the inverse map M^{-1} l times, it follows that the intersection of $M^s(N)$ and N is not empty, with $s = k - l$. To show that the recurrence occurs infinitely often, we now repeat the argument applied to the intersection of $M^s(N)$ and N.

Suppose now that, for a Hamiltonian system (1.1), we have a phase flow which is known to occupy a bounded region of the p-q plane for all times $t \geqslant t_0$. Such a situation can arise, for instance, for an autonomous system for which the Hamiltonian $H(p, q)$ is a constant on each orbit in the p-q plane, since often this constraint can be used, in turn, to infer the presence of families of bounded

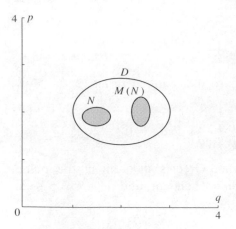

Fig. 11.4 Diagram for theorem 11.4

orbits (see, for instance, figure 11.2). Now, in general, the phase flow defines a mapping from a region at $t = t_0$ to a region at time t. By Liouville's theorem (theorem 11.2), this mapping is area preserving. Since we are also assuming that the phase flow is bounded, the conditions for theorem 11.4 to hold are satisfied and, hence, Poincaré recurrence occurs, that is, in any neighbourhood, no matter how small, there is a point $p_0 = (p_0, q_0)$ such that the solution through p_0 at time $t = t_0$ will return to this neighbourhood at some later time $t > t_0$. Further, this recurrence is repeated infinitely often. The average time for recurrence to occur is called the Poincaré recurrence time and, for some systems, is very large, being effectively unobservable, but for other systems is small enough to be readily found.

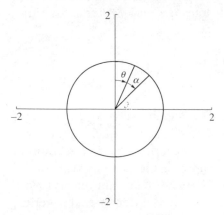

Fig. 11.5 The twist map.

As a simple example of the application of theorem 11.4, consider the **twist map**. Let C be a circle, and let θ be a polar angle on this circle (see figure 11.5). Then the twist map is

$$M(\theta) = \theta + \alpha \quad (\text{mod } 2\pi), \tag{2.10}$$

where α is a constant angle. This map simply takes a point on the circle and moves it through an angle α. It clearly preserves distance (measured along the circle) and is bounded, being confined to the circle C. Thus the conditions for Poincaré recurrence are met, where here we must use the one-dimensional version of theorem 11.4. Alternatively, we can suppose that the map (2.10) is applied to the region between two concentric circles of radii r_1 and r_2 ($r_2 > r_1$) (see figure 11.6). Then the area of the truncated sector between the angles θ_1 and θ_2 ($\theta_2 > \theta_1$) is $\frac{1}{2}(\theta_2 - \theta_1)(r_2^2 - r_1^2)$, and this is clearly conserved by the map (2.10).

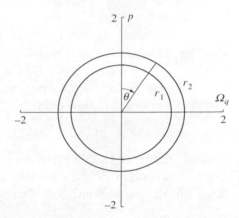

Fig. 11.6 Another version of the twist map.

Before proceeding to discuss the application of the Poincaré recurrence theorem to the map (2.10), we note that it can be regarded as a representation of the simple harmonic oscillator. This has the Hamiltonian

$$H(p,q) = \tfrac{1}{2}(p^2 + \Omega^2 q^2), \tag{2.11}$$

where Ω is the (constant) natural frequency. Hamilton's equations are

$$p' = -\Omega^2 q, \qquad q' = p. \tag{2.12}$$

Since the Hamiltonian is autonomous, $H(p,q) = E$, a constant on each orbit. In this instance E is the energy of the oscillator. Next we introduce the polar angle θ by the transformation

$$p = \sqrt{2E}\cos\theta, \qquad \Omega q = \sqrt{2E}\sin\theta.$$

Note that, with the identification $x = q$ and $y = p$, this is just the transformation to action–angle variables (I, θ), where here $E = \Omega I$, introduced in section 9.3 (see (3.5)) and discussed in more detail in section 11.4. It is readily shown that equations (2.12) reduce to

$$\theta' = \Omega,$$

or

$$\theta = \theta_0 + \Omega t. \tag{2.13}$$

This is equivalent to the twist map (2.10) with $\alpha = \Omega t$. We will show later, in

section 11.4, that using action–angle variables, a large class of nonlinear oscillators can be reduced to the twist map in a similar way.

Let us now return to a consideration of (2.10). First, suppose that α is a
rational fraction of 2π, that is,

$$\frac{\alpha}{2\pi} = \frac{m}{n} \quad \text{(for some integers } m \text{ and } n\text{)}.$$

There is no loss of generality in supposing that m and n are relatively prime.
Then, clearly, $M^n(\theta) = \theta$, and is the identity map. Hence, in this case, there is
exact recurrence. In terms of the simple harmonic oscillator (see (2.13)), this
case occurs when $t = mT/n$, where $T = 2\pi/\Omega$ is the natural period of the oscillator.

Next, let $\alpha/2\pi$ be an irrational number. Then the Poincaré recurrence
theorem implies that, given any $\epsilon > 0$, there exists a positive integer s such that

$$|M^s(\theta) - \theta| < \epsilon.$$

Indeed, theorem 11.4 implies that, given θ and any neighbourhood of θ, there
exists a θ_1 in that neighbourhood such that, for some positive integer s, $M^s(\theta_1)$
is also in that neighbourhood, that is,

$$|\theta_1 - \theta| < \tfrac{1}{3}\epsilon \quad \text{and} \quad |M^s(\theta_1) - \theta| < \tfrac{1}{3}\epsilon.$$

But

$$M^s(\theta_1) - M^s(\theta) = \theta_1 - \theta$$

and, hence,

$$|M^s(\theta_1) - M^s(\theta)| < \tfrac{1}{3}\epsilon.$$

The stated result now follows.

Further, if $\alpha/2\pi$ is an irrational number, then the set

$$S = \{M^k(\theta) : k = 0, 1, 2, \ldots\}$$

is dense on the circle C, for any θ, that is, successive iterations of the map M
applied to the point θ on C will result in the point coming arbitrarily close to
any other point on C. Note that this is clearly not the case if $\alpha/2\pi$ is a rational
number, m/n, since, in that case, $M^n(\theta) = \theta$ and the set S contains only a finite
number n of points. To prove the stated result when $\alpha/2\pi$ is irrational, first
divide the circle C into r equal intervals of length ϵ ($= 2\pi/r$). But we showed
in the preceding paragraph that there then exists a positive integer s such that

$$|M^s(\theta) - \theta| < \epsilon.$$

Further, because

$$M^k(\theta) = \theta + k\alpha \quad (\text{mod } 2\pi),$$

it follows that $M^k(\theta) \neq \theta$ for any integer $k = 1, 2, \ldots$, since $\alpha/2\pi$ is, by hypothesis, an irrational number. In particular, $M^s(\theta) \neq \theta$, and $|s\alpha| < \epsilon$ $(\text{mod } 2\pi)$. Now form the set

$$S_1 = \{M^{rs}(\theta) : r = 0, 1, 2, \ldots\}.$$

Clearly, S_1 is contained in S and, further,

$$M^{(r+1)s}(\theta) - M^{rs}(\theta) = s\alpha \quad (\text{mod } 2\pi).$$

Now, each member of the set S_1 is distinct, and successive members are a distance d apart, where $d = |s\alpha| \ (\text{mod } 2\pi)$, and $0 < d < \epsilon$. Let m be an integer such that

$$m > \frac{2\pi}{d} > \frac{2\pi}{\epsilon} = r.$$

Then, by selecting the first m members of S_1, we can ensure that any interval of length ϵ contains at least one member of S_1 and, hence, at least one member of S.

Of course, the simple twist map (2.10) is elementary since, in particular, it can be regarded as a representation of the simple harmonic oscillator. However, its significance can be extended by considering its two-dimensional analogue, the twist map on the torus. This can be obtained by considering two uncoupled oscillators with natural frequencies Ω_1 and Ω_2, and described by the polar angles θ_1 and θ_2, respectively. By analogy to (2.13) for a single oscillator, the equations are

$$\theta_1' = \Omega_1, \qquad \theta_2' = \Omega_2,$$

or

$$\theta_1 = \theta_{10} + \Omega_1 t, \qquad \theta_2 = \theta_{20} + \Omega_2 t. \tag{2.14}$$

Hence we can define the map

$$M(\theta_1, \theta_2) = (\theta_1 + \alpha_1, \theta_2 + \alpha_2) \quad (\text{mod } 2\pi), \tag{2.15}$$

where $\alpha_1 = \Omega_1 t$ and $\alpha_2 = \Omega_2 t$. Equation (2.15) defines a map on the torus (see figure 11.7(a)); equivalently it can be represented in the θ_1-θ_2 plane by the square of side 2π, provided that the opposite sides of the square are identified as being equivalent (see figure 11.7(b)). The twist map is area-preserving. Here, if α_1/α_2 is a rational number, that is, $\alpha_1/\alpha_2 = m/n$ for some relatively prime integers m and n, or, equivalently, $\Omega_1/\Omega_2 = m/n$, then the orbits are closed curves on the torus. In this case there is exact recurrence. However, if $\alpha_1/\alpha_2 \ (= \Omega_1/\Omega_2)$ is an irrational number, then the orbits are dense on the torus. This is readily shown by observing that the map (2.15) corresponds to a straight line in the θ_1-θ_2 plane $(\text{mod } 2\pi)$ (see figure 11.7(b)). If we consider the intersections of this line with $\theta_1 = \theta_{10} \ (\text{mod } 2\pi)$, then it follows that $\Omega_1 t = 2k\pi$ (for $k = 0, 1, 2, \ldots$), and so

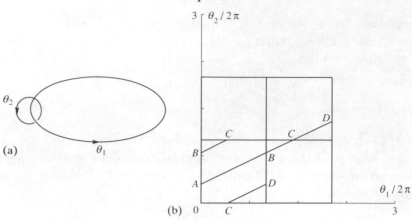

Fig. 11.7 The twist map on the torus. (a) Schematic and (b) the θ_1-θ_2 plane.

$$\theta_2 = \theta_{20} + 2\pi k \frac{\Omega_2}{\Omega_1} \quad (\text{mod}\, 2\pi;\ k = 0, 1, 2, \ldots).$$

But this is equivalent to the one-dimensional twist map (2.10). Hence, if Ω_1/Ω_2 is irrational, the intersections of the map (2.15) with $\theta_1 = \theta_{10}$ (mod 2π) are dense and, hence, the map will be dense within the square of side 2π in the θ_1-θ_2 plane.

11.3 Integral invariants and canonical transformations

We have already encountered one example of an integral invariant in section 11.2, namely, the area of the phase flow (see Liouville's theorem, theorem 11.2). Here we shall show that this is, in fact, a special case of a more general integral invariant. As in section 11.2, consider a family of solutions $(p(t), q(t))$ of (1.1), generating a phase flow in (p, q, t)-space. We picture the phase flow as forming a tube, and we let γ be a circuit around this tube (see figure 11.8). For instance, in Liouville's theorem, we considered circuits C and C_0 which lie in planes for which t is constant (see figure 11.3). Here we shall consider more general circuits γ.

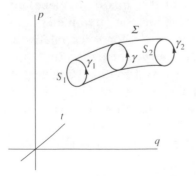

Fig. 11.8 Diagram for theorem 11.5.

Theorem 11.5: Let γ be an arbitrary circuit of a phase flow for Hamilton's equations (1.1). Then

$$\oint_{\gamma} p\,dq - H\,dt \tag{3.1}$$

is invariant.

Proof: The integral invariant (3.1) is sometimes called the **integral invariant of Poincaré–Cartan**, or **Hilbert's invariant integral**. Note that, if γ is restricted to lie in a plane for which t is constant, then (3.1) reduces to

$$\mathscr{S} = \oint_{\gamma} p\,dq,$$

where $\mathscr{S}$ is the cross-sectional area (see (2.1)). Theorem 11.5 then reduces to Liouville's theorem (theorem 11.2).

To prove the result, we use Stokes' theorem applied to the vector

$$\mathbf{F} = (0, p, -H(p, q, t)) \tag{3.2}$$

in (p, q, t)-space. For a surface S with boundary ∂S, Stokes' theorem states that

$$\iint_{S} \mathbf{n} \cdot \operatorname{curl} \mathbf{F}\, dS = \oint_{\partial S} p\,dq - H\,dt. \tag{3.3}$$

Here $\mathbf{n}$ is the unit normal to S, oriented according to the right-hand rule, and

$$\operatorname{curl} \mathbf{F} = \left(-\frac{\partial H}{\partial q}, \frac{\partial H}{\partial p}, 1 \right) \tag{3.4}$$

Now let S be the tube formed by the phase flow between two arbitrarily chosen circuits γ_1 and γ_2, which are, in turn, the boundaries of two surfaces S_1 and S_2. Thus S is a closed surface composed of S_1, S_2 and Σ, where Σ is the surface formed by the phase flows which are initiated on γ_1 and pass through γ_2. On Σ, it follows from Hamilton's equations (1.1) that

$$\operatorname{curl} \mathbf{F} = (p', q', 1),$$

which is parallel to the sides of the tube. Hence, on Σ,

$$\mathbf{n} \cdot \operatorname{curl} \mathbf{F} = 0. \tag{3.5}$$

But, since S is a closed surface, surrounding a region V, say, we can use the Gauss divergence theorem to deduce that

$$\oiint_{S} \mathbf{n} \cdot \operatorname{curl} \mathbf{F}\, dS = \iiint_{V} \operatorname{div} \operatorname{curl} \mathbf{F}\, dV = 0, \tag{3.6}$$

the last result holding because $\operatorname{div} \operatorname{curl}$ is an identically zero operation. But, from (3.5), there is no contribution to the integral over S from the side Σ and, hence,

$$\iint_{S_1} \boldsymbol{n} \cdot \operatorname{curl} F \ dS_1 = \iint_{S_2} \boldsymbol{n} \cdot \operatorname{curl} F \ dS_2,$$

where we orient the respective normals to have the same sense. But then, recalling that γ_1 and γ_2 are the boundaries of S_1 and S_2, respectively, we can use (3.3) to deduce that

$$\oint_{\gamma_1} p \, dq - H \, dt = \oint_{\gamma_1} p \, dq - H \, dt. \tag{3.7}$$

Since γ_1 and γ_2 are arbitrary circuits of the phase flow, the result is proven.

Next we show that theorem 11.5 has a converse, and that phase flows which conserve the Poincaré–Cartan integral (3.1) must satisfy Hamilton's equations.

Theorem 11.6: Consider the planar system

$$p' = f(p, q, t), \qquad q' = g(p, q, t). \tag{3.8}$$

If the phase flow for this system conserves the Poincaré–Cartan integral (3.1) for some Hamiltonian $H(p, q, t)$, then

$$f = -\frac{\partial H}{\partial q}, \qquad g = \frac{\partial H}{\partial p}. \tag{3.9}$$

Proof: Let γ_1 and γ_2 be two arbitrarily chosen circuits of the phase flow. Then we are given that (3.7) holds. As in the proof of theorem 11.5, we let S be the tube formed by the phase flow through γ_1 and γ_2 (see figure 11.8). Thus S is a closed surface composed of S_1, S_2 and Σ, where S_1 and S_2 are surfaces with boundaries γ_1 and γ_2, and Σ is the surface formed by the phase flows. The result (3.6) now holds for any twice-differentiable vector field F. As is the proof of theorem 11.2, we choose F to be defined by (3.2). Then, using (3.3),

$$\iint_{S_i} \boldsymbol{n} \cdot \operatorname{curl} F dS = \oint_{\gamma_i} p \, dq - H \, dt \quad \text{(for } i = 1, 2\text{)}.$$

But we are given that the line integrals around γ_1 and γ_2 are equal (i.e. (3.7) holds) and, hence,

$$\iint_{S_1} \boldsymbol{n} \cdot \operatorname{curl} F \ dS = \iint_{S_2} \boldsymbol{n} \cdot \operatorname{curl} F \ dS.$$

Here the unit normals $\boldsymbol{n}$ are oriented to have the same sense. Thus, in considering the integral over the closed surface S of $\boldsymbol{n} \cdot \operatorname{curl} F$, the contributions from S_1 and S_2 cancel and, hence, using (3.6),

$$\iint_{\Sigma} \boldsymbol{n} \cdot \operatorname{curl} F \ d\Sigma = 0. \tag{3.10}$$

Here F is given by (3.2), and so $\operatorname{curl} F$ is given by (3.4).

Now let us represent Σ in the parametric form:

$$p = p(u, v), \qquad q = q(u, v), \qquad t = t(u, v).$$

Then

$$\boldsymbol{n}\,\mathrm{d}\Sigma = \frac{\partial \boldsymbol{p}}{\partial u} \times \frac{\partial \boldsymbol{p}}{\partial v}\,\mathrm{d}u\,\mathrm{d}v,$$

where $\boldsymbol{p} = (p, q, t)$. Clearly, we can choose t to be one of the parameters on Σ, since Σ is the surface formed by the phase flow. Thus, we let $v = t$, so that u is a parameter which varies on circuits γ surrounding the phase flow (see figure 11.8). Then it follows that

$$\boldsymbol{n}\,\mathrm{d}\Sigma = \left(\frac{\partial q}{\partial u}, -\frac{\partial p}{\partial u}, q'\frac{\partial p}{\partial u} - p'\frac{\partial q}{\partial u} \right)\mathrm{d}u\,\mathrm{d}t$$

and hence (3.10) becomes, on using (3.4),

$$\int \left\{ \oint_{\gamma} \left[\frac{\partial p}{\partial u}\left(q' - \frac{\partial H}{\partial p} \right) + \frac{\partial q}{\partial u}\left(-p' - \frac{\partial H}{\partial q} \right) \right]\mathrm{d}u \right\}\mathrm{d}t = 0.$$

But γ is an arbitrary closed circuit on Σ and, hence, the integrand $[\,\cdots\,]$ must vanish for arbitrary choices of $\partial p/\partial u$ and $\partial q/\partial u$. It follows that

$$p' = -\frac{\partial H}{\partial q}, \qquad q' = \frac{\partial H}{\partial p},$$

which is the required result (3.9).

The fact that Hamilton's equations (1.1) are equivalent to the conservation of the cross-sectional area of the phase flow (theorems 11.2 and 11.3), or, more generally, to the invariance of the Poincaré–Cartan integral (3.1) (theorems 11.5 and 11.6), suggests that it may be useful to consider transformations of the p-q plane which preserve these invariants. These considerations motivate the following definition.

Definition: Let

$$P = P(p, q, t), \qquad Q = Q(p, q, t) \tag{3.11}$$

be a one-to-one differentiable mapping of the p-q plane to the P-Q plane. Then this is called a **canonical transformation** if it preserves area.

These transformations are also called contact transformations. Consider a region d in the p-q plane with boundary ∂d and area a. Thus,

$$a = \iint_{d} \mathrm{d}p\,\mathrm{d}q.$$

The corresponding region in the P-Q plane is D with boundary ∂D and area A, where

$$A = \iint_D \mathrm{d}P\,\mathrm{d}Q.$$

But, using the mapping (3.11),

$$A = \iint_d \frac{\partial(P,Q)}{\partial(p,q)}\,\mathrm{d}p\,\mathrm{d}q.$$

For a canonical transformation, $A = a$, and this must hold for all regions d and D. Hence,

$$\frac{\partial(P,Q)}{\partial(p,q)} = 1. \tag{3.12}$$

Alternatively, let us use Stokes' theorem to obtain the area, so that

$$a = \oint_{\partial d} p\,\mathrm{d}q, \qquad A = \oint_{\partial D} P\,\mathrm{d}Q.$$

Since $A = a$ for a canonical transformation (3.11), it follows that

$$\oint_{\partial d} p\,\mathrm{d}q - P\,\mathrm{d}Q = 0,$$

and this must hold for all closed curves ∂d. Hence the indefinite integral

$$\int^{(p,q)} p\,\mathrm{d}q - P\,\mathrm{d}Q$$

is path-independent, and is a function only of the end point. Let us denote this function by $S(q,Q,t)$. Note that we regard the independent variables of S as q and Q rather than p and q and, hence, we must assume that the mapping (3.11) can be solved to obtain p and P as functions of q and Q. $S(q,Q,t)$ is called a **generating function**, and it follows from its definition that, for fixed t,

$$\mathrm{d}S = p\,\mathrm{d}q - P\,\mathrm{d}Q.$$

More generally, if we also allow t to vary,

$$\mathrm{d}S = p\,\mathrm{d}q - P\,\mathrm{d}Q + R\,\mathrm{d}t. \tag{3.13}$$

Hence we can deduce that

$$p = \frac{\partial S}{\partial q}, \qquad P = -\frac{\partial S}{\partial Q}, \qquad R = \frac{\partial S}{\partial t}. \tag{3.14}$$

The first two of these equations now replace (3.11) in defining the canonical transformation. To ensure that there is an equivalence to (3.11), we must require that $p = \partial S/\partial q$ can be solved for $Q = Q(p,q,t)$ and, hence, we impose the condition

$$\frac{\partial^2 S}{\partial q \partial Q} \neq 0. \tag{3.15}$$

Subject to this restriction, any twice-differentiable function $S(q, Q, t)$ can be used as a generating function for a canonical transformation, the transformation being defined by the first two equations in (3.14). The third equation serves to define $R(q, Q, t)$.

We now show that canonical transformations preserve the Hamiltonian form of Hamilton's equations (1.1).

Theorem 11.7: Let the canonical transformation be defined by the generating function $S(q, Q, t)$ and the equations (3.14). Then Hamilton's equations (1.1) transform to

$$P' = -\frac{\partial K}{\partial Q}, \qquad Q' = \frac{\partial K}{\partial P}, \tag{3.16}$$

where the Hamiltonian $K(P, Q, t)$ is defined by

$$K = H + \frac{\partial S}{\partial t}. \tag{3.17}$$

Proof: We have already shown that, for phase flows in the (p, q, t)-space, the Hamiltonian form of equations (1.1) is equivalent to the preservation of the cross-sectional area in the phase flow, for sections by planes in which t is constant (theorems 11.2 and 11.3). But, by definition, a canonical transformation from (p, q) to (P, Q) preserves this area and, consequently, the cross-sectional area of the phase flow in the (P, Q, t)-space is also preserved. Hence, by theorem 11.3, the phase flow in (P, Q, t) variables is Hamiltonian.

However, to identify the form of the Hamiltonian in (P, Q, t) variables, we must use the integral invariant of Poincaré–Cartan (3.1) and theorems 11.5 and 11.6. Indeed, these theorems can be used to give a complete proof, independently of the argument in the preceding paragraph. First, let

$$K = R + H,$$

so that (3.13) becomes

$$dS = (p\, dq - H\, dt) - (P\, dQ - K\, dt). \tag{3.18}$$

Note that, from (3.18), the result (3.17) follows immediately. It remains to show that K is the required Hamiltonian. Next, form the integral invariant of Poincaré–Cartan (3.1), both for a circuit γ of the phase flow in (p, q, t)-space and for the corresponding circuit Γ of the phase flow in (P, Q, t)-space (see figure 11.9). Since

$$\oint dS = 0,$$

it follows from (3.1) that

$$\oint_\gamma p\, dq - H\, dt = \oint_\Gamma P\, dQ - K\, dt. \tag{3.19}$$

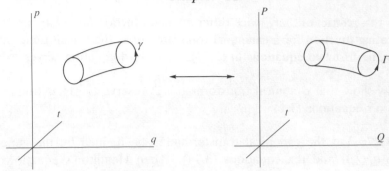

Fig. 11.9 Diagram for theorem 11.7.

But, by theorems 11.5 and 11.6, Hamilton's equations (1.1) are equivalent to the invariance of (3.1). Thus (3.19) implies that

$$\oint_\Gamma P\,dQ - K\,dt$$

is invariant in (P, Q, t)-space, and theorem 11.6 shows that the phase flow is Hamiltonian, with Hamiltonian $K(P, Q, t)$, that is, Hamilton's equations in (P, Q, t)-space are given by (3.15).

Note that, for autonomous canonical transformations (i.e. $\partial S/\partial t = 0$), the Hamiltonian is itself preserved. It is also interesting to observe that Liouville's theorem (theorem 11.2) can now be interpreted as a canonical transformation from the current coordinates (p, q) at time t to the initial coordinates (p_0, q_0) at time t_0, since theorem 11.2 shows that this transformation preserves area. In this case $P = p_0$, $Q = q_0$ and $K = 0$, so that Hamilton's equations (3.16) are

$$P' = 0, \qquad Q' = 0.$$

The generating function is defined by (3.13), and so here

$$dS = p\,dq - p_0\,dq_0 - H\,dt,$$

or

$$S = \int_{t_0}^t p\,dq - H\,dt.$$

Hence

$$S = \int_{t_0}^t L\,dt, \quad \text{where } L = pq' - H. \tag{3.20}$$

Note that L is just the Lagrangian (see (1.7)), and the generating function S defined by (3.20) is called **Hamilton's principal function**.

As an illustration of the use of generating functions and canonical transformations, consider the simple harmonic oscillator of constant natural frequency Ω, whose Hamiltonian is (see (2.11))

$$H(p,q) = \tfrac{1}{2}(p^2+\Omega^2 q^2), \tag{3.21}$$

and for which Hamilton's equations are

$$p' = -\Omega^2 q, \qquad q' = p. \tag{3.22}$$

For a generating function we choose

$$S(q,Q) = \tfrac{1}{2}\Omega q^2 \cot Q. \tag{3.23}$$

The reasons for this choice will appear in section 11.4 when we discuss action–angle variables. From (3.14) it follows that

$$p = \frac{\partial S}{\partial q} = \Omega q \cot Q, \qquad P = -\frac{\partial S}{\partial Q} = \tfrac{1}{2}\Omega q^2 \operatorname{cosec}^2 Q.$$

Solving these equations for P and Q, we find that

$$\Omega P = \tfrac{1}{2}(p^2+\Omega^2 q^2), \qquad \tan Q = \frac{\Omega q}{p}.$$

It is readily seen that (P,Q) correspond to the action–angle variables (I,θ) introduced in section 9.3 (see (3.5)), which we shall discuss further in section 11.4. Since the generating function is autonomous, the new Hamiltonian $K = H$ and, using the above expressions for P and Q, we see that

$$K = \Omega P.$$

Hence, in terms of the new variables P and Q, Hamilton's equations are given by (3.15), or

$$P' = 0, \qquad Q' = \Omega.$$

Thus P is a constant and $Q = Q_0+\Omega t$. Of course, this is just the well-known solution for the simple harmonic oscillator, where the constant P is the action, that is, E/Ω, where E is the energy. This example illustrates the principle that our interest in canonical transformations is in obtaining transformed equations which can more readily be integrated.

Although the generating function $S(q,Q,t)$ serves to define a canonical transformation, it is not always the most convenient to use in practice. For instance, the identity transformation in which $P = p$ and $Q = q$ cannot be easily written in terms of $S(q,Q,t)$, essentially because of the failure of the condition (3.15), although we note that

$$S = qQ$$

generates the canonical transformation $P = -q$, $Q = p$ with Hamiltonian

$K(P,Q,t) = H(Q,-P,t)$. A second canonical transformation with $-S$ then gives the identity.

To overcome this and other difficulties, we introduce the generating function

$$S_1 = PQ + S. \tag{3.24}$$

Then, using (3.13),

$$dS_1 = P\,dQ + Q\,dP + p\,dq - P\,dQ - (H-K)\,dt,$$

or

$$dS_1 = Q\,dP + p\,dq - (H-K)\,dt. \tag{3.25}$$

Hence, it is natural to regard S_1 as a function of q and P, so that $S_1 = S_1(q,P,t)$, and then (3.25) shows that, in place of (3.14) and (3.17),

$$p = \frac{\partial S_1}{\partial q}, \qquad Q = \frac{\partial S_1}{\partial P}, \qquad K = H + \frac{\partial S_1}{\partial t}. \tag{3.26}$$

These equations define a canonical a transformation equivalent to (3.11), provided that $p = \partial S_1/\partial q$ can be solved for $P = P(p,q,t)$ and, hence, we must impose the condition

$$\frac{\partial^2 S_1}{\partial q \partial P} \neq 0. \tag{3.27}$$

The proof that (3.26) defines a canonical transformation which transforms (1.1) to (3.16) is analogous to the proof of theorem 11.7, and follows readily from the identity (3.25). The equivalence of the canonical transformation defined by (3.14) in terms of $S(q,Q,t)$ with that defined by (3.26) in terms of $S_1(q,P,t)$ requires the further conditions that $\partial^2 S/\partial Q^2 \neq 0$, so that (3.14) can be used to find Q as a function of q and P, and $\partial^2 S_1/\partial P^2 \neq 0$, so that (3.26) can be used to find P as a function of q and Q. In terms of $S_1(q,P,t)$ the identity transformation is

$$S_1 = qP,$$

since then

$$p = P, \qquad q = Q. \tag{3.28}$$

Two other generating functions can be defined. First, let

$$S_2 = S_1 - qp,$$

so that, on using (3.25),

$$dS_2 = -p\,dq - q\,dp + Q\,dP + p\,dq - (H-K)\,dt,$$

or

$$dS_2 = -q\,dp + Q\,dP - (H-K)\,dt.$$

Hence it is natural to regard $S_2 = S_2(p, P, t)$ as a function of p and P, and then it follows that

$$q = -\frac{\partial S_2}{\partial p}, \qquad Q = \frac{\partial S_2}{\partial P}, \qquad K = H + \frac{\partial S_2}{\partial t}.$$

Next, let

$$S_3 = S - qp,$$

so that, on using (3.13),

$$\mathrm{d}S_3 = -p\,\mathrm{d}q - q\,\mathrm{d}p + p\,\mathrm{d}q - P\,\mathrm{d}Q - (H - K)\,\mathrm{d}t,$$

or

$$\mathrm{d}S_3 = -q\,\mathrm{d}p - P\,\mathrm{d}Q - (H - K)\,\mathrm{d}t.$$

Here we regard $S_3 = S_3(p, Q, t)$ as a function of p and Q, and then it follows that

$$q = -\frac{\partial S_3}{\partial p}, \qquad P = -\frac{\partial S_3}{\partial Q}, \qquad K = H + \frac{\partial S_3}{\partial t}.$$

Note that each generating function, S, S_1, S_2 or S_3, is a function of one 'old' coordinate, p or q, and one 'new' coordinate, P or Q. Hence these four generating functions define all possibilities.

11.4 Action–angle variables

We have seen in many places throughout this text that some form of polar coordinates are useful in discussing oscillations. For Hamiltonian systems, or systems which are nearly Hamiltonian, we shall now show that the most suitable way to introduce polar coordinates is through a canonical transformation from the original (p, q) coordinates to action–angle variables. Let us first consider an autonomous Hamiltonian $H(p, q)$ for which Hamilton's equations are (1.1). By theorem 11.1, the Hamiltonian itself is a constant on each orbit in the p-q phase plane. Let us suppose that the Hamiltonian is such that there is a set of orbits forming a family of bounded, closed curves in the phase plane (see figure 11.10). Such a situation occurs, for instance, for the simple harmonic oscillator whose Hamiltonian is given by (2.11) and, hence, the integrals

$$H(p, q) = E, \tag{4.1}$$

where E is a constant, define a family of concentric ellipses. Other examples are the simple pendulum, whose Hamiltonian is given by (1.17), which defines the orbits sketched in figure 6.15, and the rotating simple pendulum, whose Hamiltonian is given by (1.22), which defines orbits similar to those in figure 6.15 for low rotation rates, but, for high rotation rates, the orbits are sketched in figure 11.2. Note that, in both these cases, only a subset of the total orbits is

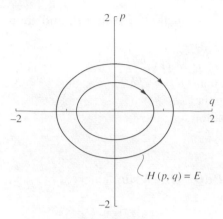

Fig. 11.10 A family of bounded, closed orbits for an autonomous Hamiltonian.

periodic. For instance, for the simple pendulum, the periodic orbits are defined by $0 < E < 2gl$ for weak rotation rates and also for the non-rotating pendulum. For high rotation rates, there are two families of periodic orbits defined by

$$-\frac{l^2}{2\omega^2}(\omega^2 - \Omega^2)^2 < E < 0 \quad \text{and} \quad 0 < E < 2gl,$$

respectively, separated by the orbit $E = 0$, which defines a saddle connection through the saddle point at the origin.

Given that there is a family of bounded, closed orbits (see figure 11.10) defined by (4.1), it is natural to seek polar coordinates such that the radial coordinate is constant on each orbit. Clearly, the Hamiltonian H satisfies this criterion, and can be used in this way. However, any function of H can also be used, and we exploit this to ensure that the new radial coordinates can be obtained by a canonical transformation from the original (p, q) coordinates. These considerations lead to the following definition.

Definition: Let (4.1) define a family of bounded, closed curves for an autonomous Hamiltonian system. Then the variables (I, θ) are called **action–angle variables** if they can be obtained by a canonical transformation from the (p, q) coordinates and satisfy the following criteria:

$$\text{(i)} \quad I = I(H), \quad \text{(ii)} \quad \oint_{H=E} d\theta = 2\pi. \tag{4.2}$$

Here we recall that E is a constant defining each periodic orbit.

To find (I, θ), we seek a generating function to define the canonical transformation. It is convenient to let this be $S_1(I, q)$, which is related to $S(q, \theta)$ by (see (3.24))

$$S_1 = I\theta + S. \tag{4.3}$$

Note that here $P = I$ and $Q = \theta$ in the terminology of section 11.3 (see (3.11)).

Now, from (3.26), we see that

$$p = \frac{\partial S_1}{\partial q}, \qquad \theta = \frac{\partial S_1}{\partial I}, \qquad K(I, \theta) = H, \qquad (4.4)$$

where we note that here the canonical transformation is autonomous, and so the new Hamiltonian K is equal to the old Hamiltonian H. Since the transformation is canonical, the new equations are also Hamiltonian, and given by (3.16). In the present notation, these are

$$I' = -\frac{\partial K}{\partial \theta}, \qquad \theta' = \frac{\partial K}{\partial I}. \qquad (4.5)$$

But, to satisfy (i) in (4.2), we must put

$$K = K(I), \quad \text{or} \quad H = H(I). \qquad (4.6)$$

Further, we assume that this relation is invertible, so that

$$\omega(I) = \frac{\partial K}{\partial I} \neq 0. \qquad (4.7)$$

We shall show below that $\omega(I)$ is the frequency of the oscillator for the orbit defined by (4.1), where we note that (4.6) ensures that I is a constant on each orbit. By virtue of (4.6) and (4.7), Hamilton's equations (4.5) become

$$I' = 0, \qquad \theta' = \omega(I), \qquad (4.8)$$

so that, as anticipated, I is a constant on each orbit, and

$$\theta = \theta_0 + \omega(I) t.$$

This relation confirms that $\omega(I)$ is the frequency.

Next, from (4.4),

$$S_1 = \int_{\substack{q_0 \\ H=E}}^{q} p \, dq, \qquad (4.9)$$

where, as the notation indicates, the integration is along the orbit defined by (4.1), on which I is a constant. Let ΔS_1 denote the change in S_1 when the orbit is traversed once. Thus, from (4.9),

$$\Delta S_1 = \oint_{H=E} p \, dq, \qquad (4.10)$$

which is just the area enclosed by the orbit. But canonical transformations preserve areas and, hence, reverting to (I, θ) coordinates,

$$\Delta S_1 = \oint_{H=E} I \, d\theta,$$

or

$$\Delta S_1 = 2\pi I. \qquad (4.11)$$

Here the second result is a consequence of the fact that I is a constant on the orbit, and of the normalization condition (ii) in (4.2). Combining (4.10) and (4.11), we see that

$$I = \frac{1}{2\pi} \oint_{H=E} p \, dq, \tag{4.12}$$

which serves to define $I = I(H)$, or $H = H(I)$ (see (4.6) and (4.7)). This completes the construction of action–angle variables, since the generating function $S_1(I, q)$ is defined by (4.9), and then the canonical transformation is defined by (4.4).

Note that $S_1(I, q)$ is not unique since, from (4.11), arbitrary integer multiples of $2\pi I$ can be added to S_1, corresponding to the number of times the orbits are traversed. However, from (4.4), p is uniquely determined by $\partial S_1/\partial q$, and we see that the non-uniqueness simply corresponds to the ability to shift θ by arbitrary multiples of 2π. However, the generating function $S(q, \theta)$, defined here by (4.3), is a periodic function of θ and, hence, is unique, since, from (4.3) and (4.11),

$$\Delta S_1 = 2\pi I + \Delta S,$$

or

$$\Delta S = 0.$$

To illustrate the construction of action–angle variables, let us again consider the simple harmonic oscillator whose Hamiltonian is given by (2.11), or

$$H = \tfrac{1}{2}(p^2 + \Omega^2 q^2). \tag{4.13}$$

The orbits are defined by the integrals (4.1) for $0 < E < \infty$, and form a family of concentric ellipses, whose axes have half-widths $\sqrt{2E}$ and $\Omega^{-1}\sqrt{2E}$. Hence the area enclosed by each orbit is $2\pi E/\Omega$. Hence, from (4.12), the action variable I is given by

$$I = \frac{H}{\Omega}. \tag{4.14}$$

Note that, since, for the simple harmonic oscillator, the Hamiltonian is just the energy, the action variable is the ratio of the energy to the natural frequency Ω. Also here (since $K = H$), the frequency $\omega(I) = \partial K/\partial I$ is just Ω, and is independent of I. Hence, as expected, the frequency of the simple harmonic oscillator is Ω. Next, the generating function $S_1(I, q)$ is defined by (4.9), where we must use (4.13) and (4.14) to obtain p in terms of I and q. Thus,

$$S_1(I, q) = \int_0^q \pm(2\Omega I - \Omega^2 \hat{q}^2)^{1/2} \, d\hat{q}. \tag{4.15}$$

Here we have put the constant $q_0 = 0$, and the alternative signs correspond to the sign of p (see figure 11.9). The integral in (4.15) can be evaluated explicitly, but we shall not find it necessary to display the result. The polar coordinate is obtained from (4.4) and we find that

$$\theta = \frac{\partial S_1}{\partial I} = \int_0^q \pm \Omega (2\Omega I - \Omega^2 \hat{q}^2)^{-1/2} \, d\hat{q}$$

or

$$q = \sqrt{\frac{2I}{\Omega}} \sin \theta. \tag{4.16}$$

Finally, p is given by $\partial S_1 / \partial q$ (4.4), and is readily obtained from (4.15), or indeed, directly from (4.13). Using (4.16), we see that

$$p = \sqrt{2\Omega I} \cos \theta. \tag{4.17}$$

Thus (4.16) and (4.17) define the action–angle variables for the simple harmonic oscillator, a result which we have already anticipated on a number of occasions (see, for instance, (3.5) of section 9.3). Note that the angle θ is measured from the p-axis in the clockwise sense, and differs from the conventional polar angle which is $\frac{1}{2}\pi - \theta$. The new Hamiltonian K is given by (4.4) and, using (4.14), we see that

$$K = \Omega I \tag{4.18}$$

and, consequently, Hamilton's equations in the action–angle variables are (see (4.8))

$$I' = 0, \qquad \theta' = \Omega.$$

Finally, the generating function $S(q, \theta)$ is obtained from (4.3), or from (3.14) on using (4.17) and (4.18). We find that

$$S = \frac{1}{2}\Omega q^2 \cot \theta. \tag{4.19}$$

In fact, we anticipated this result in section 11.3 (see (3.23)), where we determined the canonical transformation defined by the generating function S (4.19). The results obtained there are readily shown to be equivalent to (4.16), (4.17) and (4.18).

The simple harmonic oscillator is unusual in that the canonical transformation leading to action–angle variables can be explicitly determined in terms of elementary functions. In general this is not the case and, to demonstrate this, let us reconsider the simple pendulum whose Hamiltonian is given by (1.17). With the normalization $l = 1$, this is

$$H = \frac{1}{2}p^2 + \Omega^2 (1 - \cos q), \tag{4.20}$$

where we have set $\Omega^2 = g/l$, since Ω is the frequency of small-amplitude oscillations. For $0 < E < 2\Omega^2$, the integrals (4.1) define a family of periodic orbits. Now, (4.20) can be used to define p in terms of H and q, and then the action variable I is obtained from (4.12), so that

$$I = \frac{1}{2\pi} \oint \pm\{2H - 2\Omega^2(1 - \cos q)\}^{-1/2} \, dq, \tag{4.21}$$

where the alternate signs correspond to the sign of p, and the integral is around the orbit on which H is constant. Thus (4.21) defines $I = I(H)$. Since $K = H$ (see (4.4)) the frequency $\omega(I)$ is obtained from (4.7), and we find that

$$\frac{1}{\omega(I)} = \frac{1}{2\pi} \oint \pm\{2H - 2\Omega^2(1 - \cos q)\}^{-1/2} \, dq. \tag{4.22}$$

The general theory of action–angle variables now states that Hamilton's equations are (4.8) and, hence, (4.21) and (4.22) provide the essential information to solve (4.8), namely, an expression for $\omega(I)$. The integrals (4.21) and (4.22) cannot be evaluated by elementary means, although they can be expressed in terms of elliptic integrals. Indeed, it is readily shown that (4.21) and (4.22) can be reduced to

$$I = \frac{4H}{\pi\Omega} \int_0^{\frac{1}{2}\pi} \cos^2\lambda \left(1 - \frac{H}{2\Omega^2}\sin^2\lambda\right)^{-1/2} d\lambda$$

and

$$\frac{1}{\omega(I)} = \frac{2}{\pi\Omega} \int_0^{\frac{1}{2}\pi} \left(1 - \frac{H}{2\Omega^2}\sin^2\lambda\right)^{-1/2} d\lambda. \tag{4.23}$$

Next, the generating function $S_1(I, q)$ is found from (4.9), where here we use (4.20) and (4.21) to obtain p in terms of H and q. Thus,

$$S_1 = \int_0^q \pm\{2H - 2\Omega^2(1 - \cos \hat{q})\}^{1/2} \, d\hat{q}, \tag{4.24}$$

where (4.21) must be used to obtain $H = H(I)$. Finally, (4.4) is used to obtain the polar coordinate, and we find that

$$\theta = \frac{\partial S_1}{\partial I} = \omega(I) \int_0^q \pm\{2H - 2\Omega^2(1 - \cos \hat{q})\}^{-1/2} \, d\hat{q}. \tag{4.25}$$

Both (4.24) and (4.25) can be expressed in terms of elliptic functions. Altogether, the expressions (4.20) to (4.25) define the canonical transformation to the action–angle variables (I, θ) in terms of which Hamilton's equations are (4.8).

11.5 Action–angle variables: perturbation theory

The last example (4.20) of section 11.4 illustrates that, although the final form (4.4) of Hamilton's equations is appealingly simple, the canonical transformation from (p, q) to (I, θ) may be quite intricate. This observation, *inter alia*, motivates us to consider the transformation to action–angle variables as pre-conditioning for a perturbation theory. Suppose, for instance, that a planar autonomous dynamical system contains a parameter ϵ, such that, when $\epsilon = 0$, the system possesses action–angle variables (I, θ). Then it is clear that the governing equations have the form

$$I' = \epsilon F(I, \theta; \epsilon), \qquad \theta' = \omega(I) + \epsilon G(I, \theta; \epsilon). \tag{5.1}$$

Here F and G are periodic functions with respect to θ of period 2π. Assuming that the original system is analytic, it follows that F, G and ω are analytic functions of their respective arguments. We discussed systems of the form (5.1) in section 9.1 (see, for instance, theorems 9.1 and 9.2). Further, we now observe that, if the full system is Hamiltonian with Hamiltonian $H(I, \theta; \epsilon)$, then

$$H(I, \theta; \epsilon) = H_0(I) + \epsilon H_1(I, \theta; \epsilon), \tag{5.2}$$

where

$$F = -\frac{\partial H_1}{\partial \theta}, \qquad G = \frac{\partial H_1}{\partial I}, \qquad \omega = \frac{\partial H_0}{\partial I}.$$

Note also that, clearly, H_1 must be a periodic function of θ with period 2π.

As an illustration, let us consider Duffing's equation (see, for instance, (4.9) of section 6.4)

$$u'' + u = \epsilon u^3.$$

Setting $p = u'$, and $q = u$, it is readily seen that

$$p' = -q + \epsilon q^3, \qquad q' = p. \tag{5.3}$$

Clearly this system is Hamiltonian, with Hamiltonian

$$H(p, q) = \tfrac{1}{2}p^2 + \tfrac{1}{2}q^2 - \tfrac{1}{4}\epsilon q^4. \tag{5.4}$$

When $\epsilon = 0$, the system reduces to that for a simple harmonic oscillator of frequency $\Omega = 1$. The action–angle variables for the simple harmonic oscillator are given by (4.16) and (4.17) and, in terms of these, the Hamiltonian (5.4) is

$$H(I, \theta) = I - \epsilon I^2 \sin^4 \theta. \tag{5.5}$$

Since canonical transformations preserve the Hamiltonian form, it follows that (5.3) are equivalent to

$$I' = -\frac{\partial H}{\partial \theta} = 4\epsilon I^2 \sin^3 \theta \cos \theta, \qquad \theta' = \frac{\partial H}{\partial I} = 1 - 2\epsilon I \sin^4 \theta. \tag{5.6}$$

As anticipated, these equations have the form (5.1), and retain the Hamiltonian form (5.2), where here

$$H_0 = I, \qquad H_1 = -I^2 \sin^4 \theta. \tag{5.7}$$

In the context of our discussion of the method of averaging we have already discussed equations of the form (5.1) in section 9.1, and given examples of the application of theorems 9.1 and 9.2 to non-Hamiltonian systems, such as the Van der Pol equation (see (1.23) of section 9.1), and the damped Duffing's equation (see (1.25) of section 9.1). Hence here we shall confine our attention to the Hamiltonian case when F and G are given by (5.2). First let us consider the application of theorem 9.1 of section 9.1. Since $F = -\partial H_1/\partial \theta$ (see (5.2)), where H_1 is a periodic function of θ, it follows that the averaged value $\bar{F}$ of F is zero, that is,

$$\bar{F}(I;\epsilon) = \frac{1}{2\pi} \int_0^{2\pi} F(I,\theta;\epsilon)\, d\theta$$

$$= -\frac{1}{2\pi} \int_0^{2\pi} \frac{\partial H_1}{\partial \theta}(I,\theta;\epsilon)\, d\theta$$

$$= 0.$$

Let $\bar{I}$ be the solution of the averaged equation (see (1.11) of section 9.1)

$$\bar{I}' = \epsilon \bar{F}(\bar{I};0) = 0.$$

Clearly $\bar{I}$ is a constant, and theorem 9.1 of section 9.1 then implies that the solution $I(t)$ of (5.1) is such that $|I(t) - I(0)| < C_1\epsilon$ for $0 \leqslant |\epsilon t| \leqslant C_0$, provided that $|\epsilon|$ is sufficiently small, and where C_0 and C_1 are constants independent of ϵ. Significantly, we must also require that $|\omega(I)|$ is bounded away from zero.

However, the special Hamiltonian form of (5.1), namely (5.2), implies that much more can be established. Indeed, since the full system, as well as the reduced system when $\epsilon = 0$, is a planar autonomous Hamiltonian system, it follows that the orbits possess the integrals

$$H(I,\theta;\epsilon) = E,$$

where E is a constant. Assuming that these define a family of closed curves, it follows that the full system must possess action–angle variables (J,ϕ), say, such that

$$J = I + O(\epsilon), \qquad \phi = \theta + O(\epsilon).$$

Since J, the exact action variable, is a constant on each orbit, it follows that I is a constant to $O(\epsilon)$. This recovers the result established above from theorem 9.1 of section 9.1. Further, in terms of (J,ϕ), the Hamiltonian is given by

$$H(I,\theta;\epsilon) = K(J;\epsilon), \tag{5.8}$$

and Hamilton's equations become (see (4.8) of section 11.4)

$$J' = 0, \qquad \phi' = \Omega(J;\epsilon),$$

where

$$\Omega(J;\epsilon) = \frac{\partial K}{\partial J}. \tag{5.9}$$

Further, it is clear that

$$K(J;\epsilon) = H_0(J) + O(\epsilon), \qquad \Omega(J;\epsilon) = \omega(J) + O(\epsilon).$$

It remains therefore to refine these results, and to determine the respective $O(\epsilon)$ error terms.

Now, since J and ϕ are action–angle variables, it follows from the theory developed in section 11.4 that they can be obtained by a canonical transformation from the (I, θ) coordinates by means of a generating function $S_1(J, \theta; \epsilon)$ (see (4.4) of section 11.4 and note that here the old coordinates (p, q) have been labelled as (I, θ) and the new coordinates are (J, ϕ). Thus

$$I = \frac{\partial S_1}{\partial \theta}, \qquad \phi = \frac{\partial S_1}{\partial J}, \tag{5.10}$$

and the new Hamiltonian is given by (5.8), while the new Hamiltonian equations are (5.9). Further, the transformation from (I, θ) to (J, ϕ) is a near-identity transformation and, hence,

$$S_1(J, \theta; \epsilon) = J\theta + \epsilon S_1^{(1)}(J, \theta; \epsilon), \tag{5.11}$$

where we recall that $J\theta$ is the identity transformation. On substituting (5.11) into (5.10), we find that

$$I = J + \epsilon I^{(1)}(J, \theta; \epsilon), \qquad \phi = \theta + \epsilon \phi^{(1)}(J, \theta; \epsilon),$$

where

$$I^{(1)} = \frac{\partial S_1^{(1)}}{\partial \theta}, \qquad \phi^{(1)} = \frac{\partial S_1^{(1)}}{\partial J}. \tag{5.12}$$

Further, the condition (4.11) of section 11.4 that $\Delta S_1 = 2\pi J$, where we recall that ΔS_1 denotes the change in S_1 around each orbit, implies that (see (5.11) and note that $\Delta\theta = \Delta\phi = 2\pi$)

$$\Delta S_1^{(1)} = 0$$

and, hence, $S_1^{(1)}$ is a periodic function of θ, with period 2π. It follows that $I^{(1)}$ and $\phi^{(1)}$ are also both periodic functions of θ with period 2π. Further, $I^{(1)} = \partial S^{(1)}/\partial \theta$ has zero average with respect to θ and, hence, J is just the average value of I.

Finally, to determine $S_1^{(1)}$, we use the expressions (5.12) in the Hamiltonian relation (5.8). Thus,

$$H_0(I) + \epsilon H_1(I, \theta; \epsilon) = K(J;\epsilon),$$

or

$$H_0(J) + \epsilon\{\omega(J) I^{(1)} + H_1(J, \theta; 0)\} + O(\epsilon^2) = K(J; \epsilon). \qquad (5.13)$$

Here we recall that $\omega(I) = \partial H_0/\partial I$ (see (5.2)). Hence, if we let

$$K(J; \epsilon) = H_0(J) + \epsilon K_1(J) + O(\epsilon^2), \qquad (5.14)$$

it follows that

$$K_1(J) = \omega(J) I^{(1)}(J, \theta; 0) + H_1(J, \theta; 0).$$

But, since the left-hand side is independent of θ and $I^{(1)} = \partial S_1^{(1)}/\partial\theta$, where $S_1^{(1)}$ is a periodic function of θ (see (5.12)), it follows that

$$K_1(J) = \frac{1}{2\pi} \int_0^{2\pi} H_1(J, \theta; 0) \, d\theta$$

and

$$S^{(1)}(J, \theta; 0) = \frac{1}{\omega(J)} \int_{\theta_0}^{\theta} \{K_1(J) - H_1(J, \theta; 0)\} \, d\theta. \qquad (5.15)$$

Thus $K_1(J)$ is just the average of the perturbation Hamiltonian $H_1(J, \theta; 0)$ and, in effect, we have obtained an extension of the method of averaging when the system being considered is an autonomous Hamiltonian system. In particular, we now see that the exact frequency $\Omega = \partial K/\partial J$ (see (5.9)) is given by

$$\Omega(J; \epsilon) = \omega(J) + \epsilon\frac{\partial K_1}{\partial J} + O(\epsilon^2), \qquad (5.16)$$

where K_1 is defined by (5.15). Note also, however, that this perturbation procedure requires that $|\omega(J)|$ is bounded away from zero (see the expression for $S_1^{(1)}$ in (5.15)).

As an illustration, let us reconsider Duffing's equation whose Hamiltonian is given by (5.5) in terms of the approximate action–angle variables (I, θ). Thus here

$$H_0(I) = I, \qquad H_1(I, \theta; \epsilon) = -I^2 \sin^4\theta.$$

Hence $\omega = \partial H_0/\partial I = 1$ and, from (5.15) and (5.16),

$$K_1(J) = -\tfrac{3}{8}J^2,$$

so that

$$\Omega(J; \epsilon) = 1 - \tfrac{3}{4}\epsilon J + O(\epsilon^2).$$

This result can be shown to agree with the result obtained by the Poincaré-Lindstedt method described in chapter 7 (see (1.23) of section 7.1). Also, from (5.15), putting $\theta_0 = 0$, we find that

$$S_1^{(1)}(J, \theta; 0) = -\tfrac{1}{4}J^2 \sin 2\theta + \tfrac{1}{32}J^2 \sin 4\theta$$

and, hence, from (5.12),

$$I^{(1)} = -\tfrac{1}{2}J^2 \cos 2\theta + \tfrac{1}{8}J^2 \cos 4\theta, \qquad \phi^{(1)} = \tfrac{1}{2}J \sin 2\theta - \tfrac{1}{16}J^2 \sin 4\theta.$$

Next, let us consider non-autonomous Hamiltonian systems, where the explicit dependence on t is through slowly-varying coefficients which are functions of the slow variable (see (3.2) of section 9.3)

$$\tau = \epsilon t. \qquad (5.17)$$

Thus the Hamiltonian has the form $H(p, q, \tau)$. Let us suppose further that, when τ is held fixed, the system possesses action–angle variables (I, θ) obtained from (p, q) by a generating function $S(q, \theta, \tau)$. From (3.14) of section 11.3 we can infer that

$$p = \frac{\partial S}{\partial q}, \qquad I = -\frac{\partial S}{\partial \theta}. \qquad (5.18)$$

Clearly, this defines a canonical transformation from (p, q) to (I, θ), and, from (3.17) of section 11.3, the new Hamiltonian is $K(I, \theta, \tau)$, where

$$K(I, \theta, \tau) = H(I, \tau) + \epsilon \frac{\partial S}{\partial \tau}. \qquad (5.19)$$

Here the fact that (I, θ) are action–angle variables when τ is fixed ensures that $H(p, q, \tau)$ will depend only on I (and of course on τ) after using the canonical transformation (5.18). The new form of Hamilton's equations are thus

$$I' = -\frac{\partial K}{\partial \theta}, \qquad \theta' = \frac{\partial K}{\partial I},$$

or

$$I' = -\epsilon \frac{\partial^2 S}{\partial \theta \partial \tau}, \qquad \theta' = \omega(I, \tau) + \epsilon \frac{\partial^2 S}{\partial I \partial \tau}, \qquad (5.20)$$

where

$$\omega(I, \tau) = \frac{\partial H}{\partial I}(I, \tau).$$

These equations have the form discussed in section 9.3 (see (3.7) of that section), although there we discussed the more general form

$$I' = \epsilon F(I, \theta, \tau; \epsilon), \qquad \theta' = \omega(I, \tau) + \epsilon G(I, \theta, \tau; \epsilon),$$

in which the perturbation caused by the slowly-varying coefficients is not necessarily Hamiltonian.

For instance, let us consider the simple harmonic oscillator with a slowly varying frequency:

$$u'' + \omega^2(\tau) u = 0.$$

Putting $p = u'$ and $q = u$, this has the Hamiltonian (see (2.11))

$$H(p,q,\tau) = \tfrac{1}{2}\{p^2 + \omega^2(\tau)\, q^2\}. \tag{5.21}$$

When τ is fixed, this system has action–angle variables (I, θ) given by (see (4.16) and (4.17))

$$p = \sqrt{2\omega I}\cos\theta, \qquad q = \sqrt{\frac{2I}{\omega}}\sin\theta. \tag{5.22}$$

The generating function is given by (4.19), or

$$S(q,\theta,\tau) = \tfrac{1}{2}\omega q^2 \cot\theta. \tag{5.23}$$

It is readily verified that the canonical transformation (5.22) is equivalent to that defined by (5.18), on using (5.23). Next, the new Hamiltonian is obtained from (5.19), (5.22) and (5.23). Thus we find that

$$K(I,\theta,\tau) = \omega(\tau)I + \epsilon \frac{\omega'(\tau)}{2\omega(\tau)} I\sin 2\theta, \tag{5.24}$$

and so Hamilton's equations (5.20) are

$$I' = -\epsilon\frac{\omega'(\tau)}{\omega(\tau)} I\cos 2\theta, \qquad \theta' = \omega(\tau) + \epsilon\frac{\omega'(\tau)}{2\omega(\tau)}\sin 2\theta. \tag{5.25}$$

These equations agree, of course, with (3.6) of section 9.3, where we anticipated the usefulness of reformulating the equations in terms of action–angle variables.

Reverting to the general case (5.20), let us now consider the application of theorem 9.5 of section 9.3. Since here $F(I,\theta,\tau;\epsilon) = -\partial^2 S/\partial\theta\partial\tau$ and S is a periodic function of θ with period 2π, we see that

$$F(I,\tau;\epsilon) = \frac{1}{2\pi}\int_0^{2\pi} F(I,\theta,\tau;\epsilon)\,d\theta$$

$$= -\frac{1}{2\pi}\int_0^{2\pi} \frac{\partial^2 S}{\partial\theta\partial\tau}(I,\theta,\tau;\epsilon)\,d\theta$$

$$= 0.$$

Next, let $\bar{I}$ be the solution of the averaged equation (see (3.9) of section 9.3):

$$\bar{I}' = \epsilon\bar{F}(I,\tau;0) = 0.$$

Clearly, $\bar{I}$ is a constant, and theorem 9.5 of section 9.3 then implies that the solution $I(t)$ of (5.20) is such that $|I(t) - I(0)| < C_1\epsilon$ for $0 \leqslant |\epsilon t| \leqslant C_0$, provided that $|\epsilon|$ is sufficiently small, and where C_0 and C_1 are constants independent of ϵ. We must also require that $|\omega(I,\tau)|$ is bounded away from zero.

In effect, the action variable I is approximately a constant, and is said to be an **adiabatic invariant**. In section 9.3 we showed that the action is an adiabatic invariant for the simple harmonic oscillator. Here we have shown that the same conclusion holds for more general Hamiltonian systems.

Problems

1 Each of the following mechanical systems can be described by a single generalized coordinate. In each case find a suitable generalized coordinate q, obtain the Lagrangian $L(q',q,t)$ and, hence, find the Hamiltonian $H(p,q,t)$ and Hamilton's equations. For those cases where the Hamiltonian is autonomous, sketch the orbits in the p-q plane.

(a) A particle moves in the x-y plane under the influence of gravity, but is constrained to remain on the curve $y = \frac{1}{2}\alpha x^2$. Here the x-axis is horizontal and the y-axis is vertical. Assume that the constraint force needed to keep the particle on the curve does no work. [*Hint*: Let $q = x$ be the generalized coordinate, so that the vector position of the particle is $\boldsymbol{x} = (q, \frac{1}{2}\alpha q^2, 0)$. Then use (1.11) and (1.12) to obtain the Lagrangian, noting that here the potential-energy field U is gy. The Hamiltonian is then obtained from (1.6) and (1.7).]

(b) Similar to (a) above, but the particle is constrained to remain on the curve

$$y = y_0 \cosh \beta x.$$

(c) Similar to (a) above, but the particle is constrained to move on the cycloid whose parametric equations are

$$x = a(\theta + \sin \theta), \qquad y = a(1 - \cos \theta).$$

In this case use arc-length along the cycloid as the generalized coordinate, and show that the system is equivalent to a simple harmonic oscillator.

(d) A simple pendulum is constrained to oscillate in the x-y plane, and its point of support has a prescribed horizontal velocity $u_0(t)$. If θ is the angle between the vertical and the pendulum, show that the particle position is given by (see (1.14))

$$\boldsymbol{x} = (x_0(t) + l \sin \theta, l(1 - \cos \theta), 0),$$

where $x_0' = u_0$. In this case use θ as the generalized coordinate.

(e) Similar to (d) above, but the point of support of the pendulum has a prescribed vertical velocity $v_0(t)$. In this case the particle position is given by

$$\boldsymbol{x} = (l \sin \theta, y_0(t) + l(1 - \cos \theta), 0),$$

where $y_0' = v_0$.

(f) The system of (a) above is made to rotate around the vertical axis at a constant angular rate ω. [*Hint*: Let $q = x$ be the generalized coordinate, and show that the particle position is now given by

$$\boldsymbol{x} = (q \cos \omega t, \frac{1}{2}\alpha q^2, q \sin \omega t).\,]$$

2 For each of the following Hamiltonians, obtain Hamilton's equations and sketch the orbits in the p-q plane. In particular, identify and classify all the critical points.

(a) $H = \frac{1}{2}p^2 + U(q)$, where

      (i) $U(q) = \frac{1}{2}q^2 - \frac{1}{4}q^4$;   (ii) $U(q) = \frac{1}{4}q^4 - \frac{1}{2}q^2$;

      (iii) $U(q) = \frac{1}{4}q^{-4} - \frac{1}{2}q^{-2}$;   (iv) $U(q) = \frac{1}{2}q^{-2} - \frac{1}{4}q^{-4}$;

      (v) $U(q) = q^2 \exp(-q^2)$.

(b) $H = \frac{1}{2}p^2 + \alpha pq$, for both $\alpha > 0$ and $\alpha < 0$.

(c) $H = \frac{1}{2}p^2 + p\sin q$.

(d) $H = \frac{1}{2}p^2\exp(-q) + \frac{1}{2}q^2$.

3 The Hamiltonian for a single particle moving under the influence of an inverse square law is

$$H = \frac{1}{2}\left(p^2 + \frac{h^2}{q^2}\right) - \frac{k}{q},$$

where h and k are constants. In fact, this mechanical system has three degrees of freedom, but conservation of angular momentum enables the system to be described by a single coordinate q representing the radial variable in the plane of the motion. Sketch the orbits in the p-q plane, and show that, for $k > 0$ (an attractive force), there are periodic orbits representing capture of the particle for $-k^2/2h^2 < E < 0$, where the orbits are defined by $H(p,q) = E$, and all other orbits with $E \geqslant 0$ pass to infinity. For $k < 0$, show that there are no periodic orbits. Note that only that part of the phase plane where $q > 0$ is relevant to these results.

4 Show that each of the following transformations $(P,Q) \leftrightarrow (p,q)$ is canonical, and find a generating function.

(a) $P = q$, $Q = -p$;   (b) $p = \exp Q$, $P = -qp$;

(c) $P = 2\sqrt{q}e^t\sin p$, $Q = \sqrt{q}e^{-t}\cos p$;   (d) $p = q\cot Q$, $P = \frac{1}{2}(p^2 + q^2)$.

5 For a coordinate transformation $Q = Q(q,t)$, find the most general function $P = P(p,q,t)$ for which the transformation $(P,Q) \leftrightarrow (p,q)$ is canonical, and determine a generating function. [*Hint*: Here the appropriate generating function is $S_1(q,p,t)$ (see (3.26)).]

6 Show, by direct differentiation, that the canonical transformation determined by (3.14) satisfies the definition (3.12), provided that the condition (3.15) holds.

7 Show that the composition of two successive canonical transformations is canonical. [*Hint*: Use the composition rule for the Jacobians defined in (3.12).]

8 Show that the linear transformation

$$P = ap + bq, \qquad Q = cp + dq,$$

where a, b, c and d are constants, is canonical, provided that $ad - bc = 1$, and find a generating function. What is the geometrical interpretation of the condition $ad - bc = 1$? [*Hint*: If $a \neq 0$, an appropriate generating function is $S_1(q,p)$ (see (3.26)) but, if $a = 0$ and $c \neq 0$, the appropriate generating function is $S(q,Q)$ (see (3.14)).]

9 Use the canonical transformation of problem 8 to show that the Hamiltonian system with Hamiltonian

$$\frac{1}{2}\alpha p^2 + \beta pq + \frac{1}{2}\gamma q^2,$$

where α, β and γ are constants, can be transformed into a new Hamiltonian system with Hamiltonian

$$C(\tfrac{1}{2}P^2 + \tfrac{1}{2}Q^2), \qquad C(-\tfrac{1}{2}P^2 - \tfrac{1}{2}Q^2), \qquad C(\tfrac{1}{2}P^2 - \tfrac{1}{2}Q^2), \quad \text{or} \quad \pm\tfrac{1}{2}CP^2,$$

where C is a positive constant. Show that these alternative forms correspond, respectively, to the case when the eigenvalues of the matrix

$$\begin{bmatrix} \alpha & \beta \\ \beta & \gamma \end{bmatrix}$$

are both positive, both negative, of opposite sign, or such that one eigenvalue is zero. Note that this classification corresponds to the Hamiltonian being positive definite, negative definite, indefinite, or degenerate, respectively. Determine and solve the corresponding form of Hamilton's equations, and note that the first two cases correspond to a centre at which the Hamiltonian is a minimum or a maximum, the third case corresponds to a saddle point, and the fourth case corresponds to uniform motion.

10 For each of the following autonomous Hamiltonian systems whose Hamiltonian is given, find action–angle variables (where applicable), and determine a suitable generating function. In each case use the action–angle variables to obtain a solution.

(a) The simple harmonic oscillator:

$$H = \tfrac{1}{2}(p^2 + \Omega^2 q^2).$$

(b) The simple pendulum:

$$H = \tfrac{1}{2}p^2 + \Omega^2(1 - \cos q).$$

(c) The Duffing oscillator:

$$H = \tfrac{1}{2}p^2 + \tfrac{1}{2}\Omega^2 q^2 - \tfrac{1}{4}\epsilon q^4.$$

(d) A single particle moving in a potential energy field

$$U(q) = \tfrac{1}{2}\Omega^2 q^2 + \tfrac{1}{2}kq^{-2} \quad (k > 0),$$

so that

$$H = \tfrac{1}{2}p^2 + \tfrac{1}{2}\Omega^2 q^2 + \tfrac{1}{2}kq^{-2}.$$

(e) A single particle moving under the influence of an inverse square law, so that (see problem 3)

$$H = \frac{1}{2}\left(p^2 + \frac{h^2}{q^2}\right) - \frac{k}{q},$$

where $k > 0$.

[*Hint*: In (d) and (e) the following integral may be useful:

$$\int_{x_1}^{x_2} \frac{(-A^2 x^2 + 2Bx - C^2)^{1/2}}{x}\, dx = \pi\left(\frac{B}{A} - C\right),$$

where x_1 and x_2 are the zeros of $(\cdots)$.]

11 Use the perturbation method of section 11.5 to obtain approximate solutions to $O(\epsilon)$ for each of the following Hamiltonian systems, whose Hamiltonian is given. In each case, first find the approximate action–angle variables (I, θ) corresponding to the limit $\epsilon = 0$. Then find the first-order correction $K_1(J)$ to the Hamiltonian (5.15) and, hence, obtain the frequency correct to $O(\epsilon)$ (see (5.16)).

(a) $H = \frac{1}{2}p^2 + \frac{1}{2}q^2 - \frac{1}{4}\epsilon q^4$ (the Duffing oscillator).

(b) $H = \frac{1}{2}p^2 + \frac{1}{2}q^2 + \epsilon U(q;\epsilon)$.

(c) As part (b), with $U(q;\epsilon) = \epsilon^{-1}\{1 - \cos(\sqrt{\epsilon}q) - \frac{1}{2}\epsilon q^2\}$.

[This is the simple pendulum (see (1.17)) or problem 10(b).]

12 A single particle moving under the influence of an inverse square law is perturbed by a potential $\epsilon U(q)$ so that the Hamiltonian is (see problem 10(e))

$$H = \frac{1}{2}\left(p^2 + \frac{h^2}{q^2}\right) - \frac{k}{q} + \epsilon U(q).$$

Assuming that $k > 0$, use the perturbation method of section 11.5 to obtain an approximate solution to $O(\epsilon)$. Illustrate when (a) $U(q) = q^{-2}$ and (b) $U(q) = q^{-3}$ (a relativistic correction). [*Hint*: Use your result from problem 10(e) to obtain the approximate action–angle variables (I, θ) corresponding to the limit $\epsilon = 0$. Note that, in case (a), the perturbation potential is equivalent to perturbing the constant h, and use this fact to verify the result of the perturbation method.]

13 A Hamiltonian is given by

$$H(I, \theta; \epsilon) = H_0(I) + \epsilon H_1(I, \theta) + \epsilon^2 H_2(I, \theta) + O(\epsilon^3).$$

Show that if (J, ϕ) are the exact action–angle variables, then

$$H = K(J; \epsilon) = H_0(J) + \epsilon K_1(J) + \epsilon^2 K_2(J) + O(\epsilon^3),$$

where

$$K_1(J) = \frac{1}{2\pi}\int_0^{2\pi} H_1(J, \theta)\, d\theta,$$

$$K_2(J) = \frac{1}{2\pi}\int_0^{2\pi}\left\{H_2(J, \theta) + \frac{1}{2}\frac{\partial}{\partial J}\left(\frac{K_1^2 - H_1^2(J, \theta)}{\omega}\right)\right\} d\theta,$$

and

$$\omega(J) = \frac{\partial H_0}{\partial J}.$$

[*Hint*: Use the perturbation method of section 11.5 (see (5.10) and the subsequent discussion), but carry the analysis through to the second order.]

14 Use the perturbation method of section 11.5 to obtain an approximate formula correct to $O(\epsilon^2)$ for the Duffing oscillator, whose Hamiltonian is

$$H = \frac{1}{2}p^2 + \frac{1}{2}q^2 - \frac{1}{4}\epsilon q^4.$$

[*Hint*: Use the result of problem 13 after first obtaining the Hamiltonian form (5.5)].

ANSWERS TO
SELECTED PROBLEMS

Chapter 2

1 (a) $\begin{bmatrix} e^t & -e^{-t} \\ e^t & e^{-t} \end{bmatrix}$, $W = 2$.

(b) $\begin{bmatrix} 1 & -\frac{1}{2}e^{-t}(\cos t + \sin t) \\ 0 & e^{-t} \end{bmatrix}$, $W = e^{-t}$.

(c) $\begin{bmatrix} t & 1 \\ 1 & t \end{bmatrix}$, $W = t^2 - 1$.

5 $u = A\cos\omega t + B\sin\omega t + \dfrac{1}{\omega}\displaystyle\int_0^t f(s)\sin\omega(t-s)\,ds$.

6 (a) $x_1 = c_1 e^{5t} + c_2 e^{3t}$, $x_2 = 3c_1 e^{5t} + c_2 e^{3t}$.

(b) $x_1 = e^{-2t}(c_1\cos t + c_2\sin t)$, $x_2 = e^{-2t}(-c_1\sin t + c_2\cos t)$,

(c) $x_1 = e^t(c_1 + c_2 t)$, $x_2 = e^t[c_1 + c_2(1+t)]$.

(d) $x_1 = e^{-t}(2c_1\cos 2t + 2c_2\sin 2t)$, $x_2 = e^{-t}[c_1(\cos 2t - \sin 2t) + c_2(\cos 2t + \sin 2t)]$.

(e) $x_1 = (2c_2 - c_1)e^t + c_3 e^{-2t}$, $x_2 = 3c_1 e^t$, $x_3 = (c_2 + c_1)e^t - c_3 e^{-2t}$.

(f) $x_1 = c_1 + e^t[2c_2 + 2c_3(-1+t)]$, $x_2 = e^t(2c_2 + 2c_3 t)$,

$x_3 = -c_1 + e^t[3c_2 + c_3(-\frac{5}{2} + 3t)]$.

8 $x = X(t)c + \displaystyle\int_0^t X(t-s)f(s)\,ds$, where $X(t) = \begin{bmatrix} \cos t & \sin t \\ -\sin t & \cos t \end{bmatrix}$, $f(t) = \begin{bmatrix} f_1(t) \\ f_2(t) \end{bmatrix}$.

Chapter 3

1 (a) $T = 2\pi$, $X(t) = \begin{bmatrix} e^{-t} & e^{t+\sin t} \\ 0 & (2+\cos t)e^{t+\sin t} \end{bmatrix}$, $B = \begin{bmatrix} e^{-2\pi} & 0 \\ 0 & e^{2\pi} \end{bmatrix}$, $\rho = e^{\pm 2\pi}$;

$\mu = \pm 1$; $P(t) = \begin{bmatrix} 1 & e^{\sin t} \\ 0 & (2+\cos t)e^{\sin t} \end{bmatrix}$, $D = \begin{bmatrix} -1 & 0 \\ 0 & 1 \end{bmatrix}$.

(b) $T = \pi$, $X(t) = \begin{bmatrix} e^{3t}\cos t & e^{-t}\sin t \\ -e^{3t}\sin t & e^{-t}\cos t \end{bmatrix}$, $B = \begin{bmatrix} e^{3\pi} & 0 \\ 0 & e^{-\pi} \end{bmatrix}$, $\rho = e^{3\pi}, e^{-\pi}$;

$\mu = 3, -1$; $P(t) = \begin{bmatrix} \cos t & \sin t \\ -\sin t & \cos t \end{bmatrix}$, $D = \begin{bmatrix} 3 & 0 \\ 0 & -1 \end{bmatrix}$.

(c) $T = 2\pi$, $X(t) = e^t\begin{bmatrix} 2+\sin t & 0 \\ 4t - 2\cos t & 1 \end{bmatrix}$, $B = e^{2\pi}\begin{bmatrix} 1 & 0 \\ 4\pi & 1 \end{bmatrix}$, $\rho = e^{2\pi}$; $\mu = 1$ (both

repeated); $P(t) = \begin{bmatrix} 2+\sin t & 0 \\ -2\cos t & 1 \end{bmatrix}$, $D = \begin{bmatrix} 1 & 0 \\ 4 & 1 \end{bmatrix}$.

3 $\delta \approx -\frac{5}{8}\epsilon^2$, $\delta - \frac{1}{4} \approx \pm\frac{1}{2}\epsilon$, $\delta - 1 \approx \pm\frac{1}{2}\epsilon$,

6 Characteristic multipliers ρ given by $\rho^2 - 2\rho\phi + 1 = 0$, where

$$\phi = \cos(\tfrac{1}{2}\pi\delta_+)\cos(\tfrac{1}{2}\pi\delta_-) - \frac{1}{2}\left(\frac{\delta_+}{\delta_-} + \frac{\delta_-}{\delta_+}\right)\sin(\tfrac{1}{2}\pi\delta_+)\sin(\tfrac{1}{2}\pi\delta_-)$$

and $\delta_\pm = \sqrt{\delta \pm \epsilon}$.

7 $\delta \approx -\frac{1}{8}\epsilon^2 + \frac{1}{2}\epsilon\nu$, $(\delta-1)^2 \approx \frac{1}{4}(\epsilon-\nu)^2 - \nu^2$, $4\nu^2 \approx [\frac{5}{48}\epsilon^2 - (\delta-4)][\frac{1}{48}\epsilon^2 + (\delta-4)]$.

10 $-\nu$, $\dfrac{1}{T}\displaystyle\int_0^T a(t)\,dt - \nu$.

Chapter 4

1 (a) Uniformly and asymptotically stable.
(b) Uniformly and asymptotically stable.
(c) Unstable.

2 (a) Uniformly and asymptotically stable.
(b) Uniformly and asymptotically stable.

3 (a) Uniformly and asymptotically stable.
(b) Uniformly and asymptotically stable.
(c) Unstable.

5 Let $u = b\cos t + v$ and show that the linearized equation for v is

$$v'' + \nu v' + (1 + b^2 + b^2\cos 2t)v = 0,$$

which is the Mathieu equation with linear damping; now use the results of section 3.4.

7 The general solution is

$$z_1 = c_1 \mp 2c_2(c_2 + t)^{-1/2}, \qquad z_2 = \pm(c_2 + t)^{-1/2},$$

where the alternate signs correspond to $z_2(0) \gtrless 0$ and $c_2 > 0$.

8 (b) $r' = r(1-r)(2-r)$ and $\theta' = -1$. Hence $r = 1$ and $r = 2$ are both periodic solutions and, in general, $r' > 0$ for $0 < r < 1$ and $r > 2$, but $r' < 0$ for $1 < r < 2$.

10 (a) $a = \frac{2}{3}$, $b = \frac{1}{3}$, $c = 1$; uniformly and asymptotically stable.
(b) $a = 1$, $b = -1$, $c = 0$; unstable.
(c) $a = 13$, $b = 4$, $c = \frac{9}{5}$; uniformly and asymptotically stable.

Chapter 5

1 (a) Stable node.
(b) Saddle point.

(c) Stable spiral point.
(d) Unstable spiral point.
(e) Unstable improper node.

2 (a) $(0,0)$ saddle point; $(1,1)$ stable spiral point.
(b) $(0,0)$ stable node; $(2,2)$ saddle point.
(c) $(1,1)$ stable node; $(-1,-1)$ saddle point.
(d) $(1,1)$ stable spiral point; $(-1,1)$ saddle point.

4 Show that $r' = kr^3$.

5 $(0,0)$ unstable node; $(l/k,0)$ stable node; $(0,l/k)$ unstable node; (x_0,y_0) saddle point, where

$$x_0 = \frac{l(p-k)}{pq-k^2}, \qquad y_0 = \frac{l(q-k)}{pq-k^2}.$$

6 $(0,0)$ unstable node; $(l/k,0)$ saddle point; $(0,m/k)$ saddle point for $kl > mp$ and stable node for $kl < mp$; (x_0,y_0) stable node or spiral point for $kl > mp$ and saddle point for $kl < mp$, where

$$x_0 = \frac{kl-mp}{k^2+pq}, \qquad y_0 = \frac{lq+km}{k^2+pq}.$$

8 The orbits are given by

$$y + \ln|y-1| = \tfrac{1}{2}x - \tfrac{1}{4}\ln|2x+1| + \text{constant}.$$

9 $(0,0)$ stable node; $(1,1)$ saddle point. Show that $x-y = c_1 e^{-2t}$.

Chapter 6

1 Show that $r' = -rF(r)$ and, hence, that $r = r_0$ is a solution whenever $F(r_0) = 0$; for the stability determination put $r = r_0 + s$ and note that the linearized equation for s is $s' = -r_0 F'(r_0)s$. $r = 1$ (unstable), $r = 2$ (unstable), $r = 3$ (asymptotically orbitally stable).

3 Show that $r' = -\cos^2\theta\, r^3(r^2-1)$ and, hence, that $r = 1$ is a solution; for the stability determination note that $r' \gtrless 0$ as $r \lessgtr 1$, with equality occurring only when $\theta = \pm\tfrac{1}{2}\pi$.

4 (a) $(0,0)$ saddle point, index -1; $(1,0)$ stable node, index $+1$; $(-1,0)$ stable node, index $+1$; $(3,-4)$ saddle point, index -1.
(b) $(0,0)$ saddle point, index -1; $(2,0)$ saddle point, index -1; $(-2,0)$ stable node, index $+1$; $(1,3)$ stable spiral point, index $+1$. [Theorem 6.1 should also be used here to show that there are no periodic solutions, after first using the transformation $X = \ln x$ and $Y = \ln y$ in the region $x,y > 0$.]

Chapter 7

1 $\alpha_1 = 0$, $T_2 = \tfrac{1}{16}$.

2 (a) Amplitude $A = A_0 + O(\epsilon)$, where $A_0^4 = \frac{8}{5}$, period $T = 2\pi\{1 + O(\epsilon^2)\}$, stable ($\epsilon > 0$).

(b) Amplitude $A = 2 + O(\epsilon)$, period $T = 2\pi\{1 + \frac{3}{2}\epsilon + O(\epsilon^2)\}$, stable ($\epsilon > 0$).

(c) Amplitude $A = A_0 + O(\epsilon)$, where $A_0 = 2$ or 4, period $T = 2\pi\{1 + O(\epsilon^2)\}$, unstable or stable ($\epsilon > 0$).

4 (a) $T_1 = \frac{3}{8}A^2$, $T_2 = \frac{57}{256}A^4$, where A is the amplitude.

5 (a) $u = A\cos\tau + \epsilon A^2(\frac{1}{2} - \frac{1}{6}\cos 2\tau - \frac{1}{3}\cos\tau) + O(\epsilon^2)$, where $\tau = \omega t$ and $\omega = 1 - \frac{5}{12}\epsilon^2 A^2 + O(\epsilon^3)$.

(b) $u = A\cos\tau + O(\epsilon)$, where $\tau = \omega t$ and $\omega = 1 - \frac{3}{8}\epsilon A^2 + \frac{5}{16}\epsilon A^4 + O(\epsilon^2)$.

7 (a) Amplitude $A = 2 + O(\epsilon)$, period $T = 2\pi\{1 + O(\epsilon^2)\}$, stable ($\epsilon > 0$).

(b) Amplitude $A = 1 + O(\epsilon)$, period $T = 2\pi\{1 + O(\epsilon^2)\}$, stable ($\epsilon > 0$).

(c) Amplitude $A = 2\sqrt{2} + O(\epsilon)$, period $T = 2\pi\{1 + O(\epsilon^2)\}$, stable ($\epsilon > 0$).

Chapter 8

1 (a) $u = \dfrac{F_0}{\sigma^2 - 1}\cos t + \epsilon\left(\dfrac{F_0^2}{2\sigma^2(\sigma^2 - 1)} + \dfrac{F_0^2}{2\sigma^2(\sigma^2 - 1)(\sigma^2 - 4)}\cos 2t\right) + O(\epsilon^2)$,

stable for $\nu > 0$.

(b) $u = \dfrac{F_0}{\sigma^2 - 1}\cos t + \epsilon\left\{\dfrac{F_0 \sin t}{(\sigma^2 - 1)^2} - \dfrac{F_0^3}{4(\sigma^2 - 1)^3}\left(\dfrac{\sin 3t}{\sigma^2 - 9} + \dfrac{\sin t}{(\sigma^2 - 1)^3}\right)\right\} + O(\epsilon^2)$,

stable for $\epsilon\left(\dfrac{F_0^2}{(\sigma^2 - 1)^2} - 2\right) < 0$.

(c) $u = \dfrac{F_0}{\sigma^2 - 1}\cos t + \epsilon\left\{\dfrac{F_0 \sin t}{(\sigma^2 - 1)^2} - \dfrac{5F_0^3}{4(\sigma^2 - 1)^3}\left(\dfrac{\sin 3t}{\sigma^2 - 9} + \dfrac{\sin t}{(\sigma^2 - 1)^3}\right)\right.$

$\left. + \dfrac{F_0^5}{8(\sigma^2 - 1)^5}\left(\dfrac{\sin 5t}{\sigma^2 - 25} + \dfrac{3\sin t}{\sigma^2 - 9} + \dfrac{2\sin t}{\sigma - 1}\right)\right\} + O(\epsilon^2)$,

stable for $\epsilon\left(\dfrac{3F_0^4}{2(\sigma^2 - 1)^4} - \dfrac{5F_0^2}{(\sigma^2 - 1)^2} + 2\right) > 0$.

3 (a) $u = A_0\cos t + B_0\sin t + O(\epsilon)$, $\sigma^2 = 1 + \epsilon\kappa_0 + O(\epsilon^2)$,

$P_0(A_0, B_0) = \pi B_0(\kappa_0 - \frac{3}{4}R_0^2 + \frac{5}{8}R_0^4)$, $Q_0(A_0, B_0) = \pi G_0 - \pi A_0(\kappa_0 - \frac{3}{4}R_0^2 + \frac{5}{8}R_0^4)$,

where $R_0^2 = A_0^2 + B_0^2$. Response diagram defined by $B_0 = 0$ and

$G_0 = A_0(\kappa_0 - \frac{3}{4}R_0^2 + \frac{5}{8}R_0^4)$. Change of stability when $\kappa_0 - \frac{9}{4}R_0^2 + \frac{5}{2}R_0^4 = 0$.

(b) $P_0(A_0, B_0) = \pi B_0(\kappa_0 - \frac{3}{4}R_0^2 - \frac{5}{8}R_0^4)$, $Q_0(A_0, B_0) = \pi G_0 - \pi A_0(\kappa_0 - \frac{3}{4}R_0^2 - \frac{5}{8}R_0^4)$.

Response diagram defined by $B_0 = 0$ and $G_0 = A_0(\kappa_0 - \frac{3}{4}R_0^2 - \frac{5}{8}R_0^4)$. Change of

stability when $\kappa_0 - \frac{9}{4}R_0^2 - \frac{5}{2}R_0^4 = 0$.

(c) $P_0(A_0, B_0) = \pi(\kappa_0 B_0 - \frac{1}{4}\hat{\nu}A_0 R_0^2)$, $Q_0(A_0, B_0) = \pi(G_0 - \kappa_0 A_0 - \frac{1}{4}\hat{\nu}B_0 R_0^2)$, where

$\nu = \epsilon\hat{\nu}$. Response diagram defined by $G_0^2 = R_0^2(\kappa_0^2 + \frac{1}{16}\hat{\nu}^2 R_0^4)$, stable.

Chapter 9

1 (a) $u = r\cos\theta$ and $u' = r\sin\theta$, where, by the averaging method with $I = \frac{1}{2}r^2$,
   $I' = \epsilon(I - \frac{5}{2}I^3)$. Limit cycle corresponds to the critical point $I^2 = \frac{1}{5}$, stable.
 (b) $I' = \epsilon(I - \frac{1}{2}I^3)$. Critical point $I = 2$, stable.
 (c) $I' = -\frac{1}{2}\epsilon I(2 - I)(8 - I)$. Critical points $I = 2$ (unstable) and $I = 8$ (stable).

2 $u = r\cos\theta$, $u' = r\sin\theta$, where, by the averaging method with $I = \frac{1}{2}r^2$,

$$I' = -\nu I + \frac{\epsilon}{2\pi} \int_0^{2\pi} f(r\cos\theta)\sin\theta \; d\theta.$$

3 $x = r\cos\theta$, $y = r\sin\theta$, where, by the averaging method with
 $I = \frac{1}{2}r^2$, $I' = \epsilon I(1 - 2I)$. Limit cycle corresponds to the critical point $I = \frac{1}{2}$, stable.

4 $x = r\cos\theta$, $y = r\sin\theta$, where, by the averaging method with
 $I = \frac{1}{2}r^2$, $I' = \epsilon(2I - \frac{1}{2}I^2)$. Limit cycle corresponds to the critical point $I = 4$, stable.

5 (a) $u \approx R\cos(t - \Phi)$, $\sigma^2 = 1 + \epsilon\kappa$, where, by the averaging method, $R' = \frac{1}{2}\epsilon G_0 \sin\Phi$
   and $R\Phi' = \frac{1}{2}\epsilon\{G_0\cos\Phi - R(\kappa - \frac{3}{4}R^2 + \frac{5}{8}R^4)\}$.
 (b) $R' = \frac{1}{2}\epsilon G_0\sin\Phi$ and $R\Phi' = \frac{1}{2}\epsilon\{G_0\cos\Phi - R(\kappa - \frac{3}{4}R^2 - \frac{5}{8}R^4)\}$.
 (c) $R' = \frac{1}{2}\epsilon G_0\sin\Phi$ and $R\Phi' = \frac{1}{2}\epsilon\{G_0\cos\Phi - \frac{1}{2}\kappa R\}$.

Chapter 10

1 (a) $\epsilon^2 = -2\mu$, stable for $\epsilon < 0$, unstable for $\epsilon > 0$; saddle–node.
 (b) $\epsilon = 0$ stable for $\mu < 0$, unstable for $\mu > 0$; $\epsilon = \frac{1}{3}\mu$, stable for $\mu > 0$, unstable for
   $\mu < 0$; transcritical.
 (c) $\epsilon = 0$ stable for $\mu > 0$, unstable for $\mu < 0$; $\alpha\epsilon^2 = \mu$, unstable for $\alpha > 0$, stable
   for $\alpha < 0$; subcritical ($\alpha > 0$) or supercritical ($\alpha < 0$) pitchfork.
 (d) $\epsilon = 0$, stable for $\mu < 0$, unstable for $\mu > 0$; $\mu - 1 = -(\epsilon - 1)^2$, unstable for $\epsilon < 0$
   and $\epsilon > 1$, stable for $0 < \epsilon < 1$; transcritical bifurcation at $\mu = \epsilon = 0$, saddle–
   node bifurcation at $\mu = \epsilon = 1$.
 (e) Bifurcation points: $\mu = 0$, $\epsilon = 0$, transcritical; $\mu = 1$, $\epsilon = 0$, subcritical pitchfork;
   $\mu = -\frac{1}{4}$, $\epsilon = \frac{1}{2}$, saddle–node; $\mu = 2$, $\epsilon = -1$, transcritical.
 (g) Bifurcation points: $\mu = 0$, $\epsilon = 0$, transcritical; $\mu = -1$, $\epsilon = 1$, saddle–node;
   $\mu = 3$, $\epsilon = 3$, transcritical; $(\mu - 10)^2 + (\epsilon - 3)^2 = 1$ is an isolated branch with a
   saddle–node bifurcation at $\mu = 9$, $\epsilon = 3$ and $\mu = 11$, $\epsilon = 3$.

3 (a) Critical points: $x = 0$, $y = 0$; $x = \frac{1}{2}\mu$, $y = 0$; $x = 1$, $y = \mu - 2$. Bifurcation points:
   $\mu = 0$, $x = 0$, $y = 0$, transcritical; $\mu = 2$, $x = 1$, $y = 0$, transcritical.
 (b) Critical points: $x = 0$, $y = 0$; $x = \frac{1}{2}\mu$, $y = 0$; $x = 0$, $y = 1$; $x = \mu - 1$, $y = 2 - \mu$.
   Bifurcation points: $\mu = 0$, $x = 0$, $y = 0$, transcritical; $\mu = 2$, $x = 1$, $y = 0$, transcritical; at $\mu = 1$ no change in stability.
 (c) Critical points: $x = 0$, $y = 0$; $x = \frac{1}{2}\mu$, $y = 0$; $x = 0$, $y = -1$; $x = \frac{1}{3}(\mu + 1)$,
   $y = \frac{1}{3}(\mu - 2)$. Bifurcation points: $\mu = 0$, $x = 0$, $y = 0$, transcritical; $\mu = 2$,
   $x = 1$, $y = 0$, transcritical; at $\mu = -1$ no change in stability.
 (d) Critical points: $x = 0$, $y = 0$; $x^2 = 4(1 - \mu)$, $y = x + \frac{1}{2}x^2$. Bifurcation points:
   $\mu = 1$, $x = 0$, $y = 0$, supercritical pitchfork; $\mu = \frac{3}{4}$, $x = 1$, $y = \frac{3}{2}$, Hopf bifurcation.

4 (a) Let $z = x + iy = Re^{i\phi} + O(R^2)$. Then the bifurcating periodic solution is approximately given by $R^2 = 8\mu$, supercritical.

(b) $R^2 = -\frac{8}{3}\mu$, subcritical.
(c) $R^2 = 4\mu$, supercritical.
(d) $R^2 = -8\mu$, subcritical.

Chapter 11

1 (a) $H = \frac{1}{2}p^2(1+\alpha^2 q^2)^{-1} + \frac{1}{2}\alpha g q^2$.
   (b) $H = \frac{1}{2}p^2(1+y_0^2\beta^2\sinh^2\beta q) + gy_0\cosh\beta q$.
   (c) $H = \frac{1}{2}p^2 + \frac{1}{8}gq^2\alpha^{-1}$, $q = 4a\sin\frac{1}{2}\theta$.
   (d) $H = \dfrac{1}{2l^2}(p - u_0 l\cos q)^2 + gl(1-\cos q) - \frac{1}{2}u_0^2$.
   (d) $H = \dfrac{1}{2l^2}(p - v_0 l\sin q)^2 + gy_0 + gl(1-\cos q) - \frac{1}{2}v_0^2$.
   (f) $H = \frac{1}{2}p^2(1+\alpha^2 q^2)^{-1} + \frac{1}{2}\alpha g q^2 - \frac{1}{2}\omega^2 q^2$.

4 (a) $S = -qQ$.
   (b) $S = q\exp Q$.
   (c) $S = q\cos^{-1}(Qe^t q^{-1/2}) - Qe^t(q - Q^2 e^{2t})^{1/2}$.
   (d) $S = \frac{1}{2}q^2\cot Q$.

5 $P = (p/Q_q) - g_Q$, where $g(Q,t)$ is an arbitrary function. $S_1 = PQ + g$.

8 If $a \neq 0$, $S_1 = \dfrac{1}{2a}(-bq^2 + 2Pq + cP^2)$ and, if $a = 0$ and $c \neq 0$, $S = \dfrac{1}{2c}(-dq^2 + 2qQ)$.

10 (d) $I = \dfrac{1}{2}\left(\dfrac{H}{\Omega} - \sqrt{k}\right)$, $\Omega q^2 = \dfrac{H}{\Omega} - \sqrt{\dfrac{H^2}{\Omega^2} - k\cos\theta}$.

   (e) $H = -\dfrac{k^2}{2(1+h)^2}$ (note that $H < 0$ here), $\theta = \lambda - \sqrt{1 - \dfrac{h^2}{(1+h)^2}}\sin\lambda$, where

   $q = \dfrac{(1+h)^2}{k}\left(1 - \sqrt{1 - \dfrac{h^2}{(1+h)^2}}\cos\lambda\right)$.

12 (a) $K_1(J) = \dfrac{k^2}{h(J+h)^3}$.
   (b) $K_1(J) = \dfrac{k^3}{h^3(J+h)^3}$.

13 (a) $u \approx R(\tau)\cos\{t - \Phi(\tau)\} + O(\epsilon)$, $\tau = \epsilon t$, where $R_\tau = \frac{1}{2}R - \frac{5}{16}R^5$, $\Phi_\tau = 0$.
   (b) $R_\tau = \frac{1}{2}R - \frac{1}{8}R^3$, $\Phi_\tau = \frac{3}{8}R^2$.
   (c) $R_\tau = -\frac{1}{16}R(4 - R^2)(16 - R^2)$, $\Phi_\tau = 0$.

REFERENCES

These books are those referred to in the main body of the text, together with a few others which I either found useful in preparing the material for this book, or can recommend for further reading. The list is by no means exhaustive.

ABRAMOWITZ, M. AND STEGUN, I. A., *Handbook of mathematical functions*, Dover, 1965, 9th ed., 1046 pp.

ANDRONOV, A. A., VITT, A. A. AND KHAIKIN, S. E., *Theory of oscillators*, Pergamon, 1966 (English edition), 815 pp.

ARNOLD, V., *Geometrical methods in the theory of ordinary differential equations*, Springer-Verlag, 1983, 334 pp.

ARNOLD, V., *Ordinary differential equations*, MIT Press, 1973, 280 pp.

ARNOLD, V., *Methods of classical mechanics*, Springer-Verlag, 1978, 462 pp.

BIRKHOFF, G. AND ROTA, G-C., *Ordinary differential equations*, Wiley, 1978, 3rd ed., 342 pp.

BOGOLIUBOV, N. M. AND MITROPOLSKY, Y. A., *Asymptotic methods in the theory of nonlinear oscillations*, Gordon and Breach, 1961, 537 pp.

CESARI, L., *Asymptotic behaviour and stability problems in ordinary differential equations*, Springer-Verlag, 1962, 2nd ed., 271 pp.

CHOW, S. N. AND HALE, J. K., *Methods of bifurcation theory*, Springer-Verlag, 1982, 515 pp.

CODDINGTON, E. A. AND LEVINSON, N., *Theory of ordinary differential equations*, McGraw–Hill, 1955, 429 pp.

GUCKENHEIMER, J. AND HOLMES, P., *Nonlinear oscillations, dynamical systems and bifurcations of vector fields*, Springer-Verlag, 1983, 453 pp.

HAGEDORN, P., *Non-linear oscillations*, Oxford Engineering Science Series, Vol. 10, OUP, 1988, 2nd ed., 311 pp.

HARTMAN, P., *Ordinary differential equations*, Wiley, 1973, 612 pp.

IOOSS, G. AND JOSEPH, D. D., *Elementary stability and bifurcation theory*, Springer-Verlag, 1989, 286 pp.

JORDAN, D. W. AND SMITH, P., *Nonlinear ordinary differential equations*, OUP, 1977, 360 pp.

LANDAU, L. D. AND LIFSHITZ, E. M., *Mechanics*, Pergamon, 1960, 165 pp.

LICHTENBERG, A. J. AND LIEBERMAN, M. A., *Regular and stochastic motion*, Springer-Verlag, 1983, 499 pp.

MINORSKY, N., *Nonlinear oscillations*, Van Nostrand, 1962, 714 pp.

MOON, F. S., *Chaotic vibrations: an introduction for applied scientists and engineers*, Wiley, 1987, 309pp.

NAYFEH, A. H. AND MOOK, D. T., *Nonlinear oscillations*, Wiley, 1979, 704 pp.

PERCIVAL, I. AND RICHARDS, D., *Introduction to dynamics*, CUP, 1982, 228 pp.

SANDERS, J. A. AND VERHULST, F., *Averaging methods in nonlinear dynamical systems*, Springer-Verlag, 1985, 247 pp.

SPARROW, C. T., *The Lorenz equations: bifurcations, chaos and strange attractors*, Springer-Verlag, 1982, 270 pp.

STRUBLE, R. A., *Nonlinear differential equations*, McGraw–Hill, 1962, 267 pp.

THOMPSON, J. M. T. AND STEWART, H. B., *Nonlinear dynamics and chaos: geometrical methods for engineers and scientists*, Wiley, 1986, 376 pp.

INDEX

action, 216, 235, 240
action–angle variables, 217, 221, 223, 233,
    291, 303, 304, 306, 307, 308, 310, 313,
    314
adiabatic invariance, 233
adiabatic invariant, 235, 314
algebraic multiplicity, 39
asymptotically orbitally stable, 101, 110, 147,
    180
asymptotically stable, 84, 88
autonomous, 4, 98, 105, 112, 245
    equations, 216
    Hamiltonian, 283, 287, 304
    system, plane, 112, 115, 125, 142
averaged equation, 217, 219, 225, 226, 229,
    231
averaging method, 216, 223, 233, 238, 240

bifurcation
    Hopf, 247, 261
    diagram, 248, 254
    one-dimensional, 246, 255, 256
    pitchfork, 250, 256, 267, 268
    point, 246, 247
    saddle–node, 248, 249, 250, 256
    subcritical pitchfork, 250
    supercritical pitchfork, 250
    theory, 245
    transcritical, 249, 250, 256

canonical form, 36, 37, 38, 117
canonical transformation, 294, 297, 299, 303,
    307, 308
Cauchy–Euler construction, 5
Cauchy–Peano theorem, 5
centre, 120, 127, 129, 287
characteristic exponent, 49, 50, 182, 185, 194,
    202
characteristic multiplier, 49, 50, 185, 195,
    201, 228
conservative system, 165
contact transformation, 297
continuation, 15, 18
coordinate, generalized, 277
critical point, 112, 125, 145, 149, 220, 226,
    245

definite, negative, 102, 103
definite, positive, 102
dependence on parameters, 15
Duffing's equation, 166, 173, 184, 188, 191,
    197, 203, 205, 212, 213, 214, 215, 222,
    229, 231, 236, 241, 242, 309, 312

eigenvalue, 36
·eigenvector, 36
energy, kinetic, 278, 279
    potential, 278
entrainment, 210
existence, 7, 23
exponential of a matrix, 42, 53

first-order system, 3
    equivalent, 7, 32, 57
first variational equation, 92, 97, 98, 101,
    109, 116, 185, 194, 201, 227
Floquet exponents, 49, 80
Floquet theory, 47, 49
focus, 120
forced oscillations, 190, 223
fundamental matrix, 27, 32

general solution, 15, 26, 28, 30
generalized coordinate, 277
generating function, 298, 301, 302, 311, 313,
    314
geometric multiplicity, 39
Green's function, 33
Green's matrix, 31
Gronwall's lemma, 9, 21, 219

Hamiltonian, 106, 165, 276, 299
    autonomous, 283, 287, 304
    system, 165, 223, 276
Hamilton's equations, 299
Hamilton's principal function, 300
Hilbert's invariant integral, 295
Hill's equation, 56, 79
homogeneous equation, 30
homogeneous linear systems, 25
Hopf bifurcation, 247, 261
hysteresis loop, 208

326